Scanning Electron Microscopy in Archaeology

edited by

Sandra L. Olsen

BAR International Series 452
1988

B.A.R.

5, Centremead, Osney Mead, Oxford OX2 0DQ, England.

BAR -S452, 1988: 'Scanning Electron Microscopy in Archaeology'

ISBN 9780860545798 paperback
ISBN 9781407347370 e-book
DOI https://doi.org/10.30861/9780860545798
A catalogue record for this book is available from the British Library
This book is available at www.barpublishing.com

Dedicated to Don Brothwell, who foresaw twenty years ago the great contributions scanning electron microscopy has to offer archaeology.

CONTENTS

LIST OF CONTRIBUTORS

José María Bermúdez de Castro
Museo Nacional de Ciencias Naturales
Jose Gutierrez Abascal, 2
28006 Madrid
Spain

Donald J. Blakeslee
Department of Anthropology
The Wichita State University
Wichita, Kansas 67208
U.S.A.

Don Brothwell
Department of Human Environment
Institute of Archaeology
University College
University of London
31-34 Gordon Square
London, England WC1H OPY

Peter A. Bull
School of Geography
University of Oxford
Oxford, England OX1 3TB

E. Ann Butler
Department of Human Environment
Institute of Archaeology
University College
University of London
31-34 Gordon Square
London, England WC1H OPY

François Cianfarani
Department of Forensic Medicine
Faculté de Médecine La Timone
27 Bd Jean Moulin
Marseille Cedex 5
France 13 385

Donald Claugher
Electron Microscope Unit
British Museum (Natural History)
Cromwell Road,
London, England SW7 5BD

Susan Colledge
Department of Human Environment
Institute of Archaeology
University College
University of London
31-34 Gordon Square
London, England WC1H OPY

Keith Dobney
Department of Archaeology
Bradford University
Bradford, England

Richard K. Dunn
Department of Geology
The Wichita State University
Wichita, Kansas 67208
U.S.A.

Francesco d'Errico
Laboratoire de Préhistoire, U.A. 184
C.N.R.S. Institut de Paléontologie Humaine
1 Rue René Panhard
75013 Paris
France

and

Human Paleontology Laboratory
Departement of Human Paleontology and Physiology
University of Turin
Corso M. D'Azeglio 52
10126, Italy

Yolanda Fernández-Jalvo
Museo Nacional de Ciencias Naturales
Jose Gutierrez Abascal, 2
28006 Madrid, Spain

Itzhak Gedalia
Dental Research
Hebrew University Hadassah
School of Dental Medicine
Jerusalem, Israel 91 120

Dana E. Goodburn-Brown
Museum of London
London, England

Simon Hillson
Department of Human Environment
Institute of Archaeology
University College
University of London
31-34 Gordon Square
London, England WC1H OPY

Kjel Knutsson
Department of Archaeology
Uppsala University
Uppsala, Sweden S-75220

Alfred W. Magee
School of Geography
University of Oxford
Oxford, England OX1 3TB

Judy M. Maguire
Bernard Price Institute for Palaeontological Research
University of the Witwatersrand
Johannesburg 2001, South Africa

Alice Markitziu
Oral Diagnosis
Hebrew University Hadassah
School of Dental Medicine
Jerusalem, Israel 91120

Nigel D. Meeks
The British Museum Research Laboratory
Great Russell Street
London, England WC1B 3DG

Sandra L. Olsen
Department of Cell Biology and Anatomy
The Johns Hopkins University School of Medicine
725 North Wolfe Street
Baltimore, Maryland 21205
U.S.A.

V.R. O'Sullivan
Department of Anatomy
University College Cork
Cork, Ireland

C. Power
Department of Archaeology
University College Cork
Cork, Ireland

Juliet Prior
Department of Pure and Applied Biology
Imperial College of Science and Technology
Prince Consort Road
London, England SW7 2BB

Pierre-François Puech
Department of Prehistory
Musée de l'Homme de Paris
B.P. 191, 2 rue Saint-Antoine
Nîmes Cedex, France 30 008

Jennie Rose
Department of Cell Biology and Anatomy
The Johns Hopkins University School of Medicine
725 North Wolfe Street
Baltimore, Maryland 21205
U.S.A.

Friedemann Schrenk
Hessisches Landesmuseum
Friedensplatz 1
6100 Darmstadt
West Germany

David Scott
The Getty Conservation Institute
4503B Glencoe Avenue
Marina del Rey, California 90292-6537

Pat Shipman
Department of Cell Biology and Anatomy
The Johns Hopkins University School of Medicine
725 North Wolfe Street
Baltimore, Maryland 21205
U.S.A.

Bruce D. Smith
Department of Anthropology
National Museum of Natural History/National Museum of Man
Smithsonian Institution
Washington, D.C. 20560
U.S.A.

Jacqui Watson
Ancient Monuments Laboratory
Historic Buildings and Monuments Commission for England
Fortress House
23 Savile Row
London, England W1X 2HE

Tim Yates
Department of Geography
University College
University of London
Gower Street
London, England WC1H OAP

PART I:

METHODS AND TECHNIQUES

INTRODUCTION: APPLICATIONS OF SCANNING ELECTRON MICROSCOPY TO ARCHAEOLOGY

Sandra L. Olsen

This volume is the product of a conference entitled Scanning Electron Microscopy in Archaeology, held in 1986, at the Institute of Archaeology, the University of London. A gathering of over 60 scientists from 10 nations spent two days presenting their research and openly discussing their methods. It was a rare opportunity for archaeologists, conservators, and specialists from a number of other fields to share their views and exchange ideas about the potential uses of scanning electron microscopy (SEM) to a myriad of scientific problems. The results were rewarding: researchers who work on very different materials found that many of the difficulties they had encountered were shared by others and that, by interacting with each other, they could discover solutions to these problems. As a result of this meeting, those who presented papers, and others who subsequently came forward, have contributed to this volume to further disseminate information on methods of preparation of archaeological specimens and applications of SEM to the analysis of archaeological data.

Brothwell (1969) was one of the first to advocate the use of SEM in the analysis of archaeological samples. He pointed out the advantages of SEM over optical microscopy, demonstrating how it could be used to study ancient bone, dentition, textile fibers, hair, plant remains, and stone tools (Brothwell 1969: 564-566). It is clear, by the diversity of subjects covered in this volume and the increasing reference to scanning electron microscopy in recent publications, that SEM has become an important tool in the interpretation of archaeological data.

The SEM has been shown to have numerous specific applications on a wide variety of materials, but because of time, expense, and difficulties involved with specimen preparation, it is rarely used in exclusion of other analytical instruments. In fact, one of the great strengths of SEM is its complementarity with optical microscopes. Whereas the optical microscope provides a simple and inexpensive means of analyzing bulk samples, the SEM achieves a higher standard of analysis on a greatly reduced population of objects.

The well-known advantages of the secondary electron emission mode over optical microscopy apply to the study of archaeological materials as well as other objects. These include: the vast, continuous range of magnifications greatly exceeding the optical limit of around 1000 times, the increased depth of field (which is about 300 times greater than for an optical microscope), and the high resolution (practically speaking around 200 Å in the SEM). All of these features are important reasons for choosing SEM for archaeological analysis of microscopic surfaces. Even at low magnifications the high quality of scanning electron micrographs and stereo-pairs enables the researcher to demonstrate his or her findings clearly in published form.

Most archaeological studies which employ SEM are concerned primarily with the observation of surface topography and, therefore, rely mainly on the secondary electron image (SEI). Much of the work involves low magnifications between 10 and 1,000 times, although specific problems may occasionally require higher magnifications. A wide range of archaeological materials including metals, glass, faience, pottery, stone, soil particles, pigments, bone, teeth, fingernails, skin, hair, eggshell, mollusks, insects and parasites, plant remains, wood, pollen, fibers, and so on have been examined with secondary electron imaging.

The second most common use of the SEM is for determining composition, generally using energy dispersive x-ray analysis (EDAX). The advantages of using x-ray microanalysis with a SEM are numerous. It provides a relatively quick and nondestructive means of obtaining qualitative information on the constituents of a material without much specimen preparation. At this level the sample may only need to be coated with carbon, unless it is sufficiently conductive without coating. For more quantitative studies the sample may need to be sectioned and polished, but this still allows the specimen to be used for other forms of analysis. An x-ray microprobe is sometimes preferred because of its superior quantitative results, but since these instruments are generally more expensive and less common, access may be less convenient than for a SEM-EDAX. X-ray microanalysis has been used by researchers on metals, ceramics, glass, faience, bone, and other archaeological materials. Elemental mapping has been extremely enlightening in the study of metal alloys and for identifying inclusions in pottery.

Backscattered electron imaging (BEI) has seen more limited application, but has the advantage of supplying both topographic and compositional information simultaneously. It also eliminates the problem of the edge bright-up effect due to charging, a common problem with archaeological specimens that are poorly conductive. This is especially important if the edge of the object is the chief area of study as in wear analysis of stone, bone, or metal tools (Meeks, et al., 1982).

Cathodoluminescence (CL) has received little attention by archaeologists, but it has been shown to have value in the examination of quartz grains by soil analysts (Krinsley and Hyde, 1971) and for determining the likely sources of thermoluminescence signals in prehistoric ceramics (Singhvi and Zimmerman, 1979).

Auger electron emission is one of the least commonly employed SEM modes in archaeology. This may be due to a combination of limited accessibility to microscopes with Auger detectors and a lack of common knowledge among archaeologists about the principles and applications of this mode. In one case, however, Auger electron spectroscopy was used to analyze the constituents of Greek bronze arrow tips (Polak et al., 1983).

Preparation Methods for Archaeological Specimens

Working with archaeological materials may pose problems when samples are prepared for examination with a scanning electron microscope. Sometimes archaeological specimens are replaceable or available in vast quantities, as with soil samples, pollen, and certain kinds of stone, bone, or pottery collections. In these cases it is easy enough to justify the sectioning, metal-coating, or other destructive alterations necessary for optimal observation in the SEM. Very often this is not the case, however, as when the samples are unique or very rare and conservation becomes a primary concern for the researcher. In these situations alternative methods of specimen preparation must be found, such as the avoidance of coatings and adhesives, the selection of fragments of material small enough to fit into the chamber without sectioning, or the use of replicas.

Thought must be given to the kinds of analysis to be performed on the sample before any preparation is carried out. For example, if microanalysis is to be conducted, obviously the specimen should be coated with carbon rather than a metal alloy, such as gold-palladium.

In some cases further analytical techniques may need to be performed after SEM observation, requiring careful planning during preparation. On rare occasions, the sample may be later radiocarbon dated, in which case contamination by coatings and adhesives should be avoided. Whenever possible the use of coatings and adhesives should be restricted to samples that are considered expendable or replaceable; if several kinds of analysis are to be performed on a given sample, then the most destructive tests or preparation techniques should be done last.

It is not unusual for specimens to be housed and curated in a museum which does not have a loan policy for individuals or foreigners or will not allow modifications to be performed on the objects. In these situations the researcher may find it necessary to replicate the specimen in order to study it at other facilities or to avoid destructive treatment of the original.

In addition to problems pertaining to the conservation measures associated with the care and handling of rare archaeological samples, many of the difficulties encountered in preparation are specifically related to the nature of the material examined. The size, conductivity, porosity, and state of preservation of archaeological samples are the chief sources of problems with preparation. With increasingly larger SEM specimen chambers, archaeologists now have more freedom to study whole artifacts without the need to replicate or section them. The option of viewing small fragments rather than complete objects is very often feasible, since so many archaeological materials are fragmentary when recovered.

Even with a large vacuum chamber, size limitations may still be a serious problem if the material is very porous. Bone and antler, for example, are problematical since it may be difficult to evacuate the chamber if large samples are used. Placement of specimens in a desiccating chamber for two to three days prior to observation in the SEM can facilitate outgassing.

If a Charge Free Anticontamination System (CFAS) is incorporated in the SEM, then large, porous specimens can be studied. The CFAS maintains a significant vacuum differential between the column and the chamber, so that objects can be viewed in a poor vacuum. The results are significant for archaeologists, since not only can large samples be studied, but also nonconductive samples can be viewed without coating (Claugher, this volume).

If a CFAS is not available, then the conductivity of the archaeological material is important. As mentioned above, metal coating may not always be a viable procedure with rare or valuable ancient objects or those that will be displayed in museum exhibits. There are several alternatives to metal coating, however, which minimize the problem of charging considerably. Many archaeological materials are sufficiently conductive without coating if certain precautions are observed and if only low magnifications are to be used. Transparent anti-static coatings can be applied just prior to viewing to reduce charging with no visible deleterious effects. Slowing the scanning speed, reducing the accelerator voltage and minimizing the viewing time prior to photographing a particular area will all help avoid charging effects. Wrapping a metal wire around the object and twisting its ends around the specimen mount will help drain off charging, as well. It has been found that stone, bone, shell, pottery, and metals can usually be studied at magnifications under 1000 times if the above recommendations are followed.

Whenever possible, such as when samples are expendable, or when experimental imitations are to be examined, it is clearly preferable to coat the object with gold-palladium for normal imaging or carbon for microanalysis and to attach the specimen firmly to a stub with a conductive adhesive such as silver or carbon dag.

In many cases the best solution for SEM analysis of archaeological samples is to replicate the area to be studied. The most common procedure is to make a silicone rubber mould of the surface of the object. The negative replica can then be covered with a conductive metallic coating and viewed directly in the SEM, or it can be used to produce a positive resin replica for study. Considerable information has been published on replicating techniques with useful comparisons between different commercial products (Pfefferkorn and Boyde, 1974; Pameijer, 1979; Larsen, 1979; d'Errico et al., 1982/83; Rose, 1983; Bromage, 1985).

Conclusions

Although this volume does not explore the full potential of scanning electron microscopy in archaeological analysis, it is hoped that the reader will be provided with a sample from the range of possible applications. As lesser-known SEM modes, such as cathodoluminescence and Auger electron emission become more common, their use in archaeology will undoubtedly increase. Scanning electron microscopes are, themselves, becoming more readily available and affordable, so the size of samples which can be studied with SEM is rapidly expanding. The improvements in imaging through frame storage and digitalizing, as well as other computerized features, increase the quality of information obtained through SEM. One of the most important points which contributors in this volume have attempted to convey is the complementarity of SEM with other forms of analysis. It is not viewed as a substitute or replacement for optical microscopy, but rather an independent means of contributing information to archaeological research.

References

Bromage, T. 1985. Systematic inquiry in tests of negative/positive replica combinations for SEM. Journal of Microscopy 137(2): 209-216.

Brothwell, D. 1969. The study of archaeological materials by means of the scanning electron microscope: an important new field. Chapter 48 in D. Brothwell and E. Higgs (eds.), Science in Archaeology, pp. 564-566. Thames and Hudson, London. Second edition, revised and enlarged.

Claugher, D. 1987. Preparative methods, replicating and viewing of uncoated materials. In S.L. Olsen (ed.), Scanning Electron Microscopy in Archaeology. BAR International Series, London.

d'Errico, F., Giacobini, G. and Puech, P.F. 1982/1983. Varnish replicas: a new method for the study of worked bone surfaces. Ossa 9-11: 29-51.

Krinsley, D.H. and Hyde, P.J.W. 1971. Cathodoluminescence studies of sediments. Proceedings of the Fourth Annual Scanning Electron Microscope Symposium Parts I and II: 411-413. IIT Research Institute, Chicago.

Larsen, E.B. 1979. Moulding and Casting of Museum Objects. School of Conservation, Copenhagen.

Meeks, N.D., de G. Sieveking, G., Tite, M.S. and Cook, J. 1982. Gloss and use-wear traces on flint sickles and similar phenomena. Journal of Archaeological Science 9: 317-340.

Pameijer, C.H. 1979. Replication techniques with new dental impression materials in combination with different negative impression materials. Scanning Electron Microscopy 1979/II: 571-574.

Pfefferkorn, G. and Boyde, A. 1974. Review of replica techniques for scanning electron microscopy. Scanning Electron Microscopy 1974/I: 75-82.

Polak, M., Baram, J. and Pelleg, J. 1983. Auger electron spectroscopy applied to the analysis of archaeological artifacts. Archaeometry 25(1): 59-67.

Rose, J.J. 1983. A replication technique for scanning electron microscopy: applications for anthropologists. American Journal of Physical Anthropology 62(3): 255-261.

Singhui, A.K. and Zimmerman, D.W. 1979. The luminescent minerals in fine-grain samples from archaeological ceramics. Archaeometry 21(1): 73-77.1

PREPARATIVE METHODS, REPLICATING, AND VIEWING OF UNCOATED MATERIALS

D. Claugher

Introduction

Since the advent of the first SEM in 1965 design has improved and microscopes have become more versatile. Early SEMs were only capable of producing secondary electron images, but quite rapidly it was found that analytical facilities could be incorporated together with a number of alternative methods of imaging. In short the SEM became a universal tool for scientists in a large number of disciplines.

Despite the improvement of the microscope, one thing has not changed, and that is its inability to produce a first class image from poorly prepared specimens. Many people involved in the preparation of material for SEM examination give little thought to specimen preparation and this lack of thought, or perhaps lack of understanding, leads invariably to disappointing results, for which the instrument or the operator is blamed. In all fairness it may be said that there are a number of operators, who through lack of understanding are unable to get the best from their instruments, but the results from either of these circumstances is a disappointing micrograph.

Excluding the specialised techniques of analysis and metallurgy, I would like to dwell on method of getting the best from your specimen by outlining some of the techniques that have been used over the last twenty years on a variety of materials.

Specimen Preparation

Methods of preparation depend upon the material being examined, and the state in which it is found. Specimens from a marine environment require different treatment to those from a terrestrial one. Generally speaking all specimens to be examined in the SEM should be clean and absolutely dry. Drying can be achieved by desiccation over silica gel with or without vacuum, or simply by placing the specimen in a 37°C oven till it is dry. Wet or damp specimens will not stick to SEM stubs even with the most suitable adhesives, and if adhesion is obtained by some other method, the microscope will not be able to achieve operating pressure if there is vapour being given off by the specimen.

Some microscopes are able to tolerate a poorer vacuum than others, but the poorer the vacuum the poorer the resolution and the worse the quality of the micrograph.

Specimens recovered from damp or wet environments must hence be treated to remove moisture. Air drying or desiccation may cause considerable shrinkage and therefore one must consider alternative methods. The two most commonly used are freeze drying and critical point drying.

Freeze Drying

Specimens may be washed or sieved to remove adherent material and may, depending on the durability of the specimen, be subjected to ultrasonic treatment. After this initial treatment they should be thoroughly washed in at least three changes of double glass distilled water, noting that deionised water should not be used as a substitute, for glass distilled water.

After the cleaning, the material should be frozen in a minimal quantity of distilled water and placed in a previously cooled freeze drying chamber. The apparatus is then evacuated and the drying proceeds. The rate of drying is dependent upon the physical size and the density of the specimen, it is better to leave the specimen too long in the apparatus than to take it out before the drying cycle is complete.

Critical Point Drying

Specimens for critical point drying are usually "fixed" in the biological sense, that is to say that they have been treated with a chemical substance, usually formalin or alcohol, to preserve them. Archaeological fixatives or consolidating substances such as polyvinyl acetate emulsion should not be used on specimens intended for SEM examination, as the presence of this substance will obscure fine detail on the surface. Specimens that do not appear to dry well with the freeze drying method may be critical point dried. Material for critical point drying is dehydrated through ascending grades of alcohol or acetone in water. The starting solution may be as low as 10%, then the specimens are transferred through the ascending grades of 20%, 30%, 40%, 50%, and so on till 100% is reached, then two changes of the absolute solvent follows, in order to ensure that all the water has been removed. If alcohol has been used to dehydrate the material it should be treated in two changes of acetone prior to the next stage. When dehydration is complete the material is placed in the critical point drying apparatus and dried.

The two methods outlined above give satisfactory results, but you may find that some materials respond better to freeze drying and others to critical point drying; much depends upon the material and its state when found. It must be emphasised that material recovered from the sea must be thoroughly washed in distilled water prior to further treatment. Freeze drying rather than critical point drying appears to be more suitable for artifacts that have been recovered from the sea.

Assuming that you have a clean specimen, the next stage is to mount it on a stub, and this is where many mistakes are made. There are very few mountants that are suitable for this purpose. It is popularly assumed that any adhesive will suffice to hold the specimen onto the stub, but in fact nothing could be further from the truth, as there are very few adhesives that are stable inside the microscope. It was common practice some years ago to mount specimens on double sided adhesive tape, indeed some firms still sell stubs made up with double-sided tape, but these should be avoided if you value your specimens or the inside of your instrument. Double-sided tape is convenient to use but interacts with the electron beam to produce vapours that deposit on the final apertures as carbon, reducing the resolution of the instrument. Other drawbacks of double-sided tape are that it dehydrates even outside the microscope, and cracks, causing conductivity problems with the specimens.

Many other adhesives have disadvantages: the popular dags, silver and carbon, are not suitable for soft or absorbent specimens, as they tend to creep and be absorbed by the specimen. Experience has shown that for general purposes there is nothing to equal Araldite for attaching specimens to stubs. There are two types of domestic Araldite available, quick setting and normal. The normal type should be used, as the rapid or quick setting variety often does not completely cure.

Experience will show that even the smallest samples may be fixed to a stub with Araldite without creeping. Water soluble Resin "W", made by Evostick, and Durofix, made by Rawlplug, are two other suitable adhesives. Durofix may be diluted with butyl or ethyl acetate or acetone. When either of these adhesives is used it is not difficult to remove the specimens from the stubs once they have been examined.

Coating

Sputter coating has now replaced the once popular evaporative coating; it is much cheaper and more efficient.

Uncoated Material

There is now a system marketed under the name of CFAS, which enable the examination of uncoated materials in the SEM. CFAS stands for Charge Free Anticontamination System, which is probably a misleading title, as one would not necessarily associate it with a system for looking at uncoated material although it is. The CFAS system makes use of the Robinson backscattered detector which is quite different to the more commonly available and cheaper diode equivalent. The Robinson detector is a scintillator type much the same as the normal secondary detector found in all SEM's, and has a much greater sensitivity to backscattered electrons than the diode device.

When CFAS is fitted to an SEM, the gun is maintained at a higher vacuum than the specimen chamber. This differential is achieved by using the diffusion pump together with the two stage rotary pump for the gun and a second two stage rotary pump for the specimen chamber, enabling the vacuum in the chamber to be varied depending upon the material being examined.

Because it is possible to view specimens in a poor vacuum, the residual gas or air in the chamber is able to conduct away any charge that may build up, allowing observations to be made without coating at magnifications up to X10,000 for some materials. This type of set-up has been used in the EM unit of the British Museum (Natural History) for the last five years and has produced thousands of photographs of both coated and uncoated material (Figs. 1 to 6), including minerals, bryozoans, ostracods, insects, plants and archaeological artifacts.

Captions for Plate I

All specimens on this plate are fossil bryozoans, _Metrarabdotos monoliferum_ (Milne Edwards) from the Pliocene Coralline Crag of Gedgrove, Suffolk.

Fig. 1. Gold coated secondary electron image. Bar = 300 microns. Print magnification X15.

Fig. 2. Gold coated secondary electron image. Bar = 10 microns. Print magnification X750.

Fig. 3. Gold coated backscattered image. Bar = 300 microns. Print magnification X15.

Fig. 4. Gold coated backscattered image. Bar = 10 microns. Print magnification X750.

Fig. 5. Uncoated CFAS image. Bar = 300 microns. Print magnification X15.

Fig. 6. Uncoated CFAS image. Bar = 10 microns. Print magnification X750.

Note there is better detail on the low power CFAS image in Fig. 5 than on the coated secondary electron image in Fig. 3. However, at a higher magnification the coated secondary electron image in Fig. 2 is better than the CFAS result in Fig. 6.

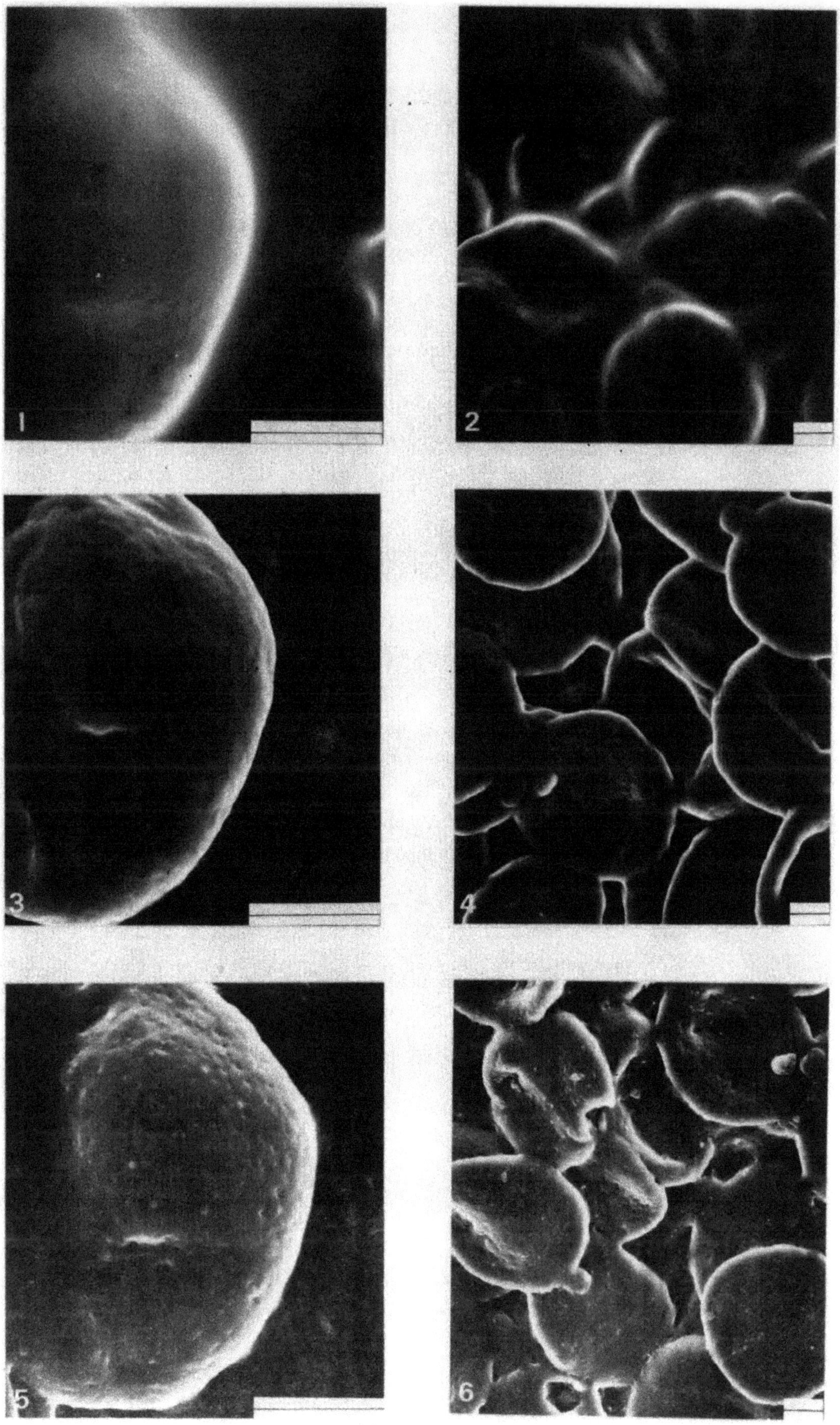

Plate 1.

Valuable type specimens that may not be coated have been examined without damage.

One other advantage of the Robinson detector is that it acts as an atomic number discriminator and may give an indication of how the various phases are distributed when looking at objects that are composed of more than one type of metal or material. A simple addition to the microscope when using this type of detector is an Image Micro Analysis System (IMAS) or a similar apparatus called a Selected Phase Imaging Module. Both of these systems enable simple qualitative analysis to be carried out on suitable specimens. It should be stressed that this only applies to inorganic materials.

A typical example of the use of the CFAS module is as follows. Microscopical examination of surface traces on bones and flints can indicate if they were fashioned intentionally by humans for a particular purpose, or if they are merely products of nature. It is impractical in some cases to coat artifacts prior to examination in the SEM and this is where the CFAS module comes into its own. Some hundreds of pieces of bone and flint have been examined by Cook (1985) and Andrews and Cook (1986) in the SEM to establish their authenticity, for the most part without coating. Results obtained show that the photographs produced using this method are of a high quality and suitable for publication.

Replicas

Replicas have been used for examining microscopical detail for well over a hundred years. Isinglass was used by mineralogists to replicate the surface details of crystals as far back as 1843 and botanists were using shellac for making replicas of leaf surfaces at the turn of the century. Since then a variety of materials have been used, including plaster of Paris, paraffin wax, electroplating, and various rubber compounds.

The advent of the electron microscope and later the SEM created a demand for replicas with a much greater resolution than was previously thought necessary, because these instruments were capable of resolving detail that was impossible to see with the optical microscope. The early transmission electron microscopists devised some very refined replicating methods for examining the surfaces of materials that could not be examined directly in the microscope. These methods produced replicas of a very high resolution, and enabled the imaging of detail down to below 10 nm. However for the general user, the methods were not always practical as many of them involved the destruction of the material that was being replicated. The advent of the SEM gave us a method of looking directly at surfaces without always needing to replicate, but the restriction imposed by the size of the chamber on the size of the object again turned user's minds towards replicating materials and methods. Pfefferkorn and Boyde (1974), Pameijer (1978), Hamilton-Waters (1983) and Bromage (1984) all examined various replicating materials with a view to them being used in the SEM, and their findings and recommendations are a good starting point for any work on this subject. The early SEM's were capable of a resolution of 20 nm, which was rarely achieved. Present day instruments are capable of achieving a resolution of 2 nm which places them in the same resolving category as a TEM, and it is necessary when making replicas to bear this in mind.

The nature of the material to be replicated often dictates to some degree the type of moulding material to be used. Tough materials such as bone, flint and ceramic, can frequently withstand harsher treatment than skin, wood or plant remains.

For tough, resilient materials cellulose acetate replica peels can be made simply by applying a strip or larger piece of cellulose acetate sheet, softened in acetone, to the surface to be replicated. When the acetone has evaporated, the cellulose acetate is peeled away from the specimen, and sputter coated and may then be examined in the SEM as a negative impression. High resolution images can be obtained using this method (Figs. 17 and 18) (Hill, 1986). Alternatively, a positive may be made using a silicone rubber compound such as Dow Corning 9161. Resolution to the order of 100 nm can be achieved with this material, and it appears to have a good dimensional stability. D'Errico et al. (1984), published some excellent examples of surface features on worked bone obtained using a nitrocellulose compound containing a plasticiser, with results very similar to those achieved by the acetate peel method. The advantage of both of these peel methods is that the results may be examined using transmitted light microscopy as well as scanning electron microscopy.

There are a great variety of moulding compounds available at the present time. Many of these are used in dentistry for taking impressions prior to making dentures or crowns, but the degree of replication of detail required for this type of moulding is often not as great as that required for SEM investigations. Some of the dental materials are expensive and the results obtained from them are poor for high resolution work, see Figs. 7 to 16.

Watson and Alvin (1976) encountered some difficulty when using Silflo, a two stage dental silicone rubber compound for reproducing details of silicified plants. Work in our laboratory with Silflo has shown that it is not suitable for high resolution moulds as it does not reproduce detail well at a microscopical level, it tends to wrinkle, and is very prone to beam damage in the microscope.

Siveter (1982) used Silcoset 105, to reproduce details of the fine ornamentation on ostracods and found that it fulfilled his requirements. We have also used this material with great success on a number of different types of specimens (Figs. 13 and 14), but have found that Dow Corning 9161 (Figs. 15 and 16) gave superior results, particularly when outgassed under partial vacuum before applying to the specimen. Reprosil is another silicone rubber, which gave poor resolution (Figs. 9 and 10). Wacker RTV-ME 622A&B is not suitable for this type of work as it is readily inhibited by the presence of contaminating material such as plasticine, a material commonly used in moulding processes. Xantopren Blau (light body) (Figs. 7 and 8) is a silicone rubber compound which gave average results. M-Polymer Wacker 4352 is not a silicone, but consists of mixed polymerisates embedded into a matrix of polydimethylesiloxanes. It is a fairly rigid material when set but does not have the dimensional stability required for

Captions for Plate 2

All specimens illustrated on this plate are spores or part of a spore of Qasimia, a new genus of marattialean fern from the Permian of Saudi Arabia.

Fig. 7. Xantopren blau. Bar = 6 microns. Print magnification X2500.

Fig. 8. Xantopren blau. Bar = 5 microns. Print magnification X1000.

Fig. 9. Reprosil light body. Bar = 6 microns. Print magnification X2500.

Fig. 10. Reprosil light body. Bar = 5 microns. Print magnification X1000.

Fig. 11. Rubber latex. Bar = 6 microns. Print magnification X2500.

Fig. 12. Rubber latex. Bar = 5 microns. Print magnification X1000.

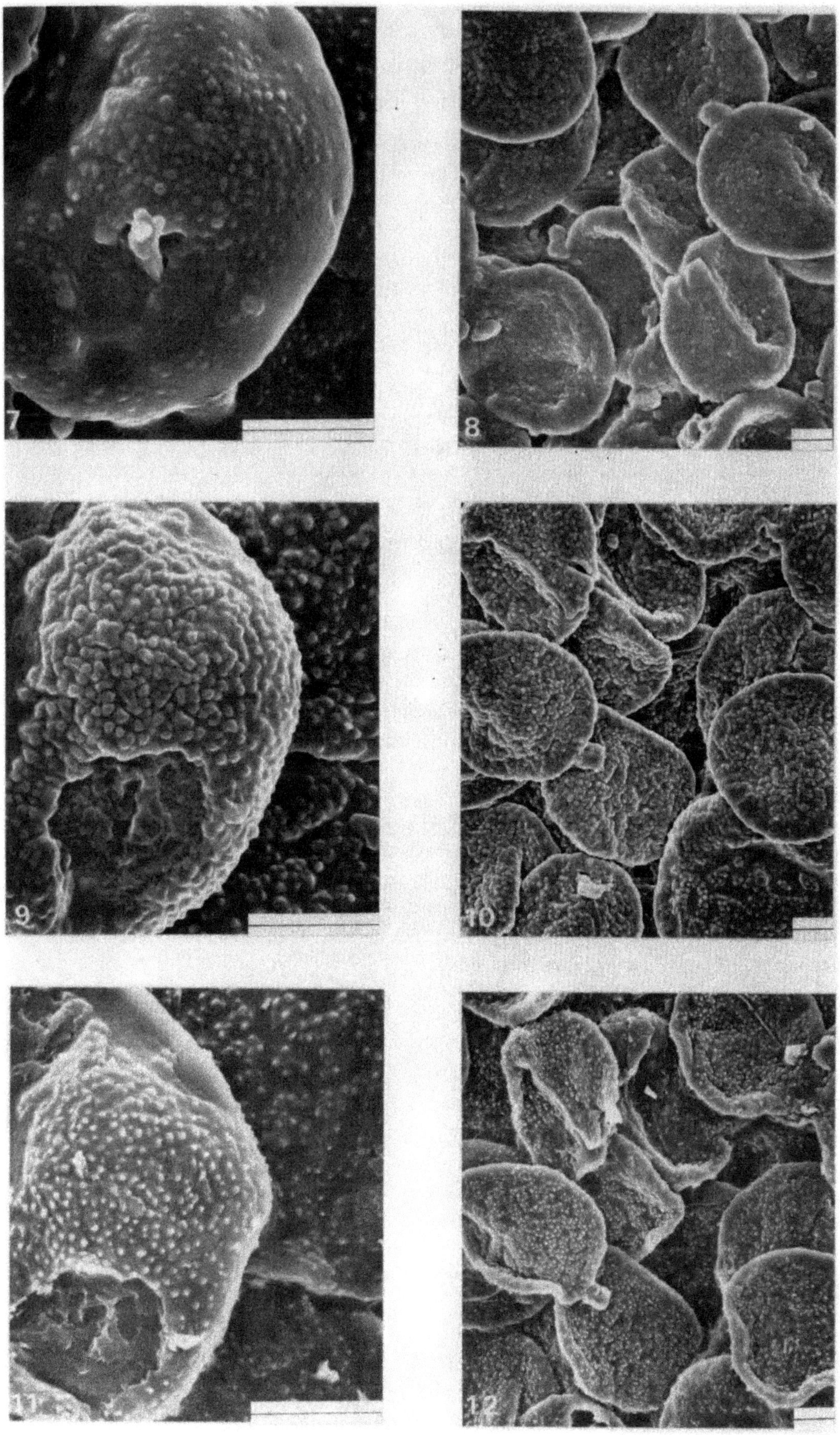

Plate 2.

Captions for Plate 3

All specimens illustrated on this plate are spores or part of a spore of Qasimia, a new genus of marattialean fern from the Permian of Saudi Arabia.

Fig. 13. I.C.I. Silcoset. Bar = 6 microns. Print magnification X2500.

Fig. 14. I.C.I. Silcoset. Bar = 5 microns. Print magnification X1000.

Fig. 15. Dow Corning 9161. Bar = 6 microns. Print magnification X2500.

Fig. 16. Dow Corning 9161. Bar = 5 microns. Print magnification X1000.

Fig. 17. Cellulose acetate peel. Bar = 6 microns. Print magnification X2500.

Fig. 18. Cellulose acetate peel. Bar = 5 microns. Print magnification X1000.

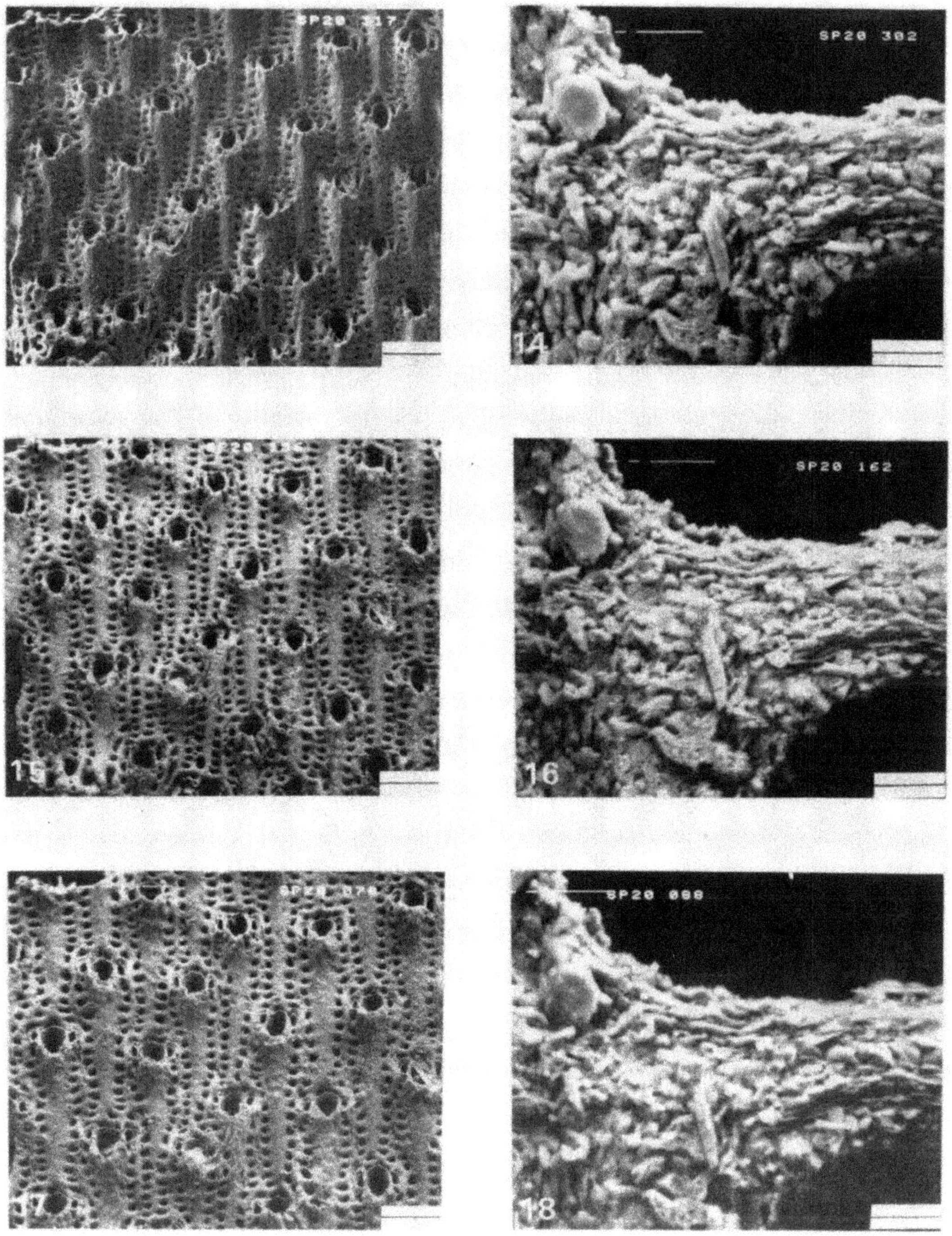

Plate 3.

microscopical work. I would stress that there are literally dozens of proprietary silicone rubber moulding compounds but very few of them are suitable for reproducing microscopical detail. Another commonly available material that has been tried is rubber latex (Figs. 11 and 12) probably the cheapest and easiest of all moulding compounds to use. It is made up in an ammonia based solution and can be applied in very thin layers which can be built up and reinforced. The resulting moulds give poor resolution but are stable in the microscope. When using this type of latex it can be pushed into the specimen with a fine jet of compressed air from a "dust off" can or a compressor.

It is suggested that Dow Corning 9161 be used with catalyst 9162 for moulding materials that are too delicate to be treated by the peel method. The moulding compound can be blown into the surface with compressed air or applied under vacuum and when the material has set it can be peeled off the specimen. Moulds may be fixed to stubs and coated but if this is proposed a small piece of filter paper should be embedded into the back of the rubber before it sets to enable the mould to be attached to the stub, as it is very difficult to stick silicone rubber compounds to metal. Moulds normally have to be coated before they are viewed in the SEM, but if one is using a CFAS apparatus they can be viewed uncoated.

If positive casts are required they may be made with Araldite, however this material frequently charges in the microscope even after coating, and it is advisable to mix a little powdered graphite with it before making the cast to eliminate any charging problems.

Appendix

Robinson backscattered detector is part of the CFAS system but it may be used independently of the system. It is a high gain detector, and it and the Selected Phase Imaging Module system are available from, Expo-Sem Ltd., Moat Farm, Church Road, Milden, Ipswich, IP7 7AF, England.

IMAS: Image and Micro Analysis System is available from Cambridge Technology, Chesterton Mill, French's Road, Cambridge CB4 3NP, England.

Solid State Diode Detector is available from K.E. Developments Ltd., The Mount, Toft, Cambridge, CB3 7RL, England. This particular detector is very good for backscattered work. It is divided into four quadrants and each can be used independently in either polarity allowing sum and difference studies to be made.

Acknowledgements

The author is indebted to, and wishes to place on record, his appreciation of the interest taken in this paper by Dr. C.R. Hill and for his permission to use Figs. 7 to 18. Thanks are also due to Dr. P.D. Taylor for the use of Figs. 1 to 6, and to the photographic staff, and members of the Palaeontological laboratory of the British Museum (Natural History).

References

Andrews, P. and Cook, J. 1985. Natural modification to bones in a temperate setting. Man 20: 675-691.

Bromage, T.G. 1985. Systematic inquiry in tests of negative/positive replica combinations for SEM. Journal of Microscopy 137: 209-216.

Cook, J. 1985. The application of scanning electron microscopy to taphonomic and archaeological problems. In Roe, D.A. (ed.), Studies in the Upper Palaeolithic of Britain and Northwest Europe. Oxford, British Archaeological Reports.

D'Errico, F., Giacobini, G. and Puech, P.F. 1984. Varnish replicas, a new method for studying worked bone surfaces. OSSA 9-11: 29-51.

Hamilton-Waters, P. 1985. A review of the moulding and casting materials and techniques in use at the palaeontology laboratory, British Museum (Natural History). The Conservator 7: 37-43.

Hill, C.R. 1986. The epidermis/cuticle and in situ spores and pollen in fossil plant taxonomy. In Spicer, R.A. and Thomas, B.A. (eds.), Systematic and Taxonomic Approaches in Palaeobotany. Systematics Association Special Volume 31: 123-136.

Pameijer, C.H. 1978. Replica techniques for scanning electron microscopy-a review. Scanning Electron Microscopy part 2: 831-836.

Pfefferkorn, G. and Boyde, A. 1974. Review of replica techniques for scanning electron microscopy. Scanning Electron Microscopy part 1: 75-82.

Siveter, D.J. 1982. Casts illustrating fine ornament of a Silurian ostracod. In Bate, R.H., Robinson, E. and Sheppard, L.M. (eds.), Fossil and Recent Ostracods, pp. 105-122. Chicester: Ellis and Horwood Ltd.

Watson, J. and Alvin, K.L. 1976. Silicone rubber casts of silicified plants from the Cretaceous of Sudan. Palaeontology 19: 641-650.

BACKSCATTERED ELECTRON IMAGING OF ARCHAEOLOGICAL MATERIAL

Nigel Meeks

Introduction

Backscattered electron (BSE) imaging in the scanning electron microscope has gained stature in recent years because of the availability of more efficient detectors that can exploit the compositional or atomic number contrast between closely similar phases in polished samples. The compositional contrast increases and clarifies the information available from a sample and therefore speeds interpretation, as well as deepening the knowledge obtained from examining the material. An additional benefit of BSE is the suppression of specimen charging in the SEM that occurs with secondary electron images, for example on the surface of antiquities due to corrosion and burial debris, and on samples that are difficult to vacuum coat, but charging is generally absent in BSE images. Topographic information is also of importance in the BSE signal coming from the specimen. However, the type of image obtained, whether compositional, topographic or mixed depends not only on the prepared condition of the sample (eg polished, etched, fractured, etc.), but also on the choice of SEM accelerating voltage and probe current, which can be particularly important for ceramic materials. The detector/specimen geometry also has an important effect on the type of image as does the size and design of the BSE detector to some extent. Therefore, detectors mounted with different geometries and working distances give images with different contributions of compositional and topographic contrasts. "Edge bright-up" effects may also be enhanced or suppressed.

The purpose of this paper is to illustrate what has been found to be the best BSE imaging conditions for a variety of archaeological materials.

Detector Geometries

The SEM used for the work in this paper is a JEOL 840 with three purpose-designed backscattered electron detectors, the geometries of which are illustrated in Fig. 1. Detector 'A' is a split solid-state annular device mounted on the final lens and close to the electron beam axis at a working distance of 39 mm from the sample (JEOL, 1980). This detector is technically in the ideal position to give maximum atomic number contrast, being located within a 40° cone region above the sample (Fig. 2). In practice, however, it is rather small in surface area for the standard long working distance. Detector 'B' is a similar, but larger diameter solid-state device mounted on the retractable optical microscope arm. In its normal position it has a working distance of 4 mm above the sample. In practice, this working distance is <u>too</u> close, as the detector elements are outside of the 40° cone region and, as a consequence, the compositional contrast is reduced and "edge bright-up" effects are enhanced. Moving the sample stage down to give 13 mm working distance cures the problem and the detector is then well positioned for good compositional contrast and low noise. In this position the detector has a significant signal to

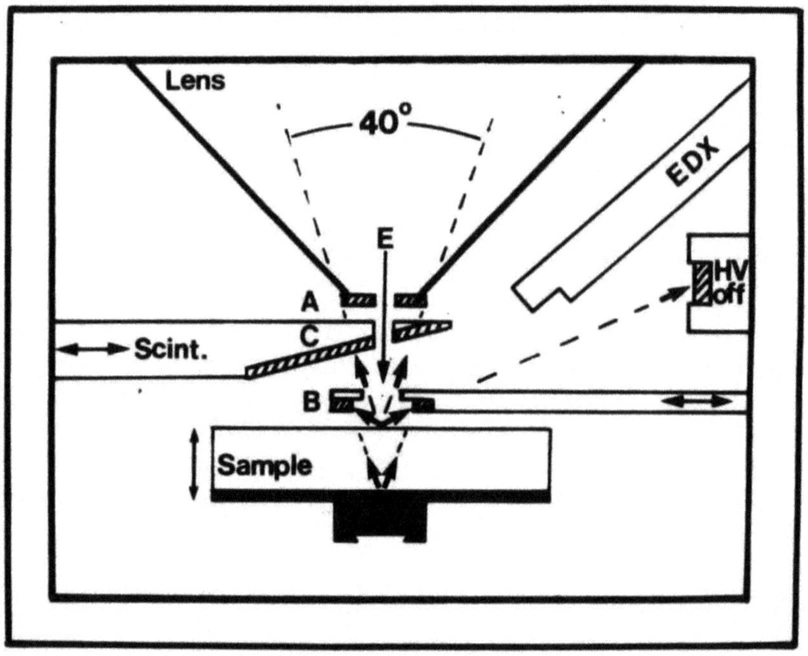

Fig. 1. Diagram of the JEOL 840 backscattered electron detector geometries.

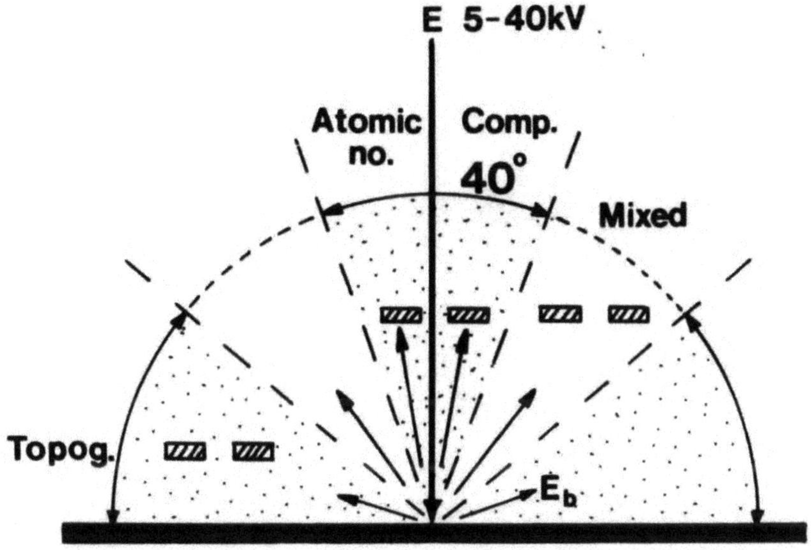

Fig. 2. Diagram illustrating the type of image seen at various BSE take-off angles.

noise and contrast advantage over the small detector 'A'. The third BSE detector, 'C', is a retractable, large surface area scintillator device (Robinson type design). but which for technical reasons has an odd geometry. It is inclined 12° from the horizontal, and because the active area is asymmetrical about the electron beam axis and because of its large size it crosses the boundary between the compositional contrast and the topography regions and collects BSE over a 100° wide cone angle. Consequently, this device always makes a topographic contribution to the compositional image. The detector can be retracted out of the electron beam axis into a position where it is excellent for topographic imaging. This device also works at 'TV' rate with low probe currents in contrast to the solid-state detectors which require excessive probe currents in this mode due to signal loss in the amplifier. However, under normal probe current conditions and at slow-scan rates the solid-state detectors can given excellent compositional images free from topography and "edge bright-up" effects.

Compositional Contrast

There are two fundamental properties of the backscattered electron radiation from the surface of a polished sample. The first property is the dependence of the BSE yield on the mean atomic number of the phases present (Fig. 3), which results in the higher mean atomic number phases appearing as brighter areas of the resulting image (Goldstein, 1981; Reed, 1975). Differences in atomic number of about 0.1 are claimed to be observable (White, 1984), and our system has been able to image a difference of 0.9 between two intermetallic compounds of copper/tin. The contrast between these two phases was difficult to observe in the unetched condition due to much higher contrast from adjacent areas of the metal sample (Hall and Lloyd, 1981). However, the chemical compositions of these two compounds were very easily distinguished by EDX analysis (ε phase 38.2% Sn and δ phase 32.6% Sn), and they were also distinguished by colour contrast in the optical microscope. This raises the need for a cautionary note that compositional imaging will not always clearly show phase differences that are present, and in practice they may be more obvious by optical microscopy and EDX analysis. However, etching does reveal phase boundaries that can be easily seen by topographic imaging in the SEM.

The second fundamental property of the backscattered electron radiation from the surface of a polished sample is that the contributions of compositional contrast and topographic contrast in the signal vary with the BSE take-off angle. This important property is often overlooked in publications but it plays an important role in the resulting image. The compositional contrast increases to a maximum as the take-off angle approaches the vertical at which point the topographic contrast is very low, and conversely the topographic contrast increases at low take-off angles where the compositional contrast is very low. The distribution of the backscattered electrons from the sample and hence the type of image seen at various detector geometries is illustrated in Fig. 2.

A dramatic demonstration of the change in compositional contrast and topographic contrast with BSE take-off angle is shown in Figs. 4 and 5. The glass spherule from the combustion of a ^{14}C charcoal sample

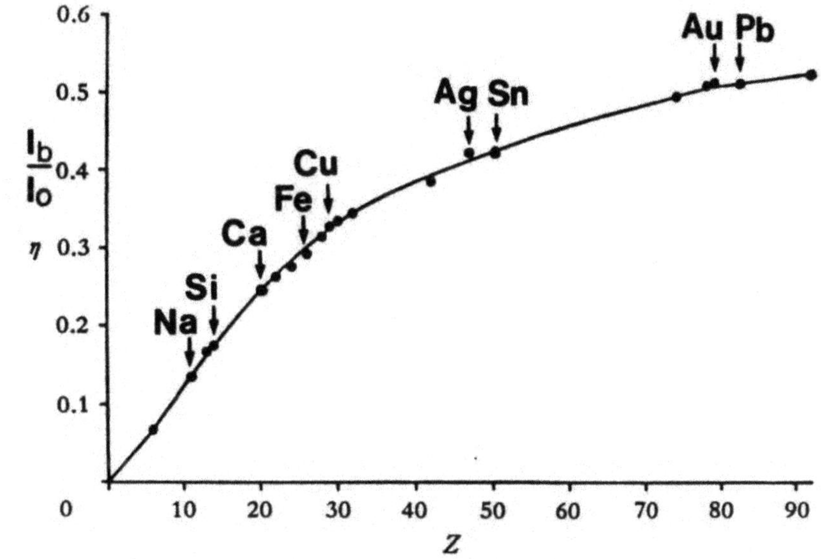

Dependence of backscattering coefficient η on atomic number Z for E_0 = 10–30 keV (after Bishop, 1966*a*).

Fig. 3. Diagram illustrating the relationship of the backscattered electron yield to the atomic number of the elements.

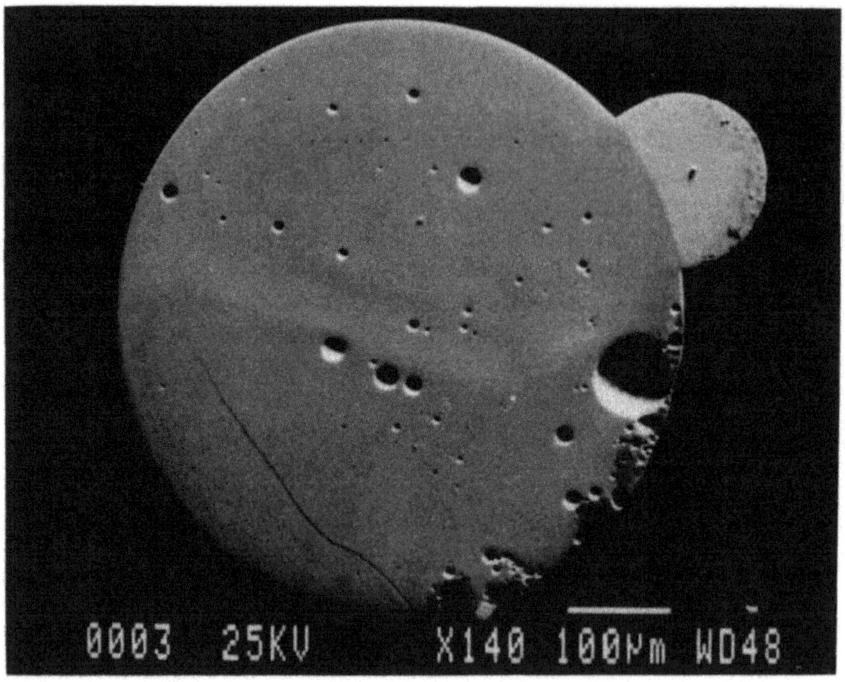

Fig. 4. Glass spherule showing the topographic BSE image taken with the retracted scintillator detector (C).

Fig. 5. Glass spherule showing the compositional BSE image taken with the solid-state detector (B) directly above the sample.

shown in Fig. 4 was taken with the retracted scintillator detector 'C' (shown in Fig. 1) and the image was dominated by topographic information, but a little compositional contrast was also seen. For comparison, Fig. 5 was taken with the solid-state detector 'B' directly above the same sample at 13 mm working distance and clearly shows the compositional variation of the calcareous convection lines in the once molten glass. Therefore, to obtain maximum atomic number contrast from a sample, the detector must be placed horizontally and symmetrically above the sample and close to the electron optical axis. It should certainly be within a cone of about 40° from the point of focus on the sample, which is found to be the region of maximum compositional contrast (Fig. 2). A detector (such as 'A') mounted in this position will effectively flatten the image and give a 'clinical' view of the compositional contrast free from "edge bright-up" effects and topography. Conversely, outside of this 40° cone region the subtle compositional contrast effects reduce due to increased BSE interactions in the sample that occur at lower take-off angles. However, these low-angle BSE give good topography and edge contrast due to the line-of-sight trajectories. For high contrast and low noise composition images, the detector should be reasonably close to the sample, provided that the detector elements remain within the 40° cone region. However, if the detector is <u>too</u> close and the detector elements are outside of the 40° cone region their compositional contrast reduces and "edge bright-up" effects dominate. This is particularly noticeable on polished ceramics which have microtopography and porosity, yet for deep-etched metal samples this effect can be unexpectedly advantageous. The ceramic sample shown in Fig. 6 which demonstrates the "edge bright-up" effect and loss of compositional contrast was taken with detector 'B' at 4 mm working distance, and was therefore too close to the sample. The same area is shown in Fig. 7 but taken at 13 mm working distance with the same detector. This dramatically shows the much increased compositional information that the sample has to offer when the correct distance is used and "edge bright-up" effects are eliminated.

The SEM accelerating voltage has an effect on the image due to beam penetration, although the backscattering coefficient does not significantly change. For compositional contrast, metal samples normally benefit from higher kV settings, eg 20 kV upwards, while ceramics and glassy material benefit from lower values in the region of 15 kV which reduces beam penetration and "edge effects" that would otherwise reduce the compositional contrast. Lower values down to 5 kV can be used to show subtle phase contrasts in some ceramics, but resolution begins to fall. In metals, crystallographic grain orientation or electron channelling contrast can also be seen more clearly at reduced kV (between 8-15 kV), again reducing beam penetration and allowing only the surface crystals to affect the backscattered (and absorbed) electron yield (Joy, 1973; Newbury, 1974; Wells, 1974; Schur, 1974).

Topographic Contrast

Topographic imaging is achieved in four ways. In the first the split annular detector ('A' or 'B' in Fig. 1) mounted directly above the sample is switched to the topography mode (ie, the signals from the two

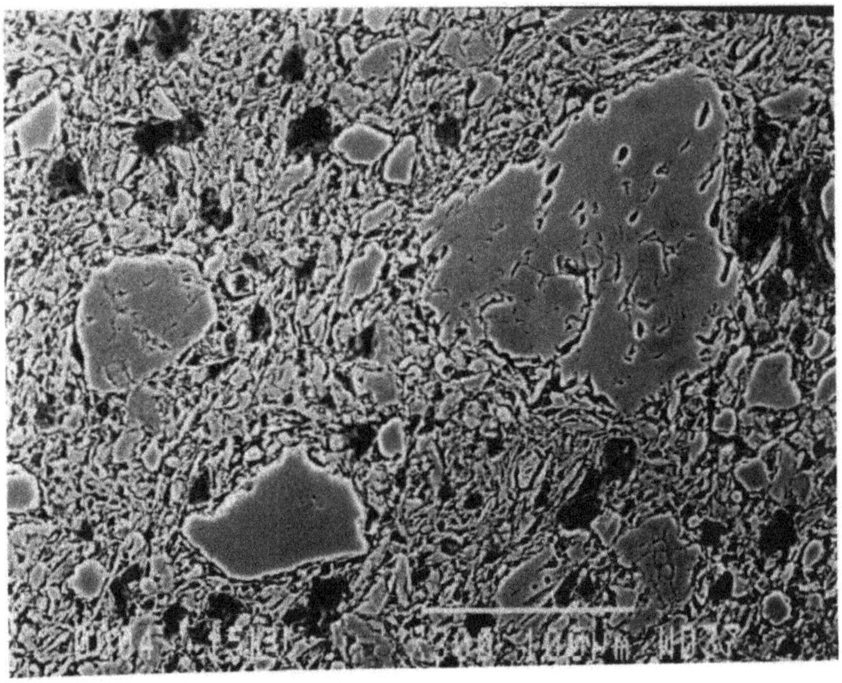

Fig. 6. Ceramic sample showing 'edge bright-up' and loss of compositional contrast when the BSE detector (B) is too close to the sample.

Fig. 7. The same ceramic sample as Fig. 6 but showing detailed compositional contrast when the BSE detector (B) is at the correct working distance.

detector elements are subtracted) but this gives a rather artificial type of image that may also lose important detail. Secondly, it is much better to move the detector ('B' or 'C') off-axis, out of the 40° cone region and use it in the composition mode (ie, summation of the two detector element signals) for line-of-sight BSE detection. In this case the contrast is much better and detail is enhanced over the first method. Thirdly, the secondary electron collector high voltage (HV) can be turned off for a similar, but more noisy image than the off-axis BSE detector, and fourthly and more unusually, certain objects or samples seen at low magnification may have a striking three-dimensional image when viewed from directly above in the compositional mode. In this way the sample is seen with a uniform 'illumination', a situation which is particularly true of gold coated samples viewed at a long working distance.

Topographic BSE imaging has rather more applications than is perhaps generally realised, and in practice an increased beam current which reduces noise, often helps to show this potential.

Sample Preparation

Resin mounted polished sections and glass mounted thin sections are prepared in the normal way for optical microscopy; the better the polish the better the results are for both techniques. SEM examination of metals by BSE before etching is an advantage in that the subtleties between phases, diffusion zones and coring are visible, and is essential for EDX analysis in order to ensure that no metals have been dissolved from the surface. Etching can follow compositional BSE imaging and EDX analysis. It is often best if deeper etching than for normal optical microscopy is used for SEM examination in order to enhance grain boundary detail. Carbon coating is used for all polished samples in resin mounts, because the compositional contrast clearly shows through the carbon coat. Some three dimensional samples for topographic imaging, for example silicone rubber moulds, can be gold coated to increase the BSE yield. Gold coating effectively suppresses atomic number contrast from the sample but enhances topography.

Tapered sections are polished through the surface layers of some archaeological objects if samples cannot be taken. With tapered sections, the object is polished at a low angle to the surface, often on a convex edge. The surfaces of uncleaned corroded objects can be successfully examined by BSE techniques because charging effects are suppressed.

Examples of Archaeological Material Viewed by Backscattered Electrons

To illustrate the range of options available for BSE imaging of archaeological material, examples are given for a variety of materials that have been recently examined.

Metals. An unprepared bronze Frankish strap-end of the 5th century AD is seen at low magnification in the BSE compositional mode in Fig. 8. The surface of the object clearly shows the deep decorative pattern, but in addition displays areas of wear through the light-coloured

Fig. 8. The unprepared surface of a Frankish strap-end shown in the composition mode.

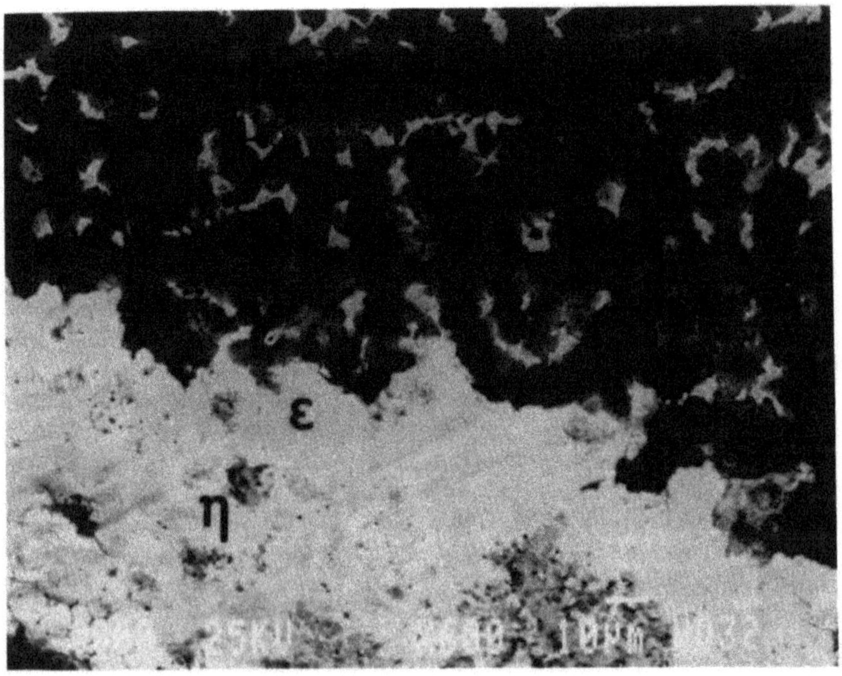

Fig. 9. Detail of the worn surface of the Frankish strap-end showing the dendritic core metal and the two intermetallic compound layers (eta and epsilon) of the tinned surface.

surface layers to the core metal below. These areas of wear form natural tapered sections through the surface layers. Fine dendrites of the cast metal can be seen in these areas. Fig. 9 shows a view at higher magnification, in which the dendrites are seen to be totally corroded, but still retain their original cored structure. The low-tin primary dendrites are dark and the higher-tin outer dendrite layers are light. Additionally, α + δ eutectoid islands are seen in the interdendritic high-tin regions. Thus a classical bronze microstructure is clearly shown on an unprepared object by compositional BSE imaging. The two light surface layers in Fig. 9 (at the edges of the worn tapered section) appear as two distinct grey levels which, when analysed, were found to be the compositions of the intermetallic compounds η and ε. These layers are typical of a bronze that has been tin plated by melting tin onto the surface of the bronze.

Tapered sections of, or samples from metal objects often show great detail in polished, but unetched, section when viewed by compositional contrast. In the forged coin of silver-clad copper alloy shown in Fig. 10, setting the BSE contrast to give a full range of grey levels emphasizes the layered structure of the coin. The resin mount has low BSE yield and is therefore blacked out, while the three-layered structure of the coin is clearly seen. The white outer layer consists of silver overlying a grey intermediate layer and a dark copper alloy body metal. The body metal has light spots showing the presence of small lead globules, and black corroded areas and residual coring are also visible. A second micrograph at higher magnification (Fig. 11) is shown with expanded contrast that has blacked out both the plastic mount and the copper core. This exaggerates the contrast between the two surface layers and shows the light silver surface layer clearly bonded to the silver/copper eutectic solder region which joins the silver to the copper core.

Etched metals can show their structure very clearly but the image type depends on where the detector is placed. The deep etched high-tin bronze shown in Fig. 12 is taken in the composition mode with the annular split detector ('A') mounted at 39 mm above the sample. The etched structure shows <u>no</u> topography at the grain boundaries, but clearly shows the lighter, high-tin $\alpha + \delta$ eutectoid regions adjacent to the darker, cored, low-tin bronze α phase. Casting porosity is seen as dark areas due to the low BSE yield from these regions, although the large pores do show some detail within. Also note that several tiny globules of lead are distributed throughout the sample and show up as white spots. Compare this with the same area in Fig. 13 which is viewed by the larger diameter annular detector ('B') at 4 mm working distances, with the detector elements well outside of the 40° cone region. In this case the image is very clearly topographic although the detector is switched to the compositional mode yet retains some compositional information. The topography is emphasised by the bright edge effects on the eutectoid adjacent to its own shadow which is cast on the bronze. Note also how the detail is lost in the porosity due to the low angle of the BSE, producing the shadow effect in the holes. Additionally, the lead globules are now invisible due to the "edge bright-up" and reduction of available atomic number contrast. Much detail is visible in the bronze body metal in this image. Figure 14 shows the same sample taken with the large scintillator detector ('C')

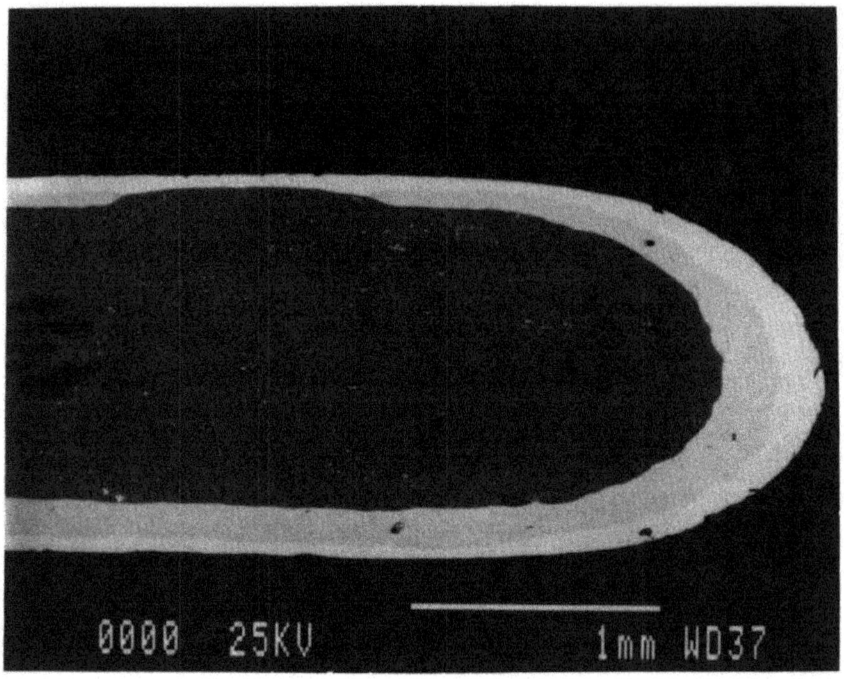

Fig. 10. Forged silver-clad coin showing the dark core metal, the light surface silver layer and the mid-grey fusion zone between.

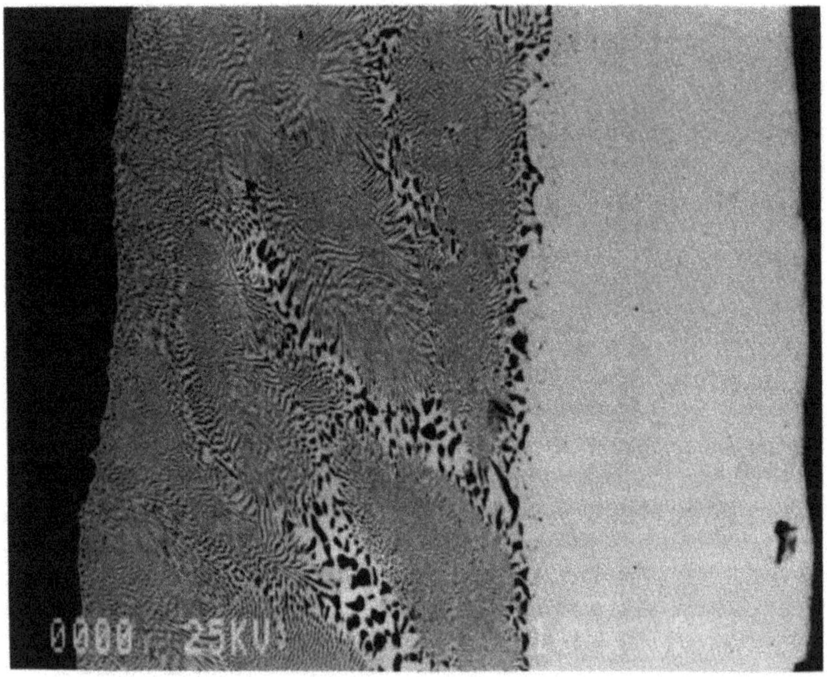

Fig. 11. Detail of the eutectic solder microstructure between the light silver layer and the dark copper alloy core.

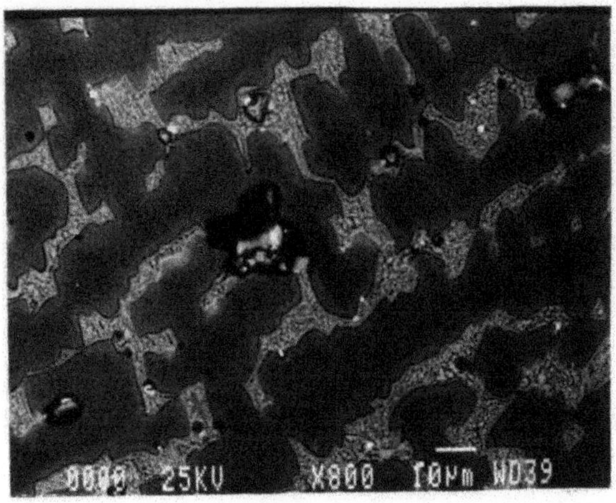

Fig. 12. High-tin bronze in the composition mode at long working distance, detector 'A'. The etched sample appears flat, and tiny lead globules can be seen.

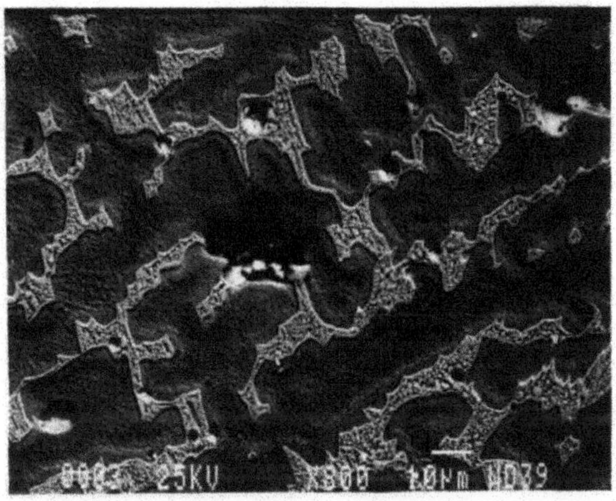

Fig. 13. High-tin bronze in the composition mode at short working distance, detector 'B'. Note the topographic type of image and the fine detail in the etched body metal.

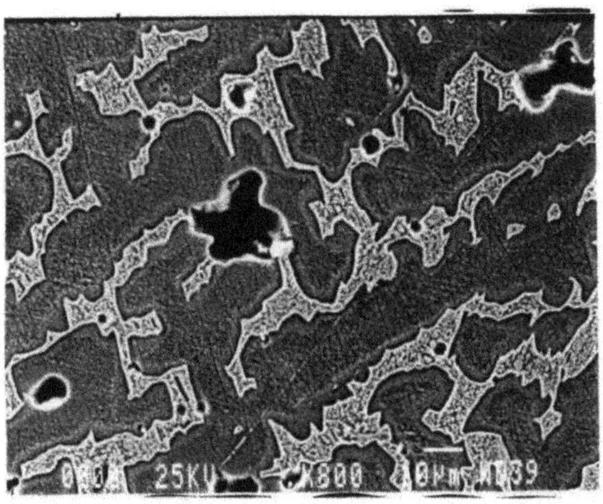

Fig. 14. High-tin bronze taken with the large area scintillator detector in the composition mode. Note the mixed topographic and compositional information in the image.

directly above the sample and is clearly a combination of atomic number contrast and topography.

The combination of atomic number contrast and grain orientation or electron channelling contrast is shown in Fig. 15, taken with detector 'B' at 13 mm working distance. In this polished cross-section of a Bronze-Age spiral gold torc from Shropshire the four arms clearly show the patchwork of large annealed grains. The ends of the arms are burred over and the gold grains are very small, indicating that a lot of cold working was followed by annealing to soften the prepared rod before coiling the torc. This textural detail in the image is purely the effect of grain orientation contrast, as the gold is a homogeneous alloy (Au 80.3%, Ag 16.2%, Cu 3.5%). In two regions between the arms, however, there is a solder infill of the following composition: Au 77.6%, Ag 13.9%, Cu 8.5%. The solder has a lower mean atomic number than the gold torc metal and is therefore darker. The solder region shows a dendritic structure with grain orientation contrast and coring superimposed. Higher magnification of one arm with a soldered region is shown in Fig. 16. The diffusion zone between the gold and solder is visible due to atomic number or compositional contrast, while the grains of the torc appear by means of grain orientation contrast, revealing the growth of grains across the compositional boundary. (A similar image is obtained from the specimen current signal, but this technique is not discussed in the paper.)

Niello of the ternary metal sulphide type has a distinctive microstructure that shows clearly in BSE compositional imaging and can be observed in situ on favourable objects if no sample can be taken. The ternary copper/silver/lead sulphide niello shown in Fig. 17 is an experimentally prepared sample that shows the light dendrites of lead sulphide, the silver/copper sulphide eutectic forming the matrix, and the dark streaks of covellite, CuS.

Ceramics and glass. There is an enormous range of archaeological material composed of silica-based or mineral type matrices that can be very effectively viewed in polished section by compositional contrast. These include ceramics, glass, slag, and refractory materials, faience, Egyptian blue, natural minerals and corrosion products. Vitrification and diffusion zones within these materials can be clearly observed as well as the mineral phases that are fused in the matrices. The BSE techniques often complement transmitted light petrographic analysis of thin-sectioned material, and the same samples are normally examined in the SEM after carbon coating. The matrices are often very complex, but compositional contrast imaging presents the component phases in a clear and detailed manner. Additionally, the 'edge effects' and beam penetration effects are eliminated, along with the micro-topography that often accompanies these polished materials.

It is found that for the best compositional information the sample should be viewed at a relatively long working distance to ensure that the detector is well within the 40° cone region, thereby minimising the 'edge effects' and maximising atomic number contrast. Lower accelerating voltage of around 15 kV limits beam penetration, enhancing resolution. Theoretically, there is more contrast between the phases in this type of matrix at lower accelerating voltages, even as low as 5 kV in some cases. The probe current (ie spot size) should be large

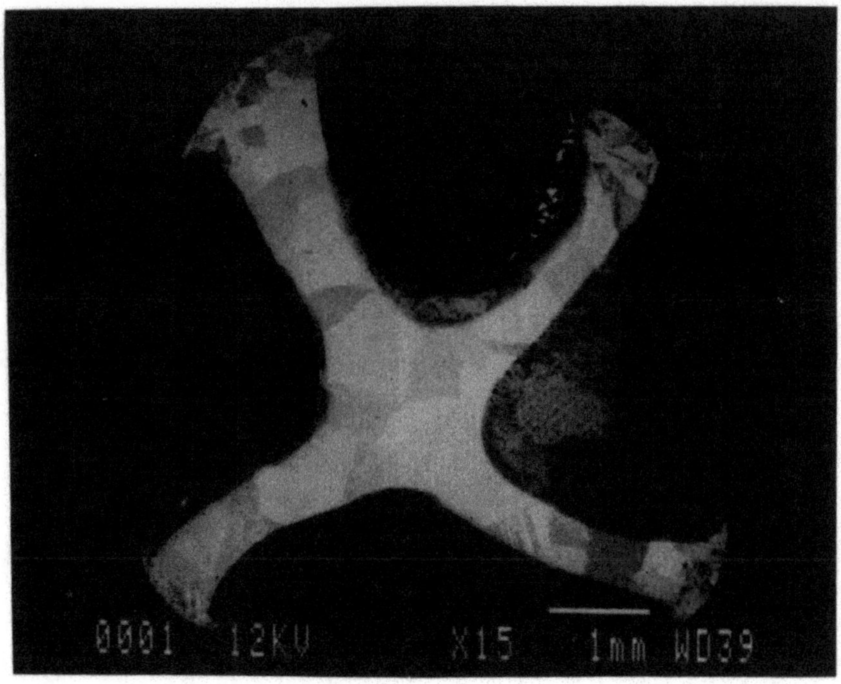

Fig. 15. Grain orientation contrast seen in the gold torc section. Solder infill is seen in two areas between the arms on the right.

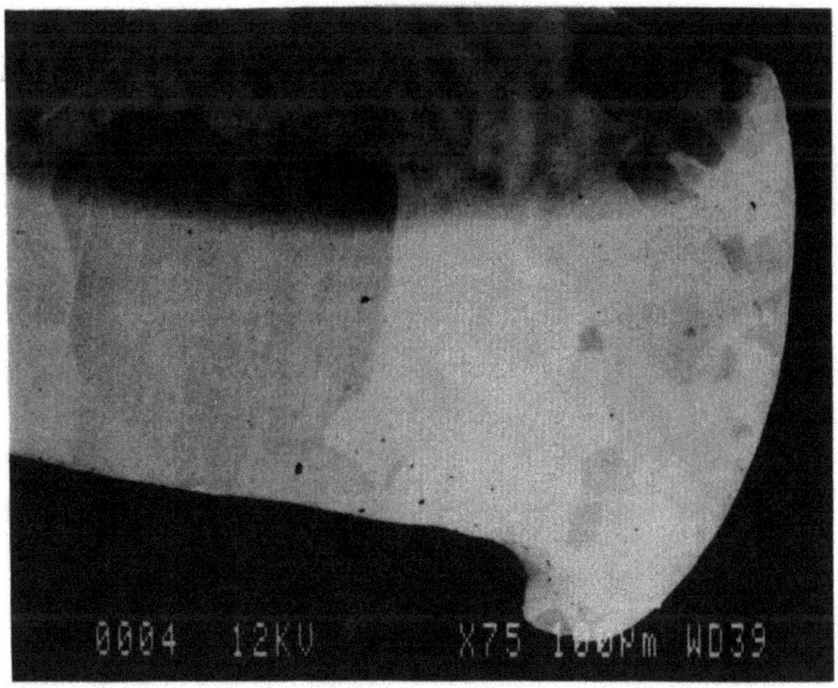

Fig. 16. Detail of one of the torc arms showing the grain orientation contrast which reveals fine grains in the heavily worked end region. Compositional contrast is seen between the solder and gold arm, and as dendritic coring in the solder.

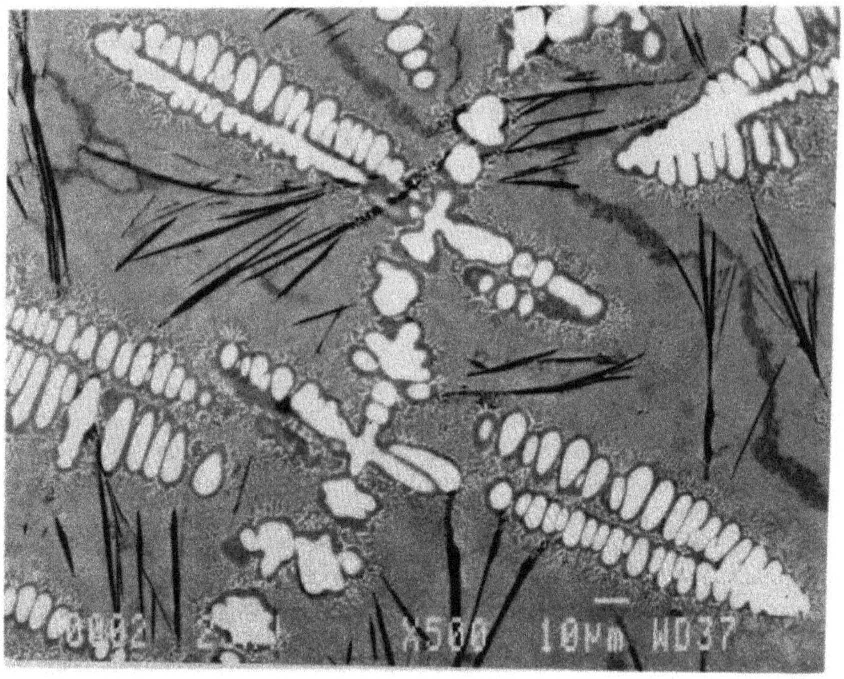

Fig. 17. Ternary sulphide niello. Primary dendrites of lead sulphide are seen in the matrix of eutectic copper/silver sulphide. The black streaks are covellite, CuS.

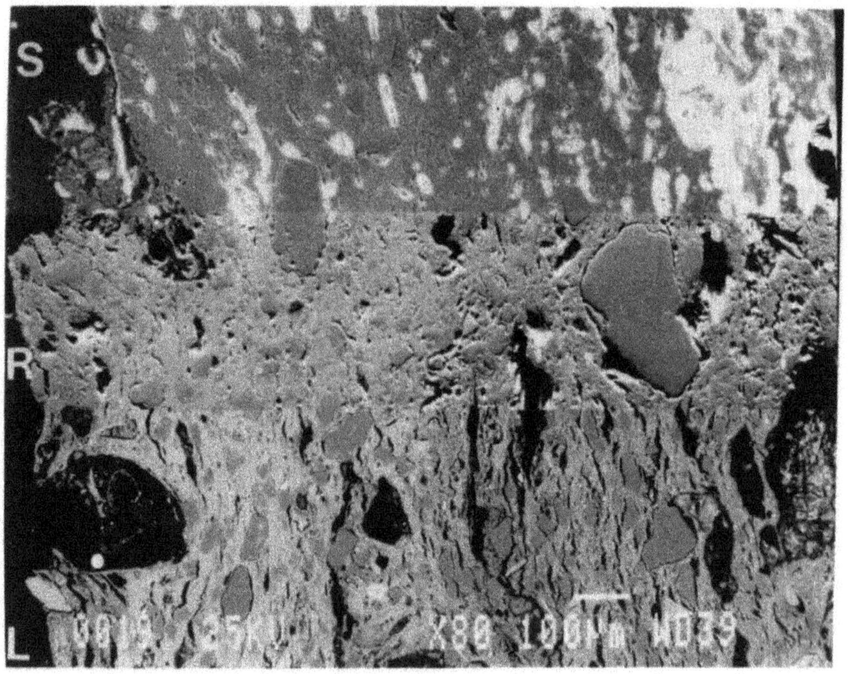

Fig. 18. Ceramic sample shown in split image:-

S = secondary electron image showing gross charging effects

R = scintillator detector (C) BSE image showing mixed composition/topography image

L = solid-state detector (A) BSE image showing the best compositional contrast.

enough to maximise contrast and minimise noise, but not large enough to affect resolution. Under these conditions, clear and detailed images are obtained at both low and high magnification images of the mineral distribution of ceramics and of the vitrification structures across furnace fragments.

Figure 18 shows a typical polished ceramic sample from a well vitrified crucible fragment. The micrograph shows three different images recorded sequentially by switching to different detectors. The top (marked S) is a secondary electron image which demonstrates the problems of low phase contrast and severe charging in the porosity. The centre (marked R) is a backscattered electron image using the large surface area scintillator detector ('C') above the sample. This shows some improvement in atomic number contrast, but topography is a noticeable feature, due to the odd detector geometry discussed above. The bottom (marked L) is taken using the small annular solid-state detector ('A') mounted directly above the sample on the final lens of the SEM. It is ideally placed, being both symmetrically around, and very close to the electron axis. Note both the dramatic increase in the compositional information between the quartz grains and the glassy matrix, and the very fine and subtle information which is seen within the glassy matrix. A round globule of iron sulphide is visible as a small white blob on the left. Also, a subtle gradation is seen in the glassy matrix from slightly lighter on the left to slightly darker on the right due to the presence of a higher K_2O concentration on the left. In the lower image (L) there is no topographic information or 'edge effects', but rather just a clean presentation of the structure.

Figures 19 and 20 are low magnification micrographs of the section through a furnace lining, from Rio Tinto in Spain, taken using the solid-state detector ('B') at 13 mm above the sample to give increased signal and low noise, compared to the smaller detector 'A' at longer working distance. Figure 19 shows the fine structure of the low-fired outer surface of the wall with quartz and crushed slag used as grog. Figure 20 shows the highly vitrified inner surface of the wall of the furnace, diffusion zones of glassy material, and lead-rich glass having been incorporated into the surface layers. This compositional contrast technique is now used by us almost exclusively to image polished sections of ceramics and glassy materials in the SEM.

Other materials. Silicone rubber moulds taken from the surface of antiquities to demonstrate tooling can be effectively imaged by BSE methods. The mould shown in Fig. 21 is taken from a drilled edge on a cast bronze figure on one of the bronze Basse-Yutz flagons in the British Museum and was compared with similar moulds from the second flagon. It is imaged by the retracted scintillator detector 'C' which gives high relief contrast to the gold-coated mould. This is compared to Fig. 22, showing the same sample viewed with the split solid-state detector 'A' above the sample and switched to the topography mode. The former gives a far more 'real' image with much better detail than the latter. For example, the low profile turning lines on the sample stub in the upper left hand corner of Fig. 21 are totally missing in Fig. 22. A secondary electron image by comparison also shows lack of detail in the low-profile scored lines on the mould.

Fig. 19. Furnace wall showing very fine detail compositional contrast of the low-fired ceramic, at low magnification.

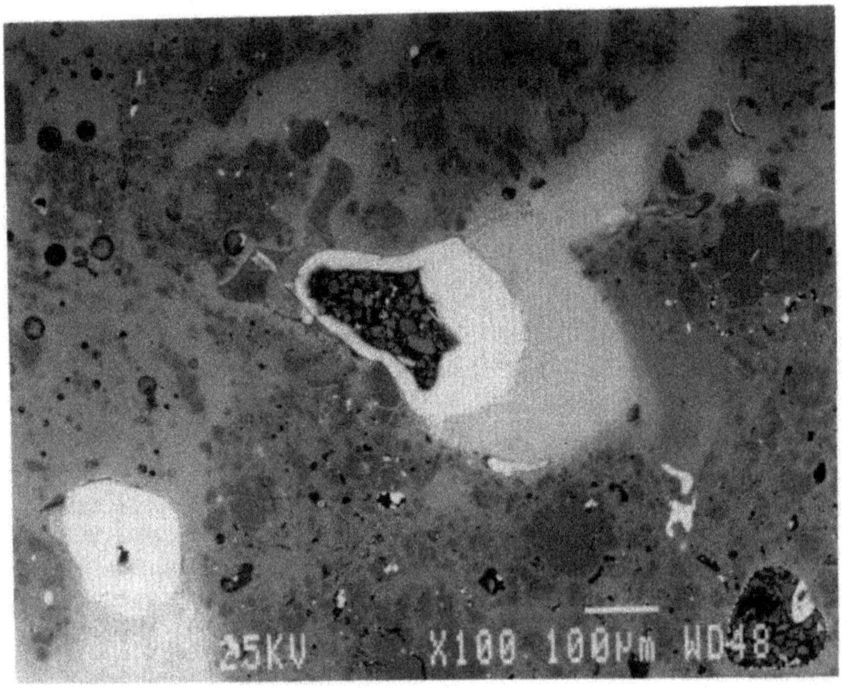

Fig. 20. High-fired region of the furnace wall showing diffusion zones in the glassy matrix and intrusions of lead-rich glass.

Fig. 21. Topographic image of a gold coated silicone rubber mould taken with the retracted scintillator detector, showing high relief contrast with good detail.

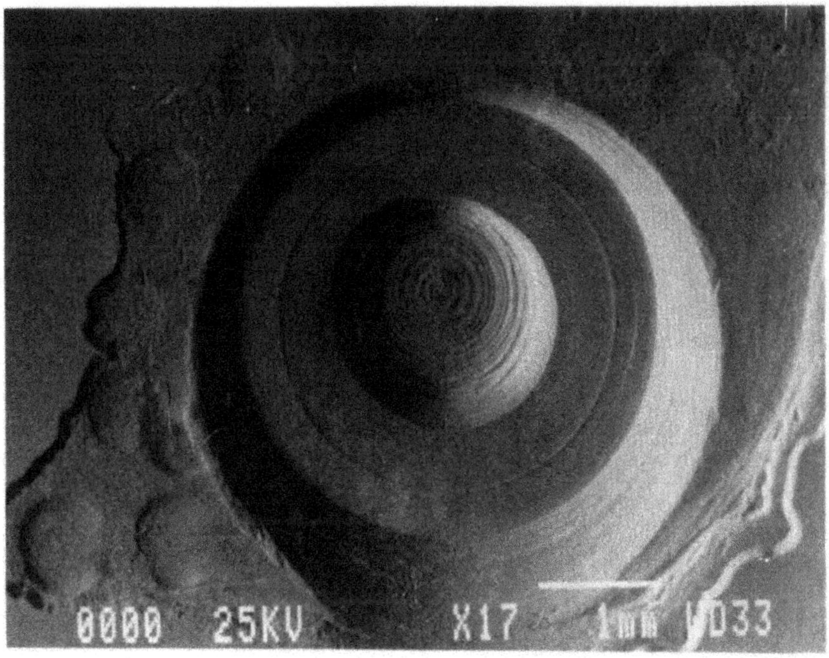

Fig. 22. Topographic image of the silicone rubber mould taken with the solid-state detector above the sample. Note the loss of fine detail.

Charcoal or wood sections can sometimes be difficult to examine if specimen charging occurs within the vessels. BSE imaging from directly above the sample in the composition mode suppresses charging and effectively blacks out the vessel pores, leaving a clear and detailed section. This gives a 'drawing' quality to the section and emphasizes the vessel size distribution that forms the sequence of spring and summer growth and winter dormancy. Figure 23 shows a fracture section through a piece of hearth charcoal from a ^{14}C sample. The roughness of the carbon-coated fracture section is beneficially flattened in this composition mode image. The negative form of the image may be preferred and is shown in Fig. 24.

Conclusions

Atomic number or composition contrast imaging is now a routine practice for examining polished sections of all types of material. Indeed, it is often the sole technique used. The basic rules regarding the geometry of the detector with respect to the sample, BSE yield with respect to probe current, and the effects of the accelerating voltage on the resulting image of the sample are straightforward. Bearing these rules in mind, the SEM may be in a less than ideal condition for BSE imaging with the fixed parameters dictated by EDX analysis. Therefore, for best results, BSE imaging needs to be decoupled from microanalysis, although compromise conditions can be satisfactory, particularly when the type of material under examination is known. With unknown samples it is best to initially use larger probe currents (than for EDX analysis), a long working distance and, when possible, fast scan rates for general survey work to make the compositional contrasts more obvious. This is where 'TV' rate detectors have a real advantage over 'slow-scan only' detectors.

BSE topographic imaging is perhaps less generally used in archaeology than the secondary electron mode, but much useful information can be obtained by this means. From our viewpoint it is secondary electron imaging, the traditional staple diet of SEMs, that is now out of fashion.

One of the remarkable things about the compositional contrast technique is that it is so complementary to optical microscopy. Apart from carbon coating for the SEM, sample preparation is essentially the same for both types of examination. The high quality polished specimens required for light microscopy are ideal for compositional imaging in the SEM. Use of the compositional contrast technique has revolutionised our capability for studying archaeological material.

Acknowledgements

The following Research Laboratory staff are thanked for their contributions to this paper: Dr. M.S. Tite for useful comments, Susan La Niece for providing Figures 10 and 11, and Tony Milton for printing the micrographs. The scintillator detector used is manufactured by Cambridge Technology, and the solid-state detectors used are JEOL production-standard devices.

Fig. 23. Fracture section of a charcoal sample viewed with the solid-state detector in the composition mode. Note how the uneven fracture section is apparently flattened. The annual growth rings are clearly seen at low magnification, in the horizontal direction, while the medullary rays are the vertical white lines on the micrograph.

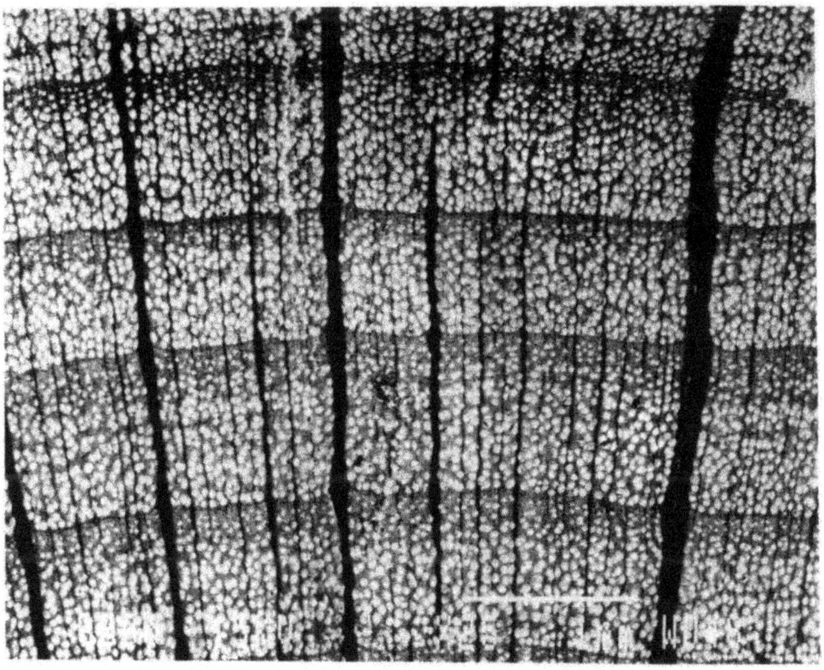

Fig. 24. Negative image of the charcoal sample.

References

Goldstein, J.I. et al. 1981. Scanning Electron Microscopy and X-Ray Microanalysis. London: Plenum Press.

Hall, M.G. and Lloyd, G.E. 1981. The SEM examination of geological samples with a semiconductor backscattered electron detector. American Mineralogist 66: 362-368.

JEOL, 1980. Backscattered electron image and electron optics control system. JEOL News 18e: 1: 15-18.

Joy, D.C. 1973. Crystallographic effects from imperfect crystals. In Scanning Electron Microscopy: Systems and Applications. The Institute of Physics Conference Series 18: 288-293.

Newbury, D.E. 1974. The origin, detection and uses of electron channelling contrast. In Scanning Electron Microscopy, Proceedings of the 7th Annual SEM Symposium, pp. 1048-1057. IIT Research Institute, Chicago.

Reed, S.J.B. 1975. Electron Microprobe Analysis. Cambridge: Cambridge University Press.

Schur, K., Blaschke, R. and Pfefferkorn, G. 1974. Improved conditions for backscattered electron SEM-micrographs of polished sections using a modified scintillator detector. In Scanning Electron Microscopy, Proceedings of the 7th Annual SEM Symposium, pp. 1004-1010. IIT Research Institute, Chicago.

Wells, O.C. 1974. Scanning Electron Microscopy. New York: McGraw Hill.

White, S.H., Shaw, H.F. and Huggett, J.M. 1984. The use of backscattered electron imaging for the petrographic study of sandstones and shales. Journal of Sedimentary Petrology 54: 2: 487-494.

PART II:

ARCHAEOMETALLURGY

SEM AND MICROPROBE ANALYSIS: COMPLEMENTARY METHODS OF INVESTIGATION

David A. Scott

Introduction

Scanning electron microscopy, especially with energy dispersive analysis facilities, represents one of the most useful techniques for the visual examination of ancient metallic specimens. At the present time, some scanning electron microscopes are available with sample chambers sufficiently large to enable complete objects such as hand guns or fragments of very large objects to be accommodated. For carrying out completely non-destructive analysis of the unprepared surface of metallic artefacts, this facility is extremely useful.

The kind of samples, on the other hand, which can usefully be examined by electron probe microanalysis, or electron microprobe analysis (EPMA) as it is otherwise known, are more limited in scope and size. The ideal sample for EPMA study is a polished section which can be coated with a thin film of carbon. Carbon coating is usually necessary in EPMA analysis because high accelerating voltages are used in order to obtain sufficient x-ray emission to give good analytical data. The accelerating voltages may be typically 20kv or 30kv. The function of the two methods of analysis are essentially different: the SEM is primarily designed as an imaging machine with analytical facilities of secondary importance whilst with EPMA, analysis takes priority over imaging. With modern EPMA machines, such as the JEOL Superprobe 733, the quality of the image obtained may, nonetheless, be extremely good, although the degree of specimen movement in the chamber is usually more restricted than that which the SEM permits. Older EPMA machines only allow the polished sample to be moved in its position relative to the electron beam backwards and forwards or up and down whilst retaining the same geometrical position as far as the beam is concerned. Limited sample size and movement, combined with an emphasis on analysis has meant that the electron microprobe has not received particularly widespread attention. This is compounded by the price of the instrument which, at 1986 prices, may cost $400,000. EPMA has found most favour with geologists and metallurgists and is less suitable for organic materials which may be damaged by the beam, even if carbon coated.

A wide range of ancient materials may however be examined by EPMA, such as ceramics, slag, metallurgical waste products, rock and mineral samples, corrosion products, core materials as well as metallic artefacts themselves. Investigations to be followed up by EPMA analysis often start as thin sections, or as polished sections for reflected light optical microscopy. Having examined the material carefully in reflected light microscopy mode, analyses of areas of interest can be obtained by EPMA. There are usually the following facilities on the EPMA instrument:

(1) energy dispersive analysis.
(2) crystal spectrometers for wavelength analysis.
(3) line scanning across the specimen surface using either wavelength or energy dispersive modes.

(4) elemental mapping of selected areas of the specimen using either wavelength or energy dispersive modes.
(5) analysis programmes which provide print-outs in terms of oxides based on fixed ratios between the elements sought and oxygen.
(6) storage of elemental standard data with reference to cobalt.
(7) back-scattered and secondary electron imaging: in more recent machines this may be quite satisfactory for obtaining good images, although the SEM is generally superior.

Elements below sodium in the periodic table are generally difficult to analyse by EPMA, but using crystal spectrometers even oxygen may now be analysed with some confidence.

It is standard for many metallic elements to be profiled and the information stored for future analyses by reference to a cobalt standard which is carbon coated at the same time as the specimens for study are coated. Without coating, even very conductive samples may build up a surface charge, especially when they are mounted and polished in a resin mount, as most archaeological samples are. A dab or streak of silver paint touching both the specimen holder and specimen may also be necessary, although some care must be taken that this does not mask off areas of interest in the sample. The EPMA is equipped with good optical microscopes which allows visual contact with the coated sample. This is most necessary, particularly where selected areas of inclusions, corrosion products, or surface features are to be examined. Carbon coating affects the visual contrast between areas of the specimen which may be of different composition, so it is therefore important that the carbon coating is not excessively thick. It is also important that the area selected by optical examination is centred exactly for the position of the electron beam, otherwise when the sample is presented to the beam the area analysed is not the same as that selected optically. Spot analyses may now be computer driven, as may be the wavelength spectrometers which were formerly rather time-consuming to use.

Neither the SEM or EPMA provide the perfect solution to the problems of inclusion characterisation in archaeological material. If the inclusion or feature is only some 2-3 microns across then it is very difficult to exclude matrix excitation as well. An example of this problem is seen when attempting to identify the precipitation of AgAu rich phases in gold-copper-silver alloys which are just beginning to show signs of two phase composition. The shape and size of thin plates or particles, especially when they differ only slightly from the matrix composition make them very difficult to analyse. Larger inclusions, measuring perhaps 10 microns across present less trouble. As an example, let us look at the nature of the mineral inclusions in a placer gold nugget from the area of the Rio de la Miel, a western tributary of the Rio Magdalena on the Caldas-Antioquia border, Colombia.

Analysis of the gold matrix by EPMA gave the following data: Au: 85.1% Cu: nd Ag: 11.7 Fe: nd Pt: nd Total: 96.8%. With the EPMA, spot and line scans for gold and silver can be carried out to see if any elemental variation occurs within the gold matrix; here the placer nugget was fairly uniform in composition. The mineral inclusions within the gold were found to be rich in lanthanum, cerium, neodymium, samarium, gadolinium and phosphorus. The inclusions are therefore monazite, and a special analysis programme operating at 30kv on the EPMA could be used to obtain quantitative information. The print-out from this analysis is given in Table 1.

The EPMA is less useful for the study of inclusions which are associated with casting remains, such as core material. This is because the material itself may contain organic components, such as carbonised plant material which is not well suited to EPMA study. An example of material of this kind is given in Figure 1. The SEM secondary electron images are from a fragment of core material composed of clay and charcoal. The object itself was a debased gold alloy cast by the lost-wax process: a Tairona necklace piece from the Sierra Nevada de Santa Marta in the North of Colombia. The SEM reveals interesting detail of the core composition; in Figures 1a and b can be seen a fragment of a woody cell structure containing recognisable remains of vestured pits, typical of tropical hardwoods, although without reference material identification cannot be taken any further at present. Microchemical analysis, employed here to determine the percentage of carbon present in the material, gave a result of 19.66%, fairly typical for some of these lost-wax casting cores.

The study of corrosion products and corroded metallic fragments can be carried out in both the SEM and the EPMA: both methods of examination may be very useful. An example of the EPMA results for elemental mapping of a polished cross-section of a debased gold pan-pipe from the Municipio of Pupiales, Department of Narino, Southern Colombia, can be seen in Figure 2. Corrosion has left some of the grains of the material isolated in a sea of corrosion corresponding to the depressed areas in the BSE image shown in Figure 2a. As a result of corrosion the gold content towards the surface region of this piece has been slightly enhanced; note also the slightly segregated distribution of the silver in this gold-copper-silver alloy (composition about 25.9% gold; 62.4% copper and 4.0% silver by EPMA analysis). A further example of the use of the EPMA for the study of corroded metallic materials is given in Figures 3 and 4. Figure 3 shows a sketch of a cross-section of a bronze rod from Iran (Scott, 1985). Figure 4 shows the copper and tin distributions as they vary across the corroded rod. The information revealed by line scans of this kind, slowly moving the specimen across the path of the beam, can be extremely useful in attempting to investigate surface composition of gilded or silvered or tinned surfaces, or for investigation of corrosion effects. In Figure 4 can be seen a number of different zones of structure of the rod which can be described briefly as follows:

Zone 1: bronze metal grains containing about 9.5% tin grading into a region of corroded metallic grain remnants.
Zone 2: a region of massive tin precipitation.
Zone 3: compact cuprite layers interspersed with some tin precipitation.
Zone 4: the original surface of the rod.
Zone 5: the outer corrosion crust which only contains very small traces of tin.

Table 1. Monazite Inclusions in Placer Gold Nugget from Colombia

A typical EPMA print-out for a quantitative analysis carried out on a Cambridge Mark 5 X-Ray Microanalyser at the Dept. of Geology, University College, London. The fit index for the data and the apparent concentration are shown in the first part of the print-out. After ZAF corrections have been applied, and an oxygen content to the monazite assigned, then the quantitative print-out is given below, both for the percentage of the element concerned and for the oxide. Although the figures are quoted to three decimal places, results for analyses carried out by EPMA are really only accurate to one decimal place.

FIT INDEX = 2.86

ELMT	APP. CONC.	ERROR (WT. %)	
LA	7.820	0.190	
CE	22.300	0.256	
PR	2.655	0.256	
ND	14.486	0.199	
SM	2.169	0.157	
EU	0.384	0.176	
GD	0.796	0.138	
DY	0.047	0.114	2 SIGMA
P	12.951	0.056	
CA	0.083	0.024	
FE	0.046	0.054	2 SIGMA
MN	0.032	0.089	2 SIGMA
TI	0.013	0.045	2 SIGMA

NUMBER OF OXYGEN ATOMS = 1 (1 2 3 ZAF)

30.00 KV TILT=0.00 ELEVATION=75.00 AZIMUTH=0.00 COSINE=0.00

SPECTRUM: 1037.A4

LAST ELEMENT BY STOICHIOMETRY

ELMT	ZAF	% ELMT	%OXIDE	FORMULA
LA	0.901	8.691	10.193	0.039
CE	0.945	23.710	27.771	0.107
PR	0.946	2.807	3.285	0.013
ND	0.942	15.419	17.984	0.067
SM	0.924	2.347	2.722	0.010
EU	0.903	0.426	0.493	0.002
GD	0.774	1.029	1.187	0.004
DY	0.865	0.055	0.063	0.000
P	1.041	12.477	28.593	0.254
CA	1.129	0.074	0.103	0.001
FE	0.942	0.049	0.062	0.001
MN	0.942	0.000	0.000	0.000
TI	1.002	0.000	0.000	0.000
O	0.303	25.372	0.000	1.000
TOTAL		92.456	92.456	

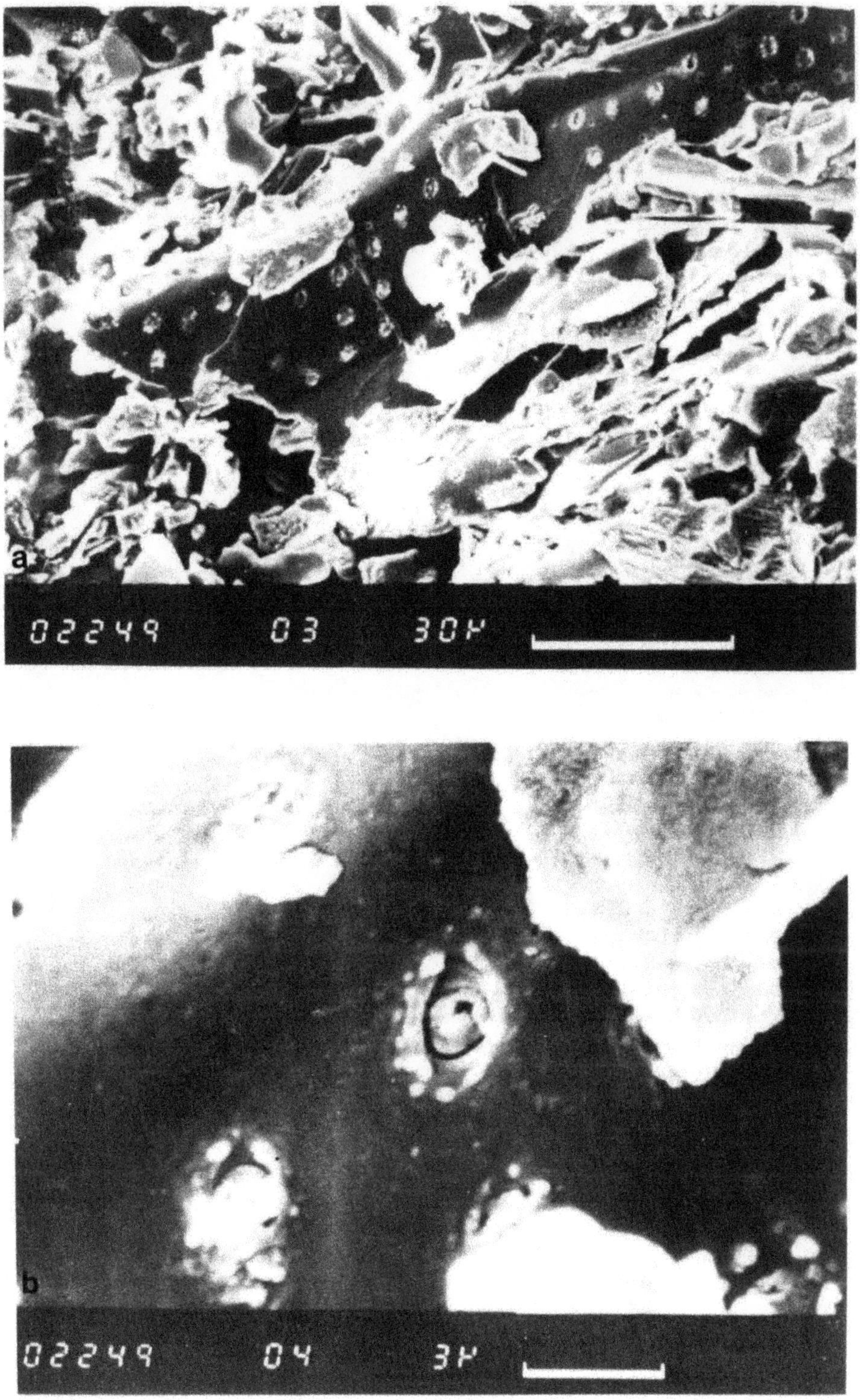

__Fig. 1.__ SEM secondary electron images of a fragment of casting core from a Tairona necklace bead. a. Low magnification view showing extensive remains of carbonised plant material used in the fabrication of the core before the wax was applied for modelling of the shape. Scale = 30 microns. b. High magnification view of the pits evident in the large wood structure of Figure 1a. The pits in the wood here can be seen to be vestured, typical of tropical hardwoods. Studies such as this can provide important clues to manufacturing technique and local preferences in the fabrication of lost wax casting cores. Scale = 30 microns.

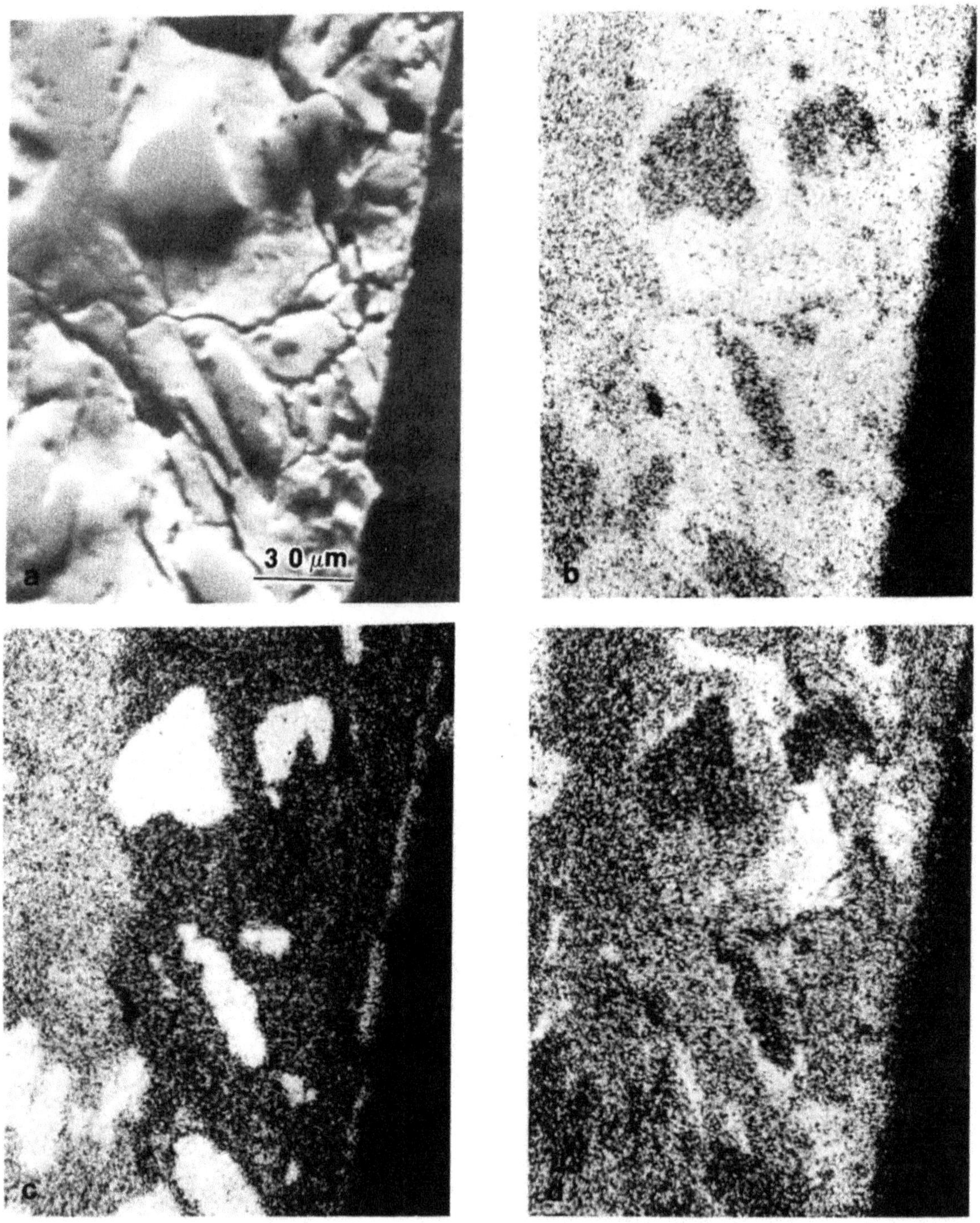

Fig. 2. Elemental maps for gold, copper and silver across part of a debased gold pan-pipe from Nariño, Colombia. a. Back-scattered electron view showing surface topography with sound metal grains occurring as isolated elevated areas. b. Elemental map for gold showing some enrichment towards the outer surface as a result of corrosion. Note the uncorroded metal grains appear less rich in gold, because here copper has not been lost. c. Elemental map for copper, showing thin layer of copper rich corrosion to the outer surface and several well-preserved metal grains in the corroded material which correspond to the elevated regions shown in Figure 3. 2d. Elemental map for silver showing some segregation in the corroded matrix. The matrix itself is now mostly cuprite containing both gold and silver. The silver is corroded here to silver chloride.

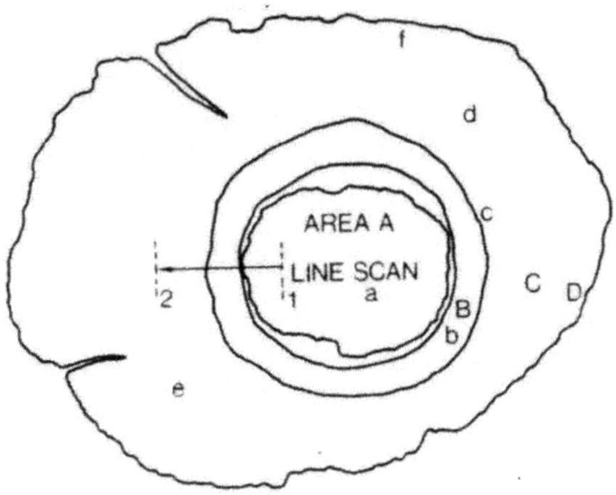

Fig. 3. Sketch of the cross-section of object AI 862, the bronze rod fragment from Iran, showing the location of the electron microprobe line scan for copper and tin: a. bronze metal grains containing 9.5% tin, b. compact layers which are principally cuprite and tin oxide in segregated areas, c. original surface of the object, d. some regions of high Ca content (soil mineral inclusions), e. only trace amounts of tin in the outer corrosion crust, f. outer areas contain some Sn, Cl, Ca, Fe and Cu immediately adjacent to the surface.

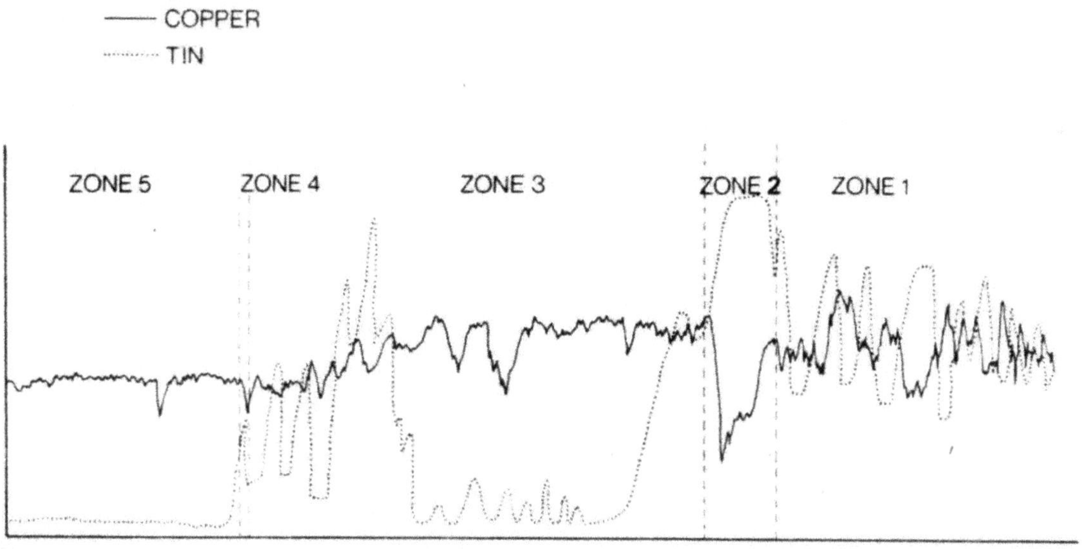

Fig. 4. Sketch showing the principal zones to be found in the cross-section of object AI 862. The nature of the different regions is explained in the text. _______ = copper, - - - - = tin.

The other elements mentioned in Figure 3 were also determined by EPMA analysis using individual spot determinations in different regions. Despite difficulties with spot analyses, these are often very useful for characterisation of the composition of the original metal before corrosion, simply by analysis of uncorroded metallic grains, and it is also the most commonly used technique for the study of rocks and minerals in thin section, as well as archaeometallurgical waste products.

These brief examples show some of the limitations and advantages of EPMA analysis: for further details concerning the technique the reader is referred to the very readable account provided by Long (1977) although it should be borne in mind that the scope and sophistication of the probes currently available are far in advance of that described.

References

Cope, L.H. and Warren, R. 1972. Comparison of EPMA with other methods for determining the bulk composition of ancient coins. In Methods of Chemical and Metallurgical Investigation of Ancient Coinage, pp. 237-249. Royal Numismatic Society: Special Publication No. 8, London.

Hallerman, G. and Picklesimer, M.L. 1969. The influence of the preparation of metal specimens on the precision of electron probe microanalysis. In A.J. Tonsimis and G. Marton (eds.), Electron Probe Microanalysis, pp. 197-225. New York: Academic Press.

Long, J.V.P. 1977. Electron probe microanalysis. In J. Zussman (ed.), Physical Methods in Determinative Mineralogy, 2nd edition. London: Academic Press.

O'Boyle, D. 1973. Metallurgical applications of the microprobe. In Proceedings of the 8th National Conference on Electron Probe Analysis, pp. 411-429. New Orleans: Electron Probe Analysis Society of America.

Tite, M.S. 1972. Methods of Physical Examination in Archaeology. London and New York: Seminar Press.

Scott, D.A. 1985. Periodic corrosion phenomena in bronze antiquities. Studies in Conservation 30: 49-57.

METALWORKING TOOLS AND WORKSHOP PRACTICES: INTERPRETATION OF WORKED METAL SURFACES VIA SILICONE RUBBER MOULDS

Dana E. Goodburn-Brown

Introduction

This paper focuses on the recovery of technological information which is stored on the surfaces of metal artefacts. It is a non-destructive technique of analysis, whereby a silicone rubber mould (Dow Corning Silastic 9161) is taken from the surface of an artefact to be studied. It is this mould, rather than the artefact itself, which is then sectioned and manipulated under the microscope. The magnifications utilized span a range from x20 - x1000 and are dependent on the level of question being asked about the surface texture. The metalworking tools, type of materials being used in the workshop, and even wrought and annealing processes used in the production of an artefact may be examined. The term 'workshop' here relates to the hypothetical surroundings of a metalworker during fabrication processes, and specifically refers to the entire collection of tools and materials involved in the process of manufacture.

Toolmarks

The first level of interpretation is concerned with tool marks; that is, the identification of tool type and ultimately the recognition of individual marks or "signatures". Tool types may be distinguished by the manner in which they are used in manufacture; for example, a SCRIBER is often used as a pencil, its sharp point breaks through the metal surface and leaves a characteristic burr. A TRACER is used with a hammer and depresses metal; whereas a GRAVER is hand-held and actually cuts away metal from its path. A PUNCH also depresses metal-but it is applied with a single blow, whereas a tracer is hammered along in succession in order to create linear decoration.

By examining the characteristics of a toolmark, the type of tool used to create the mark may be determined. This kind of study is best carried out indirectly, via silicone rubber moulds of artefacts' worked surfaces. Aside from the recording of fine detail, the use of a silicone rubber mould in this work has two major advantages. Firstly, it facilitates an extensive technological examination of an artefact, taking place quite removed from the artefact itself; and secondly, it reproduces a raised image of the toolmark, rather than the recessed feature which is observed when examining an artefact first-hand.

The value of technological research carried out on silicone rubber moulds is attested to by the works of Larsen (1985), Shorer (1984), Lowery, Savage and Wilkins (1971). Stated succinctly: "indirect examination, since it presents the bottom and sides of the groove in relief uncamouflaged by colour differences in stain or patina, is almost always more informative, and we would hesitate to attribute many types of line without it" (Lowery et al., 1971:176).

Tool types may usually be distinguished with the use of a hand lens (x10) or an optical microscope; although the greater resolution and depth of field available with the SEM may be required in some circumstances. It is in the search for identifying characteristics of individual tools that the SEM is of great value. Scars or nicks on the face of a tool would leave details which may be identified via silicone rubber moulds. Benner Larsen (1984) speaks of toolmarks as being "fingerprints" or the "signatures" of the individual tools used. In his study of Scandinavian Iron Age metalwork, using the SEM and silicone rubber moulds, he has been able to verify that the same tool was used to decorate different pieces of jewelry recovered from a single grave and has also observed the use of an individual tool on objects recovered from more than one grave.

In a preliminary study which explores some aspects of late Pre-Columbian metalworking techniques, an unusual tracer type has been observed on a decorated gold alloy disk from Southern Peru (Brown, 1984). The tool appears to have a broken or jagged edge, and it is this kind of characteristic detail which may be used to verify the multiple use of a single tool (Fig. 1).

While the recognition of tool types is relatively easy, the identification of an individual tool's signature is complicated by variables in the manufacturing process. Factors such as the angle of application, force of hammer blows, and the sharpening and/or wearing down of a tool may hinder comparative studies. The build-up of surface deposits on a worked piece may also obscure the required level of detail for tool signature recognition. It should be noted that metal artefacts should be carefully examined in order to ascertain the research value, as well as the condition, of the various surfaces, prior to taking any silicone rubber moulds.

Surface Texture: (Imprints)

Another level of technological interpretation concerns itself with the identification of surface imprints which reflect the fabric of other materials used in the workshop. In the preliminary study described above (Brown, 1984), an attempt was made to determine the type of surface (e.g., leather, wood, stone, or metal) upon which such artefacts were worked.

A series of Peruvian gold alloy plume holders and beakers depicting faces with large protruding noses, was examined. It has been suggested that such items were manufactured on wooden formers (Easby, 1966). Silicone rubber moulds were taken from the interior surface of the artefacts and examined with the SEM. Traces of wood grain structure were found to be present.

As the SEM analysis suggests that a wooden former was used in the production of the artefacts, a former was carved and an experimental plume holder was produced. Experimental work serves both to support and augment observations obtained from the objects themselves. The surface texture of the silicone rubber mould taken from the interior of the experimental plume holder shows a characteristic resemblance to

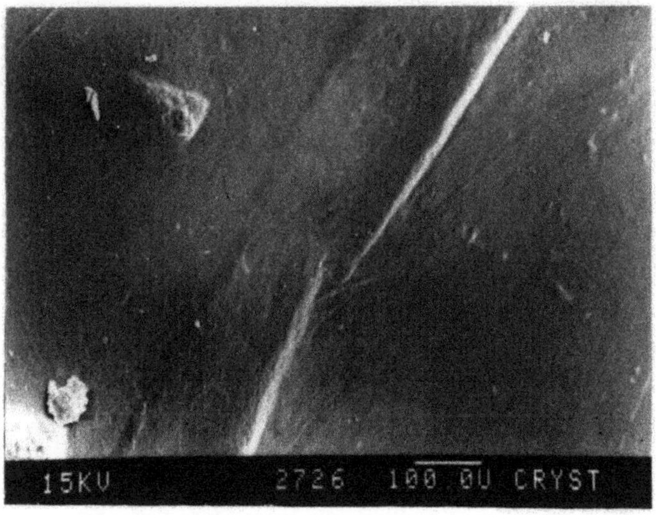

Fig. 1. Silicone rubber mould, taken from a Pre-Columbian gold alloy disk, illustrating toolmarks from a jagged edged tracing tool. (Brown, 1984).

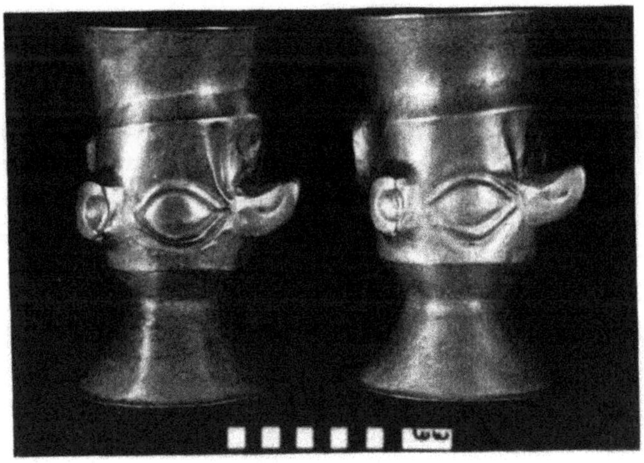

Fig. 2. Pair of Pre-Columbian gold alloy vases (private collection), which have been manufactured using a wooden former (Brown, 1984).

that of the textures on the moulds taken from corresponding features of the plume holders and beakers studied.

Examination under magnifications greater than x50 provides the only reliable means of surface texture analysis; although even here, one must be aware that the surface under observation reflects the entire manufacturing process, as well as corrosion and perhaps cleaning/restoration processes subsequent to fabrication. For example, the metal sheet which was used in this experimental work was 'thinned' through steel rollers before the forming process was begun. When examining the resulting overall texture, care had to be taken not to confuse the striations of wood grain (Fig. 3). with the characteristic striations resulting from modern 'rolled' sheets (Fig. 4).

Microstructure

The third level of interpretation deals with the metal structure by observing the surface which reflects the metalworker's manipulation of the material's physical properties through wrought and annealing processes.

During SEM analysis carried out on silicone rubber moulds of some Peruvian gold alloy artefacts, it was noted that the silicone rubber was capable of reproducing the crystalline structure of an artefact's surface. Preferential corrosion at the grain boundaries had occurred; consequently, the silicone rubber picked up this microcrystalline change in 'texture' (Fig. 5). In response to this observation, an analysis of the etched surface of an experimental axe was undertaken, in collaboration with Peter Shorer, a conservation technologist working in London.

Shorer provided a silicone rubber mould taken from the surface of a cast 5 tin-bronze axe, which he had hammered and annealed 5 times (working from the cutting edge towards the hafting hole). Samples were cut from the silicone rubber mould from four areas, each one depicting variously worked areas of the bronze surface. Sample 1 represents dendritic structures "as cast"; samples 2 and 3 represent the surface at intermediate stages of working; and sample 4 represents the surface at its final work-hardened state.

Each sample (approx. 1 cm sq.) was carbon-coated and examined with the SEM at various magnifications: x100, x150, x500, x1000. The x1000 series is illustrated here (Figs. 6-9). The conclusions drawn from this preliminary examination rely on the assumption that changes in the microstructure from the "as cast" state through the intermediate and final working stages may be detectable at the surface of an artefact, using silicone rubber and the SEM. Further work, with a larger sample of material, may establish the accuracy of interpretations of such manufacturing processes.

In order to better understand the changes observed in the microstructure at the surface, and to enable the possible correlation of such changes with various stages in the working of metal, a more controlled examination is required. A proposed study would involve the casting and working of a bronze bar; silicone rubber moulds would be

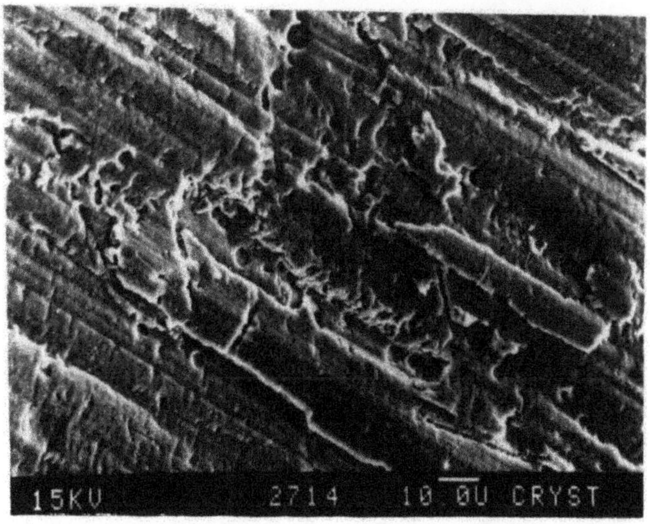

Fig. 3. Wood grain structure evident on silicone rubber mould taken from interior surface of projecting nose of Pre-Columbian vase, see Fig. 2 (Brown, 1984).

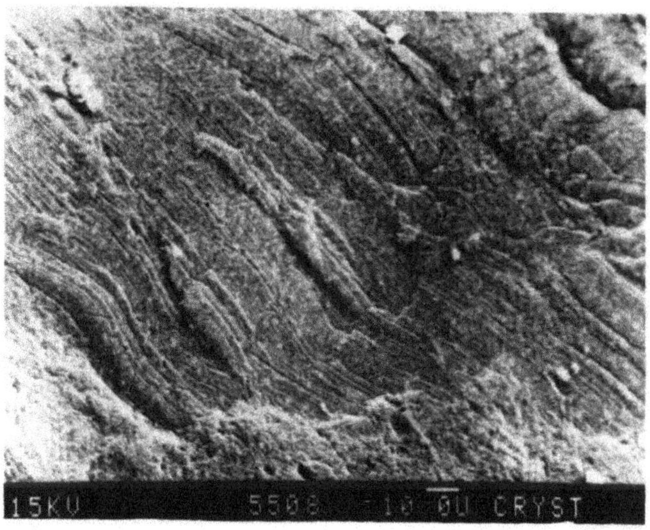

Fig. 4. Silicone rubber mould taken from copper sheet used in experimental work. Parallel striations, (raised features in this case), reflect the action of thinning the sheet through steel rollers (Brown, 1984).

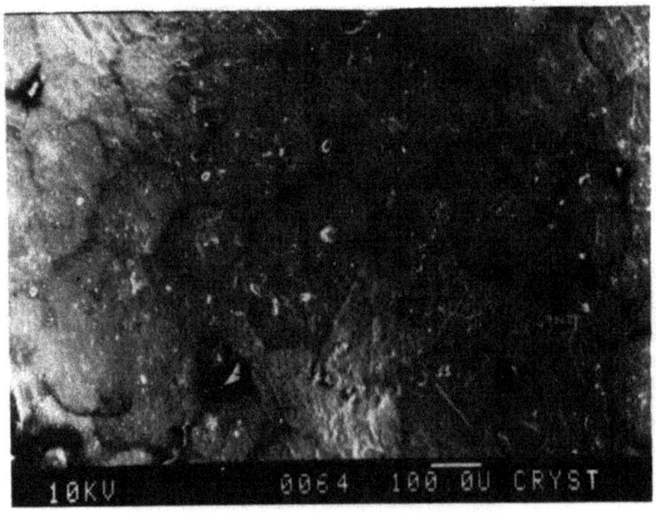

Fig. 5. Grain boundaries visible on a silicone rubber mould taken from the surface of a Pre-Columbian gold alloy disk (Brown, 1984).

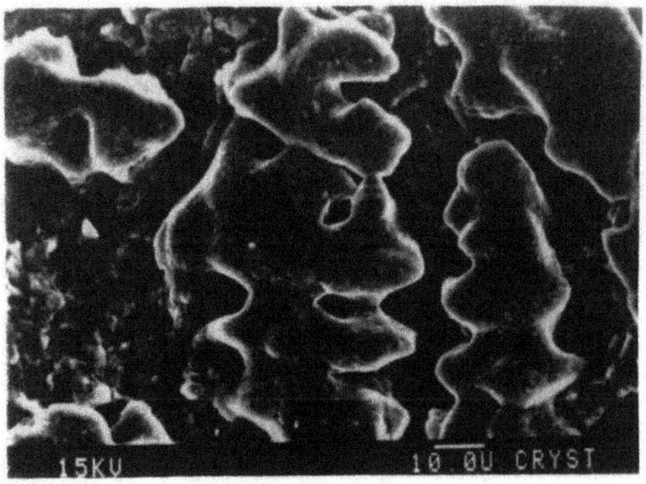

Fig. 6. Microstructure of an experimental bronze axe, as evidenced on a silicone rubber mould showing a portion of the axe 'as cast'.

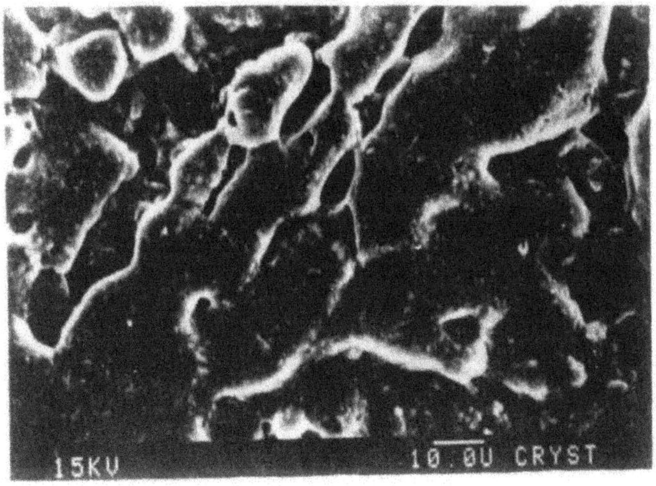

Fig. 7. Microstructure of the same experimental bronze axe, as evidenced on a silicone rubber mould showing an intermediate stage of hammering.

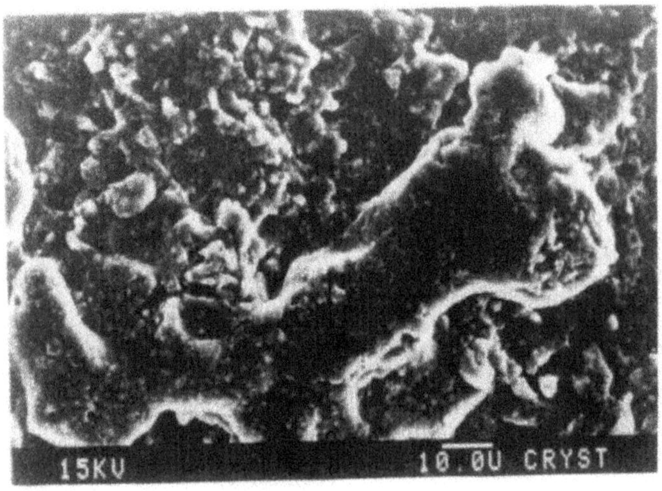

Fig. 8. Microstructure of the same experimental bronze axe, rubber mould showing a later stage of hammering.

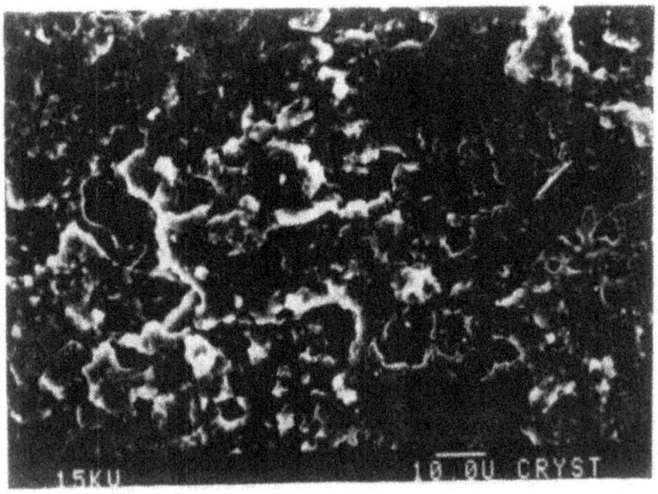

Fig. 9 Microstructure of the same experimental bronze axe, showing its final work-hardened, cutting edge.

taken both from the etched surface as above, and in section, at each stage of working (i.e. 'as cast', annealed, hammered, annealed, etc.). In this manner, a reference set may be compiled for the particular bronze alloy cast, as changes in the microstructure at the surface may be compared with the changes observed at the corresponding cross-sections. Case studies would then be undertaken in order to test the applications of such examination on archaeological metals.

Conclusions

The surface level analysis described here may aid in the clarification of various phases of the manufacturing process. Tool types can be identified with greater accuracy and this may lead to a better understanding of ancient metalworking technology and the recognition of tool kits from the archaeological record. The use of an individual tool on one or more objects may be determined by the recognition of matching toolmark "signatures". It is possible to gain an understanding of the methods employed in cutting metal sheet, wire production, abrasion and finishing techniques, in addition to the forming and decorative stages which have been described above. The surface interpretation of hammering and annealing stages has only received preliminary testing to date. The significance of this technique lies in the non-destructive nature of the process; as compared with cross-section analysis, which is presently used to retrieve this type of data.

Acknowledgements

I would like to give special thanks to those who have both helped and inspired my work: Benner Larsen, Phillip Lowery, Richard Savage and Peter Shorer. Thanks also to: Damian Goodburn, Peter Kvietok and Sandra Olsen, for reading my manuscript. However, none of the above are in any way responsible for any errors which may still remain in this text.

References

Brown, D. 1984. A Reconnaissance of Some Peruvian Goldworking Techniques. Dissertation for the Diploma in Conservation, University of London, Institute of Archaeology (unpublished).

Easby, D. 1966. Early metallurgy in the New World. Scientific American 24: 4: 73-81.

Larsen, B. 1985. Vaerktoejsspor Og Overfladestruktur. Metoder Til Identification Af Vaerktoejsspor Og Overfladestrukturer Pa Arkaeologisk Materiale. School of Conservation, Royal Danish Art Academy. Copenhagen, 208 pp.

Larsen, B. 1985. The Gundestrup Cauldron, identification of tool traces. In The Third Nordic Conference on the Applications of Scientific Methods in Archaeology 5: 561-574, Helsinki, Finland.

Larsen, B. 1984. Tracing the tools by their traces. Identification and documentation of toolmarks, illustrated through punch-decorated objects from the Early Germanic Iron Age. KUML 1982-83: 169-180.

Lowery, P.R., Savage, R.D.A. and Wilkins, R.L. 1971. Scriber, graver, scorper, tracer: notes on experiments in bronzeworking technique. Proceedings of the Prehistoric Society 37: 167-182.

Shorer, P.H.T. 1984. Interpretation through reconstruction of a rusted 10th century Viking iron sliding key. ICOM Preprints, 7th Triennial Meeting, Copenhagen 2: 84.22: 33-43. ICCROM, Rome.

THE IDENTIFICATION OF ORGANIC MATERIALS PRESERVED BY METAL CORROSION PRODUCTS

Jacqui Watson

In Britain, organic objects are often preserved from non-waterlogged excavations solely as traces in the corrosion products of adjacent metalwork. Corroding iron, copper and lead produce salts that protect organic materials from micro-organism activity and preserve recognisable microstructures. Such characteristic structures can frequently be identified by visual examination if well-preserved, but the use of optical microscopy alone in some instances may produce incorrect identifications, as minerals and alteration by fungi suggest diagnostic features which were not actually present on the original organic material. In such cases, the scanning electron microscope can often be employed to confirm the identification of an organic material.

The potential of these mineral-preserved traces was first recognised by L. Biek (1963). Since then, the Ancient Monuments Laboratory (AML) has built on the work of E. Crowfoot and C. Keepax (1975), so that we are now able to identify wood and sometimes leather to species level, and distinguish between bone, horn and ivory (Cronyn et al., 1985).

Mineral Preservation of Organic Materials

In order for organic material to be preserved, it must be in contact with metalwork that is corroding rapidly and the salts produced must cover the organic material before it is consumed by micro-organisms. The corrosion products on the metalwork are usually a variety of iron oxides, basic copper carbonates and basic lead carbonates.

The ideal environment for the preservation of organic material is, therefore, one which is aggressive to metals, sometimes converting them totally to minerals. Burials often provide this environment, probably because of the proximity of a decomposing body (Janaway, 1985).

Organic materials are preserved by metal salts in one of three ways: coating, replacement, or impregnation.

Coating. This is the most common form of mineral preservation. In this case, the original organic material is lightly coated with metal salts which inhibit micro-organism activity. The morphological structure of the organic material is thus preserved and can sometimes be identified using conventional means, such as transmitted light microscopy on thin-sections after the salts have been removed with a sequestering agent. This technique is mainly applicable to wood or fibres preserved by contact with copper or lead corrosion products that are often in the form of large crystals (Fig. 1) that obscure the surface detail. If the metal is allowed to continue corroding at a rapid rate after excavation, the formation of large crystals will disrupt and deform the preserved organic material. The organic

1

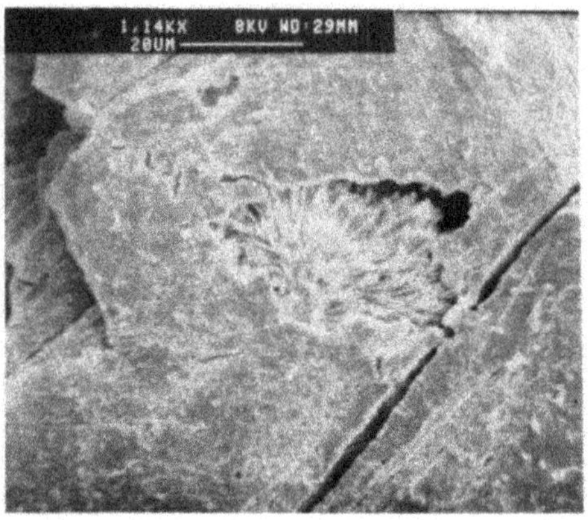

2

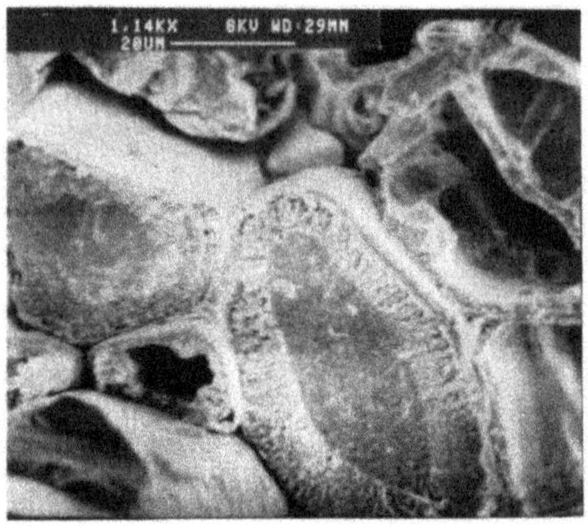

3

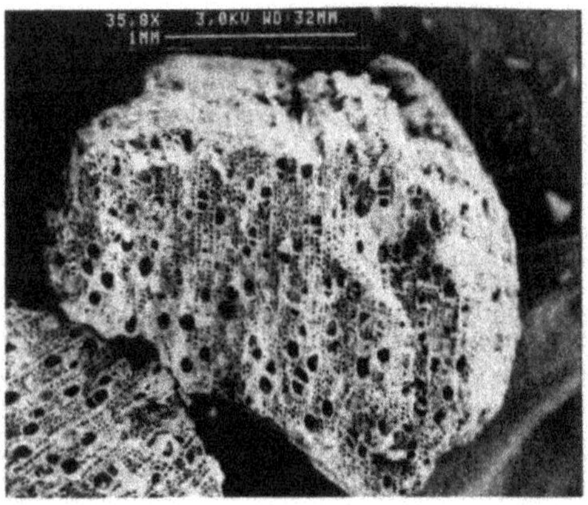

4

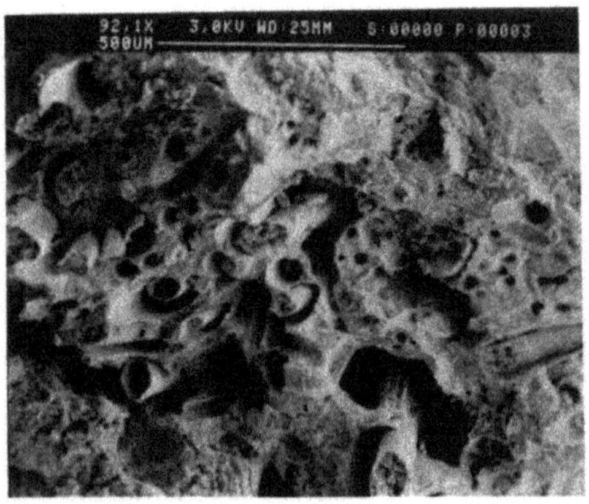

Fig. 1. Mineral preserved wood with large crystals of copper corrosion products.

Fig. 2. Cross section of wood preserved by copper salts that have been laid down in layers of differing densities.

Fig. 3. Cross section of *Salix* sp. or *Populus* sp. preserved by a light coating of iron oxides.

Fig. 4. Cross section of wood consolidated with iron oxides, but with the cell wall still intact.

material itself can still warp and collapse out of shape if allowed to dehydrate too quickly after excavation. Under these circumstances, it may be very difficult to identify the type of organic material, not to mention the species.

Replacement. This is a less common type of preservation than coating. It is frequently observed in iron corrosion products, but can also be seen to a lesser extent with copper salts. Replacement takes place when the organic material becomes consolidated with corrosion products. The original structure then decays, leaving a negative impression of the organic material in the mineral. Sufficient morphological characteristics may be preserved to allow identification of the material, despite the fact that only a negative impression is preserved.

In the case of wood, very fine surface detail can be preserved, so that standard microscope keys can be used. With textiles and fleece, the individual fibres have been encapsulated with corrosion products, forming moulds of the fibre casts which may exhibit enough surface detail to make identification possible even when the originals have gone.

Replacements in copper salts do not have the same fine detail as those in iron oxides and identification can prove more difficult. In Figure 2, it is possible to see that the copper salts have been laid down in two layers of differing densities, with the more granular one next to the cell wall. Obviously, this granular layer is incapable of reproducing the fine details of features in the cell wall, such as the pit chambers and apertures often seen in the replacements in iron oxides.

It is quite common for organic material such as wood and textiles to be preserved by both coating and replacement on the same sample.

Impregnation of porous organic material by metal salts. This is the main type of preservation for bone, antler, ivory, horn and leather. These materials appear unaltered microscopically, but are fragile and powdery. Preservation by impregnation can occur with both iron and copper salts. Iron oxide-impregnated organic materials tend to be more powdery than those preserved by copper corrosion products.

Sample Selection

Preservation by metal corrosion products is non-discriminating, so it is essential that the occurrence of preserved organic material is correctly interpreted before destructive techniques are used to take samples for identification of the metal. Any organic material lying adjacent to corroding metalwork may be preserved whether it is part of the object, a piece of organic material that the metal object is lying in or above, or relates to the context in which the metal object has become incorporated. From contexts such as urban rubbish pits, many kinds of organic material may be preserved that are important in interpreting the context, but which have no direct relationship to the object upon which they are preserved.

Metalwork from graves may have different layers of preserved organic material. By studying these it may be possible to reconstruct the original organic objects buried within the grave. For this reason the identification of mineral-preserved organic material has been most beneficial in the interpretation of cemetery remains, where, with the aid of grave plans, it has been possible to reconstruct most organic objects from traces preserved on different pieces of metalwork.

It is essential that the specialist sees the objects with the mineral-preserved organic material in situ, rather than being presented with small fragments. With out-of-context samples, the specialist will only be able to supply the identification of the organic material, at best. Important information regarding technology may be lost totally, if key features, such as stitches, the selvage of a mineral-preserved textile, or evidence for dowels in joinery are accidentally taken for samples. Additionally, it is difficult, if not impossible, to reconstruct large organic objects from traces on various small pieces of metalwork.

The ethics of sampling also need to be raised in cases when all the mineral-preserved organic material has to be removed from the metalwork in order for it to be identified. If this cannot be avoided, the identification report should be supported by a full photographic record, including micrographs of any preserved structures so that further study and interpretation can be made if necessary at a later date.

For further details on the categories of material preserved on metalwork see Edwards (AML report No. 4124 and forthcoming).

Sample Preparation

Using the SEM one can observe the surface of the sample in detail. Specimens have to be mounted with fractured surfaces, since cutting can smear the detail or cause the sample to be reduced to powder. When identifying wood, the samples should be mounted in three planes: cross-section, radial longitudinal, and tangential longitudinal. However, when dealing with replaced wood it is often difficult to obtain a cross-section, so one often has to try to find enough features to identify the material from the two longitudinal sections. For other materials one longitudinal section and a cross-section are usually required.

In this study, samples were mounted on brass or aluminum stubs with silver dag and were coated with either evaporated gold/palladium or sputtered gold.

Using a Cambridge S200, which has a large vacuum chamber, it has been possible to examine whole objects, a very valuable facility when little organic material is preserved and sectioning would remove the entire evidence. Even so, where possible, sectioned samples were mounted in order to look at the internal structure.

Although it was often not necessary to coat specimens with a conductive layer of metal, coating was done whenever possible to insure better resolution for viewing delicately preserved features.

Identification of Wood

The anatomical features of wood preserved in metal corrosion products can be used to identify the wood species by consulting an atlas (Schweingruber, 1982) or a microscope key (Brazier and Franklin, 1961). The use of a suitable reference collection can aid in establishing accurate identification.

As mentioned earlier the diagnostic features need to be assembled from three different planes. The following lists provide an indication of which sections exhibit the different features. On the whole this is applicable to both coated and replaced samples. A glossary of the terms used can be found in Butterfield and Meylan (1980).

1. Cross-sections (Transverse sections, or T.S.):

   a. Pore arrangement, usually visible in samples preserved under light coatings of metal salts (Fig. 3).

   b. Resin canals, if present, in conifers.

   c. Any unusual features of the rays, such as noding at the annual ring boundary.

   d. Distribution of axial parenchyma.

When wood is heavily impregnated with metal salts, but retains cell walls intact, it is almost impossible to make a taxonomic identification because of the lack of topographical detail (Fig. 4). In wood preserved by replacement, it is impossible to obtain cross-sections for study because the mineral breaks up into small splinters since there is no cell wall to maintain the pore arrangement.

2. Radial longitudinal sections (R.L.S.):

   a. Perforation plates in vessels (however, note that scalariform perforation plates in replaced samples are more easily recognised in T.L.S.).

   b. Spiral thickening in vessels or tracheids (this can also be seen in T.L.S.). In unaltered and lightly coated wood this can be observed as raised lines on the cell wall (Fig. 5), but in replaced material it appears as depressions in the cast (Fig. 6). This is an important diagnostic feature and its presence should be confirmed by using the SEM. In incident light microscopy, optical birefringence caused by the mineral component may give the illusion of spiral thickening. Etching of the cell wall by fungal hyphae can also be misinterpreted as a spiral thickening, especially as the hyphae tend to follow the

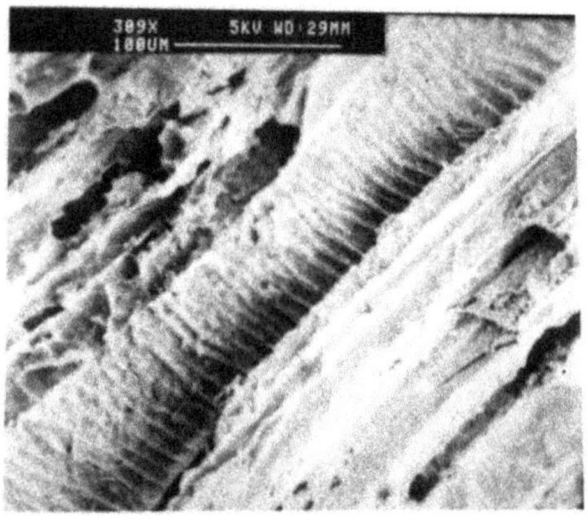

5

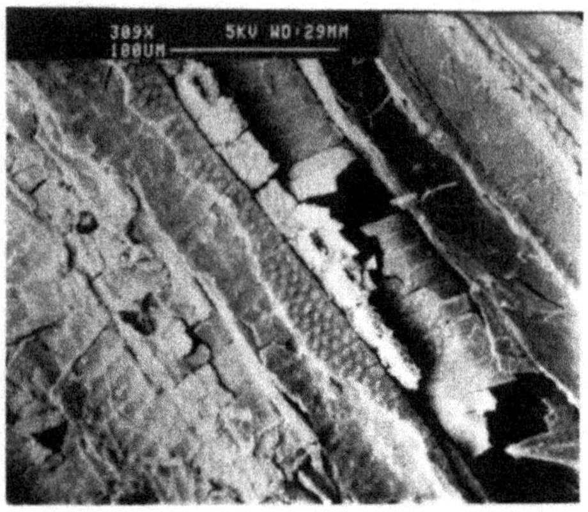

6

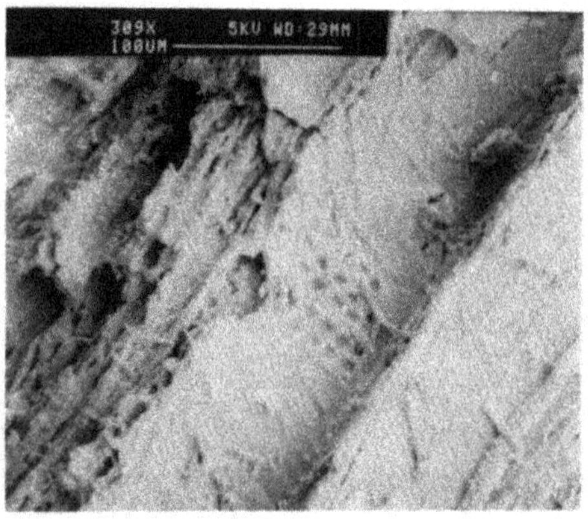

7

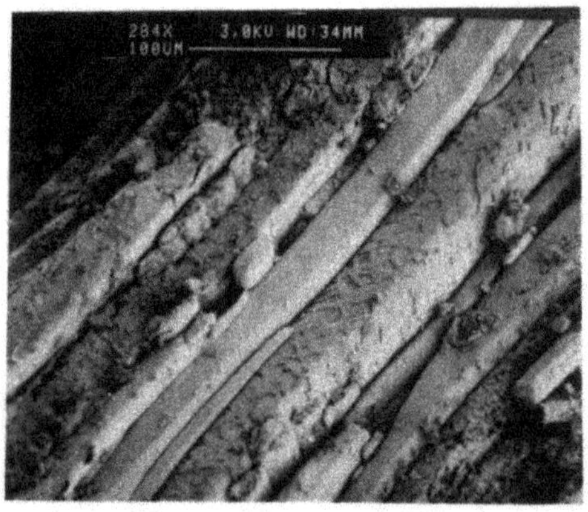

8

Fig. 5. Section of vessel with spiral-thickening, from _Tilia_ sp. preserved by a light coating of iron oxides.

Fig. 6. Spiral thickening in iron oxide replaced wood from _Tilia_ sp. The original cell walls have dissolved leaving casts of the pit chambers in relief.

Fig. 7. Section through a vessel which has been colonised by fungi. The wood, _Salix_ sp. or _Populus_ sp., has been preserved by a coating of iron oxides.

Fig. 8. Iron oxide preserved wood (either _Salix_ sp. or _Populus_ sp.). The vessel has been attacked by fungi, and the etching of the original cell wall can be seen in relief.

spiral orientation of cellulose in the cell wall (Fig. 7). By SEM examination this will appear as troughs in the cell wall, and in the replaced examples the etching will be in relief (Fig. 8). This mirrors spiral thickening.

c. Pitting in tracheids can be indicative of certain species, but should not be considered as diagnostic.

d. Cross field pits: the most important feature in the identification of conifers.

e. Longitudinal features of ray cells.

f. In replaced samples, the R.L.S. may be the only section from which one can establish whether the wood is ring or diffuse porous.

3. Tangential longitudinal sections (T.L.S.):

a. Cross-section of ray cells and their arrangement among the wood fibres.

b. Perforation plates in replaced vessels, especially scalariform perforation plates where the voids may be counted to obtain the number of bars in the original plate (Figs. 9 and 10).

c. Axial resin canals in conifers.

d. Axial parenchyma.

In order to confirm the species of the preserved wood, it is essential that one looks at all three sections. However, it is possible to <u>suggest</u> the wood species from two sections; for hardwoods the T.S. and T.L.S. are most indicative, whereas for conifers the T.S. and R.L.S. are required.

Other Materials

Various types of animal tissue such as bone, antler, ivory, horn, leather, pelts, and hairs are preserved by contact with corroding metalwork. All these materials have a porous matrix which can be impregnated with metal salts, preserving the organic structure. As yet, there are no examples of replacement of these materials with the exception of hairs and fibres (see below).

Animal tissues are much more difficult to identify than wood, and it is only possible to distinguish the material used rather than to identify the species of animal from which it was derived.

Based on the work initiated in the AML, further study on distinguishing between the different types of hard animal tissue was undertaken as a student dissertation topic by G. Moraitou (1983). In this work she discusses the morphology and degradation of these

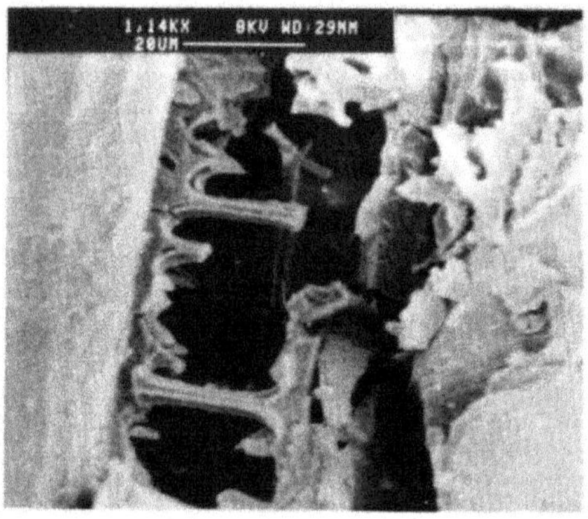

9

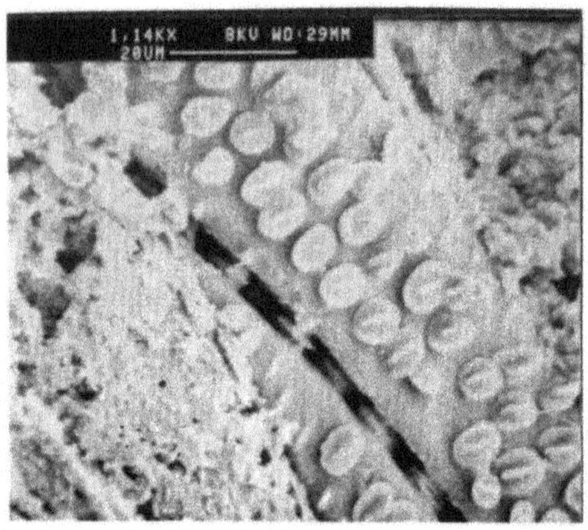

10

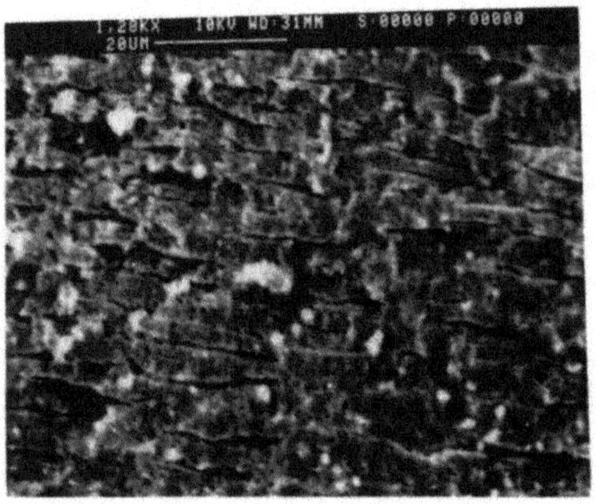

11

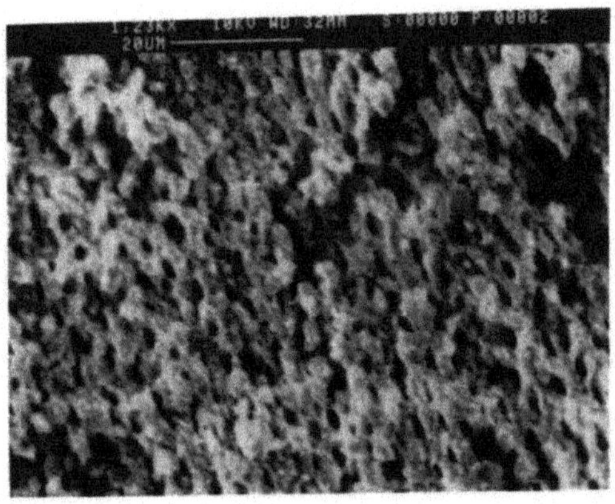

12

<u>Fig. 9.</u> Scalariform perforation plate in RLS of <u>Alnus</u> sp., preserved by a coating of iron oxides.

<u>Fig. 10.</u> Scalariform perforation plate in <u>Alnus</u> sp. replaced by iron oxides and viewed in TLS.

<u>Fig. 11.</u> Radial section of elephant ivory exhibiting undulating lines of dentinal tubules.

<u>Fig. 12.</u> Tangential section of the openings of dentinal tubules in ivory (same sample as Fig. 11).

materials and makes suggestions on the diagnostic features that may be used.

Bone and Antler. Bone can sometimes be recognised by the presence of Haversian canals, both in longitudinal and cross-sections. It is very difficult to distinguish between normal bone and antler in small samples preserved on metalwork and further research in this field is needed.

Ivory. Technically, the term ivory should only apply to elephant tusks, but it is commonly also used to describe artifacts made from other large mammalian teeth, such as walrus, hippopotamus, narwhal and wart hog.

In well-preserved specimens it is easy to distinguish between the different types of dentine (Penniman, 1952), but in mineral-preserved examples all that usually remains is a fine powder with no obvious grain orientation. In the latter condition, the use of the SEM can confirm that the material is dentine by the presence of dentinal tubules seen at magnifications of 500-1000x. Figure 11 shows a radial section with dentinal tubules arranged in undulating lines, while Figure 12 shows their cross-section. It is almost impossible, however, to distinguish between the different types of "ivory" on the basis of such features.

Horn. Unlike the previous three materials, horn is composed solely of an organic component - keratin. Through burial, horn breaks up into fine, rippled layers which probably correspond to the original layers developed by the animal, although this structure is not easily distinguished in fresh horn. When preserved by metal corrosion products, the layered structure is maintained and is usually recognised by the rows of parallel ridges, which can resemble wood grain at low magnifications (Fig. 13). By using the SEM, though, it is often easy to distinguish between horn and wood by the apparent lack of morphological features in horn at any magnification.

In identifying mineral-preserved horn, comparisons were made with degraded horn, rather than fresh material, since that preserved by minerals appeared to have been altered before impregnation with metal salts.

Leather and Pelts. At present leather can only be identified if a grain pattern of the hair follicles survives, and often such detail has been removed by abrasion when the leather article was in use. The structure in the cross-section of skin often remains, however,and further research may provide a means to identify leather when the grain pattern no longer survives.

The hairs of pelts may be identified by well-preserved moulds in metal corrosion products (Fig. 14), after the original hairs have deteriorated. The moulds usually exhibit negative bract patterns which are diagnostic of different animals (Fig. 15). However, some experience and access to a reference collection are required before one can use this feature alone to identify the species of animal involved (Appleyard, 1978).

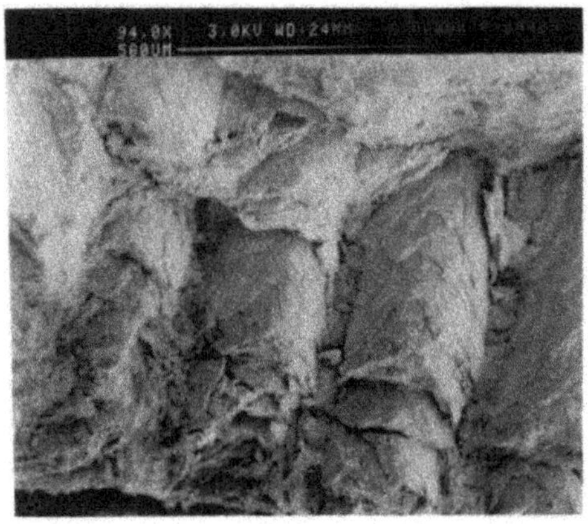

13

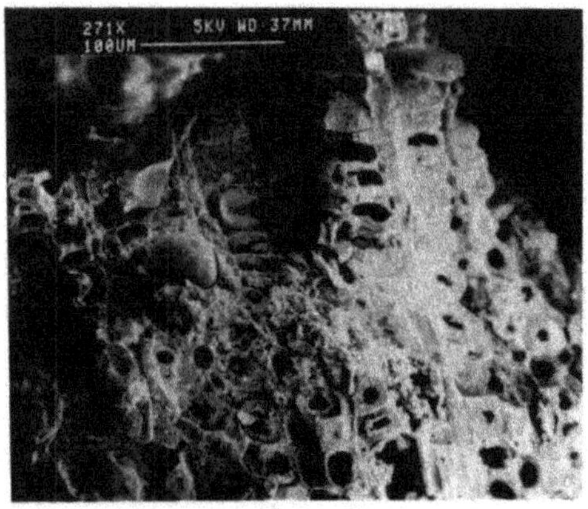

14

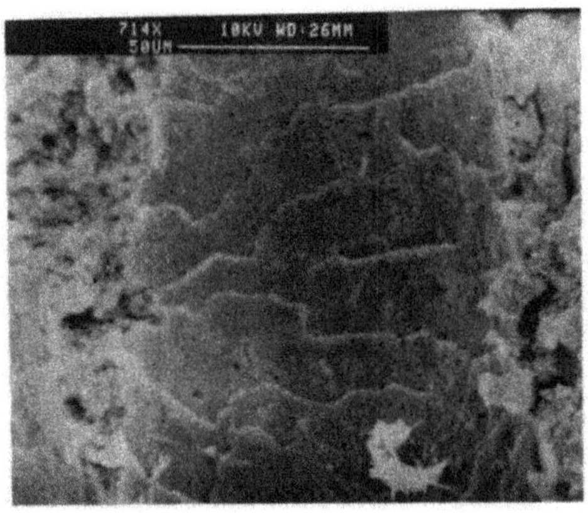

15

<u>Fig. 13.</u> Structure of iron-oxide preserved horn.

<u>Fig. 14.</u> Cross-section of iron preserved fleece. The residual hairs have shrunk and preserve less surface detail than the casts.

<u>Fig. 15.</u> The impression of a hair in iron corrosion products, with the scale pattern of the original clearly preserved.

Textiles and Fibres. Corrosion products can encapsulate the individual fibres in a textile and simultaneously allow the weave to be distinguished macroscopically. Quite surprisingly, the overall weave pattern on a mineral-preserved textile may remain fairly free of distortion.

In lightly-coated samples, fibres can be extracted from the corrosion by using sequestering agents such as EDTA, and be identified using transmitted light microscopy. With textiles preserved on heavily corroded objects, it may be easier to use SEM techniques to identify fibres from their moulds in the corrosion products. For the identification of animal fibres see Appleyard (1978), and for vegetal fibres see Catling and Grayson (1982).

Summary

Under certain conditions organic materials are preserved by corroding metalwork and the resultant structures can be identified. Most of this work can be undertaken using established keys for the unaltered material. However, there is scope for more work in distinguishing between bone and antler, and looking at the possibility of identifying leather from cross-sections rather than the rarely preserved grain patterns.

Using the SEM rather than an optical microscope enables the specialist to make accurate identifications of mineral-preserved organic material. In combination with the study of technological details, this work forms the basic record of perishable raw materials utilised in ancient metalworking periods.

Acknowledgements

The work presented in this paper was undertaken at the Ancient Monuments Laboratory of the Historic Buildings and Monuments Commission for England, and the author would like to thank the Commission for its continued support of this research.

References

Appleyard, H.M. 1978. Guide to the Identification of Animal Fibres. 2nd edition. Leeds: WIRA.

Biek, L. 1963. Archaeology and the Microscope. London: Lutterworth Press.

Brazier, J.D. and Franklin, G.L. 1961. Identification of hardwoods: a microscope key. Forest Products Research Laboratory Bulletin 46: London: H.M.S.O.

Butterfield, B.G. and Meylan, B.A. 1980. Three-Dimensional Structure of Wood. 2nd edition. London: Chapman and Hall.

Catling, D. and Grayson, J. 1982. <u>The Identification of Vegetable Fibres</u>. London: Chapman and Hall.

Cronyn, J., Pye, E. and Watson, J. 1985. The recognition and identification of traces of organic materials in association with metal artifacts. In Phillips, P. (ed.), <u>The Archaeologist and the Laboratory</u>, pp. 24-27. London: CBA Research Report No. 58.

Edwards, G., forthcoming. Guidelines for dealing with material from sites where organic remains have been preserved by metal corrosion products. In <u>Evidence Preserved in Corrosion Products</u>. London: UKIC Occasional Paper. Also available as an AML report, No. 4124.

Janaway, R.C. 1985. Dust to dust: the preservation of textile materials in metal artefact corrosion products with reference to inhumation graves. <u>Science and Archaeology</u> 27: 29-34.

Keepax, C. 1975. Scanning electron microscopy of wood replaced by iron corrosion products. <u>J. Archaeol. Sci.</u> 2: 145-150.

Moraitou, G. 1983. <u>Structure, Deterioration, and Identification of Bone, Antler, Horn, and Ivory</u>. Unpublished B. Sc. dissertation, Univ. London.

Penniman, T.K. 1952. Pictures of ivory and other animal teeth, bone and antler. Oxford: <u>Occasional Paper on Technology</u>, 5, Pitt Rivers Museum, Univ. Oxford.

Schweingruber, F.H. 1982. <u>Microscopic Wood Anatomy</u>, 2nd edition. Swiss Federal Institute of Forestry Research.

PART III:

POTTERY

IDENTIFICATION OF SUBSOIL INCLUSIONS IN POMONA CERAMICS

Donald J. Blakeslee and Richard K. Dunn

Introduction

This paper illustrates the use of the SEM in conjunction with thin-section analysis in resolving a problem involving prehistoric ceramics. Previous descriptions of Pomona ceramics from the eastern Great Plains of North America disagree on the nature of the tempering material used. Recent reports emphasize the wide range of materials seen -- up to 25 different combinations in one study -- and various analysts report conflicting results from the same assemblages.

The study reported here combined SEM examination of freshly broken sherd edges with optical examination of ground edges and thin sections. The results indicate that Pomona ceramics represent an unusual adaptation to a particular environment, the Osage Cuestas physiographic province. Frequently, the clays were used without the addition of temper and with a minimum of preparation. Subsoil structures, including peds, concretions and clay-lined voids mimic a variety of temper types, and most sherds contain a combination of these subsoil inclusions. The nature of the clay and the presence of the inclusions precluded the necessity of purposefully adding temper. Lack of aging of the clay to remove organics mandated a low firing temperature.

Background

The Pomona variant (originally the Pomona focus) is an expression of the Plains Village tradition of central North America. Sites of this complex date to A.D. 950-1400. They are found in the Osage Cuestas physiographic province of eastern Kansas and western Missouri (Fig. 1). The sites consist of the remains of lightly-built wattle and daub houses with associated rock-filled hearths, storage pits and activity areas (Blakeslee and Rohn, 1986; Brown, 1984; Rohn and Daniel, 1984; Witty, 1967, 1981). The contents of the sites are indicative of a mixed economy that included horticulture, hunting and gathering.

The Pomona variant is bordered by other units of the Plains Village tradition. To the northeast is the Steed-Kisker complex, in the Kansas City area. The ceramics of this unit, heavily influenced by the Mississippian styles of the east, are usually tempered with crushed shell. The Nebraska phase, which lies north of the Pomona range, also has some shell-tempered pottery, but the bulk is tempered with crushed granite. To the northwest is the Smoky Hill phase, also with some shell tempering, but with a majority of grog (crushed pottery) temper. South of the Pomona area are units of the Caddoan tradition, whose pottery is usually tempered with crushed bone.

The various cultural groups mentioned, and others of the same time period, have a tendency to grade into one another near their boundaries. Transitional sites, in which attributes of ceramics and other artifacts are blended, are common. Thus, it is not surprising to learn that some Pomona sites yield vessels with shell, crushed granite,

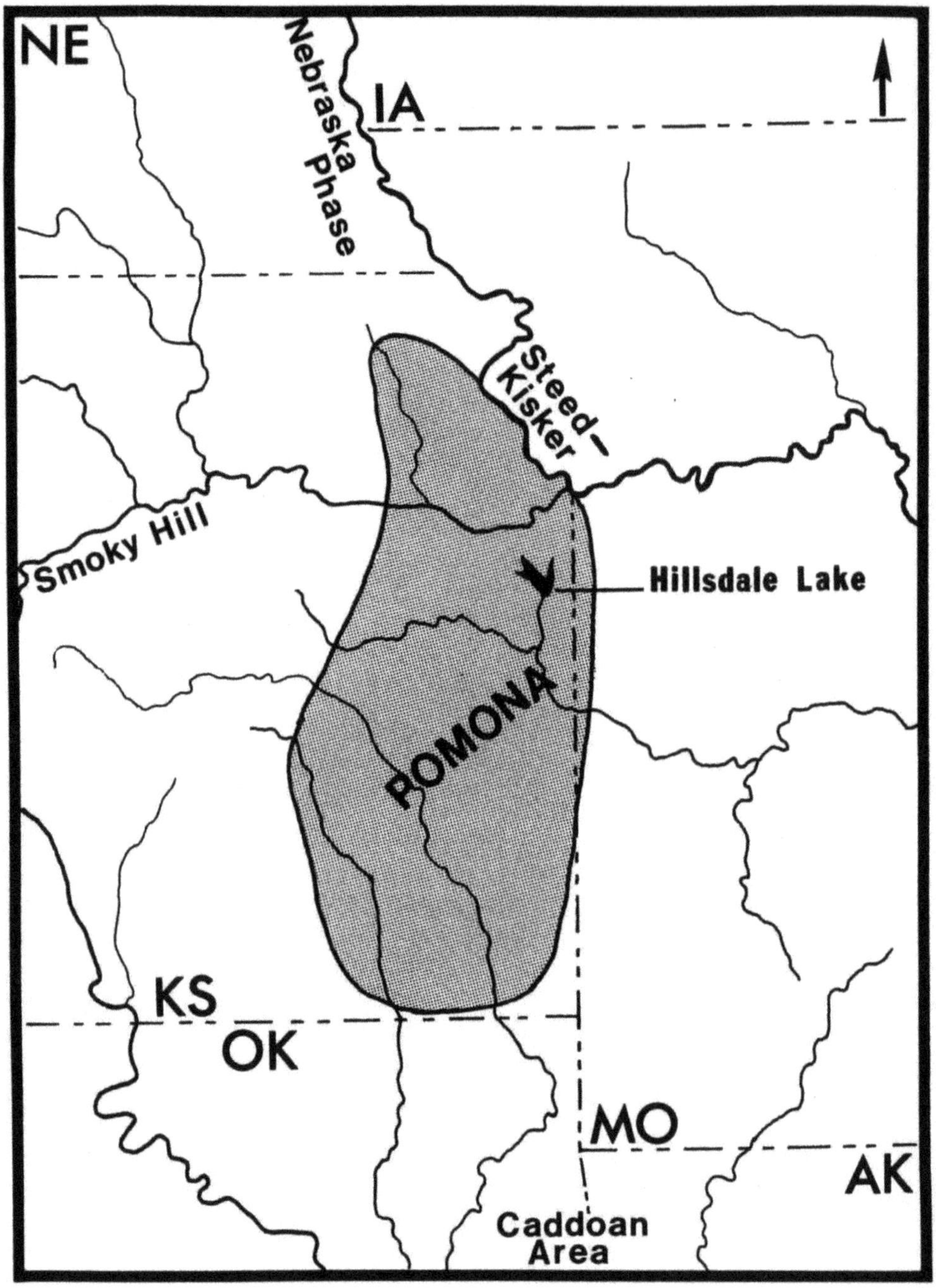

Fig. 1. Location of the Pomona variant and contemporary complexes.

grog, or bone temper (Witty, 1967: 4-5). What sets Pomona apart from the others is an apparent lack of a predominant temper type in its core area.

Pomona pottery is a simple, generally undecorated, utilitarian ware. Jars are small and globular, usually with a high, nearly vertical rim. They were formed by lump-modelling, followed by shaping and thinning using the paddle and anvil technique. In many instances, the anvil appears to have been the potter's hand. The paddle was loosely wound with cordage (in some cases, netting), which has left impressions on the exterior surfaces of the jars. The vessels were fired at relatively low temperatures. Evidence for low firing temperatures includes the softness of the sherds (averaging just above 2 on Mohs' scale of hardness) and gray cores visible in sherd cross-sections indicating underfiring. Experimental vessels made from Osage Cuesta subsoils and fired in a small surface campfire were harder than Pomona sherds. A ceramist who examined a set of Pomona pottery came to the same conclusion about firing temperature, independently (Wright, cited in Wilmeth, 1970: 27). The low firing temperatures yielded ceramics that were readily abraded and eroded after deposition in the ground. As a result, most sherds from Pomona sites have more or less rounded edges. The highly porous sherds, once deposited, are subject to the same forces that form the surrounding soil, and various substances may be added to or leached from them.

The Pomona variant is limited in its distribution to the Osage Cuestas physiographic province. This region consists of rolling hills and east-facing escarpments (Fenneman, 1931). The vegetation consists of tallgrass prairie communities. The distribution of trees is limited to gallery forests along streams, and to groves on escarpments and on the few hilltops where the bedrock is sandstone.

All streams except the largest rivers are entrenched, and the region is south of the area that was glaciated during the Pleistocene. The land surface is thus relatively old, with highly developed soil horizons. The climate is continental, with a well-marked wet season in the spring and dry summers.

The soils are generally acidic, and bone is seldom preserved in the archaeological sites. The region is dominated by mollisols that have an argillic (clay rich) horizon. The Woodson soil that is widespread in the uplands is heavy, and the zone from 10 to 30 cm below the surface is frequently recorded as having a clay texture (Soil Conservation Service, 1979: 87). This is the finest of a series of eleven texture classes for soil and is indicative of an extremely clay-rich zone (Soil Conservation Service, 1979: 46). This illuvial horizon (zone of clay deposition by soil formation processes) is plastic enough to form pottery without extensive aging or processing of the clay. Woodson and other clay-rich soils in the region contain numerous ferrous-manganese oxide concretions and blocky to prismatic peds.

The Problem

The literature regarding Pomona pottery is far from clear and consistent. The original description and definition of Pomona ware is by Wilmeth (1970: 26-28). He describes the paste as "a fairly fine but impure clay, [that] was probably used without much working, often resulting in a rather lumpy appearance." He also suggests, however, that the tempering material was grog (crushed sherds), which he described as small inclusions of the same texture and hardness as the paste.

Later descriptions of tempering materials do not correspond closely to Wilmeth's. Witty (1981: 80) states that Pomona ware contains "sparse inclusions of indurated clay, weathered shale and occasional crushed burned bone fragments." Other descriptions make frequent mention of indurated clay or shale, and many combinations of different temper types have been recorded. Rohn and Daniel (1984: 14, 67), on the other hand, agree with one of Wilmeth's assertions, that Pomona pottery is essentially untempered but with inclusions of iron oxide, quartz sand, indurated shale platelets, and occasional pieces of organic matter.

In a recent survey of Pomona archaeology, Brown (1984) recorded 25 different temper combinations. Primary ingredients listed by him include sand, sherd (grog), limestone, bone, shell, granite, calcite and no temper. His shell temper category includes both those sherds which retain shell particles and those which exhibit flat parallel-sided voids, presumably left by shell particles leached from the sherd by acidic groundwater. Some authors prefer to use the term "cell temper" in the absence of proof of the origin of these features (Bell and Gilmore, 1936: 318).

Since some Pomona sherds contain shell temper and because the soil environment is frequently acidic, it is likely that at least some of the flat cells seen in Pomona sherds were produced by leaching of shell. Other sources are possible, however. Witty suggests that most derive from flaky, indurated clay used as temper. "Many sherds weather to have the appearance of the 'cell temper'. This is formed by the flaky clay inclusions dissolving out of the paste" (Witty, 1967: 2). Why clay flakes which did not dissolve when added to a moist paste should leach out after firing is not explained.

Still other sources for the cells are possible. When the sherds from the Hillsdale Lake (Fig. 1) were examined with a low power binocular optical microscope, several lines of evidence suggested multiple origins. Some sherds contained both empty cells and a few cells that retained remnants of shell temper. On other sherds the voids were entirely empty, with none of the sherds from a whole site exhibiting any shell particles. In some rare cases, a few of these empty cells were surrounded by areas of reduced (blackened) paste. This implies that at least some contained combustible organic material that took up the available oxygen at the time the pot was fired. Still other sherds had flat cells lined or completely filled with lighter colored material. As this material did not react with hydrochloric acid, it cannot be shell temper.

Sometimes, this lighter material resembles clay, and it is probably what Witty identified variously as indurated clay and weathered shale. Leaching, however, would proceed from the exterior of a soluble particle toward the center, not the reverse. Hence the cells that are lined with this material cannot result from the dissolution of it, and Witty's explanation of the origin of the empty cells by this process appears weak.

Prior to the Hillsdale Lake project (Blakeslee and Rohn, 1986; Rohn and Daniel, 1984: 67), none of the descriptions of Pomona pottery correctly identified the ferrous-manganese oxide concretions which this study shows are common inclusions in the paste. These derive from local subsoils. Those observed varied in size from 0.1 mm to 1.9 mm in diameter and range in color from black to red. Presumably, the concretions, which occur densely in some sherds, were often incorrectly identified in earlier reports as some kind of temper. On heavily eroded sherds, they are difficult to distinguish from grog unless a freshly broken edge is examined.

The variety of substances found in Pomona ceramics and the varying interpretations of them combine to make uniform reporting difficult. There is a great potential for both inter- and intra-observer error, even in the face of the most careful analysis. In the report of the Hillsdale Lake project inter-observer error was eliminated by assigning a single technician to identify all of the pottery from the various sites, while intra-observer error was minimized by using templates for all discrete traits (Blakeslee and Rohn, 1986: 143-144). All of the sherds larger than 7 mm in diameter from each site were examined with a binocular microscope. Similarly, Brown (1984) analyzed samples from various localities solely by himself in order to eliminate inter-observer error from his survey.

Where comparisons of the analyses of the two observers are possible, the results are shockingly different. Blakeslee and Rohn (1986: 734-735) report 35% untempered, 42% sand tempered, 9% shell temper, and 4% grog temper in the 386 sherds from site 14MM213. Brown (1984: 156) examined a sample of 113 of these sherds and found 7.1% untempered, 0% sand tempered, 0.9% grog tempered, and 91.2% shell tempered. Similarly, Rohn and Daniel (1984: 13-16) found all but four of the 32 sherds from site 14MM21 to be untempered, with three examples of bone and one of shell temper. Brown (1984: 156), on the other hand, found 77.8% of a sample of nine sherds to be shell tempered, and one each with no temper and a combination of bone and grog temper.

Obviously, there is a serious problem here, only a small part of which can be attributed to sample size. One problem has to do with whether all empty cells, regardless of number and size, are identified as the remains of shell temper. Another is the question of whether various particles are purposefully added or are part of the paste of an otherwise untempered vessel. Still another is the determination of whether a sherd with an angular, lumpy appearance on a freshly broken edge has grog temper or merely a poorly mixed paste.

The study presented here allows resolution of most of these problems and additionally provides an explanation of the nature of Pomona ceramics. First, however, it is necessary to review the nature of ceramics and the materials available to the Pomona people to manufacture them.

Clays, Nonplastics, and Subsoils

Ceramics are complex substances that usually include both plastic and nonplastic materials. The plasticity of the wet clay allows vessels and other objects to be shaped. Nonplastics must frequently be added, however, to reduce plasticity. The more plastic a clay is, the greater the shrinkage during drying and firing and hence the greater the resulting potential for cracking during these processes. The presence of nonplastics "counteracts shrinkage and facilitates uniform drying, thus reducing strain and lessening the risk of cracking" (Shepard 1954: 25).

Nonplastics can be added purposefully or may already be present in the clay. In the former instance, they are called temper; in the latter, inclusions. Some clays will dry and fire satisfactorily without the addition of temper. Frequently these are residual clays that contain a substantial number of inclusions (Shepard, 1954: 53).

Generally, ceramics are formed from either of two basic categories of clay. Residual clays are those that remain in contact with their parent material of igneous rock. Such clays often retain other minerals from the parent rock, i.e., inclusions (Shepard, 1954: 11). Secondary or sedimentary clays have been transported from their point of origin by water. Frequently, the inclusions are removed during this process, leaving a pure clay (Shepard, 1954: 11). Hence, because of inclusions already present, residual clays are more likely to be used without temper than are sedimentary clays.

There is a third potential source of clays that was only rarely used for ceramics. It is so rare as a source that it is not discussed by Shepard (1954) in her comprehensive volume, Ceramics for the Archaeologist. This is subsoil clay formed during the process of soil development. While many soils have an alluvial origin, clay deposits within them may not be produced by the same forces that shape the alluvial clays discussed by Shepard. Chemical and mechanical processes slowly remove clay particles from the A horizon, or eluvial zone of a soil, and deposit them farther down in the soil, in the illuvial zone. In some soils, enough clay accumulates that the illuvial zone is termed an argillic horizon (Soil Conservation Service, 1975: 19).

Not all soils end up with clay-enriched argillic zones, but in some, the concentrations are remarkable. A soil that has only eleven percent clay in its composition near the surface may be 50 percent clay only 18 to 36 cm below the surface (Soil Conservation Service, 1975: 22). Factors involved include time, climate, soil acidity and type of clay. Old soils (i.e., on old, stable land surfaces) have had more time for the clay to be leached from the eluvial horizon and to be redeposited in the illuvial horizon. A climate with a well-defined dry season seems to enhance the process (Soil Conservation Service, 1975: 19-21), while carbonates in the soil tend to inhibit the process. Montmorillonite clays are transported more rapidly than other clay minerals (Soil Conservation Service, 1975: 19-21).

Other processes that also occur in soils are pertinent here. All soils evolve a "structure" over time, producing a patterned set of solid particles and voids (Brewer, 1964: 13). Included among the solid particles are peds, naturally occurring aggregates of soil particles (Soil Conservation Service, 1975: 474). These can be platy, columnar, granular, blocklike or polyhedral in shape. Sometimes they are delimited by thin clay films on their surfaces (Soil Conservation Service, 1979: 46).

Peds differ from clods, lumps formed by disturbance, and from concretions, which are cemented together chemically. Concretions do occur in some subsoils. Those formed in alkaline soils are calcareous, while those that form in acidic soils are cemented with iron and/or manganese oxides. The latter type consists of black to reddish spherical pellets of varying size.

Also to be found in subsoils are organic materials and several kinds of voids. Some of the latter are formed by plant roots, others by burrowing insects, and still others, such as drying cracks and spaces between peds, are the result of mechanical and chemical processes in the soil (Brewer, 1964: 179-196). In certain environments, subsoil voids become lined with fine clay or other materials. The resultant coatings are called cutans (Brewer, 1964: 206-231).

The extent to which pottery made form subsoil clays will retain these features depends primarily on how much the clay is processed prior to the manufacture of ceramic objects. Grinding the clay will destroy peds and concretions and will remove the voids. Thorough kneading of the clay in the absence of grinding will destroy peds and unlined voids but not concretions or some cutans. Illuviation cutans, formed from clay particles in suspension, consist of parallel alignments of tightly packed clay particles. As a result of their structure, they are far less soluble in water than the randomly-oriented clay particles in the general soil matrix. They are, therefore, resistant to destruction when the subsoil clay is mixed with water to create a ceramic paste. Experimental mixing of unground subsoil clays shows that some dissolve readily in water, while others must be soaked for long periods before all the peds dissolve.

Aging (adding water to clay and letting it stand for a period of days to weeks) will remove large organic matter through decomposition. Most of the organic material is thus converted to gases (carbon dioxide and methane) which escape into the air. This process is usually performed so that the gasses created by combustion of organic materials will not bloat the clay when the vessel is fired. When it bloats, clay actually foams up, leaving a scoria-like appearance. Aging should also weaken or destroy the ped structure as the clay cutans slowly dissolve.

Hypothesis

A possible explanation for the unusual nature of Pomona ceramics is that the Pomona people used the local subsoil clays to manufacture their vessels. The environment in which they lived contained unusually rich subsoil clays because of a combination of factors, including a stable old land surface, a dry season, and acidic soil conditions. The clay-rich subsoils were apparently used without any preparation of the clay other than the addition of water and some minimal preliminary kneading. In this way, the ferrous-manganese oxide concretions were almost always intact, and peds and cutans have occasionally been preserved in the pottery. The presence of numerous inclusions meant that no addition of temper was necessary. Lack of aging required that the pottery be fired at relatively low temperatures.

It is hypothesized here that Pomona pottery was normally formed from untempered subsoil clays and fired at low temperatures. While it is definitely the case that some Pomona sites yield sherds tempered with shell, grog, bone, or granite, these can be attributed to contact with neighboring contemporary societies. This hypothesis implies that many of the identifications of nonplastic materials in Pomona ceramics as various sorts of temper are incorrect and that these are, in fact, inclusions derived from the subsoil sources of the paste.

Sherds in which the paste retains blocky or prismatic peds will be similar in appearance to those with grog temper. Concretions also can be mistaken for grog. This is so because some concretions have nearly the same texture as the soil from which they were formed, and sherd temper is also usually similar in texture to the surrounding paste. Because grog is angular in outline, however, it can be distinguished from concretions, which tend to be rounded. Some pieces of grog are derived from the outer or inner surfaces of vessels which may retain diagnostic features that allow them to be distinguished from angular peds.

Voids present in the subsoil may be preserved in the sherds, but usually in distorted form. The paddle and anvil technique of manufacture will flatten most previously existing voids to a lenticular shape. Exceptions may be voids lined with relatively insoluble materials, those containing rigid objects, and voids in peds that do not dissolve completely. Flattened subsoil voids and voids left by the combustion of leaf fragments may be mistaken for spaces left by leached shell temper. Other combustible or leachable materials may leave voids of varied shapes.

Argillic cutans lining various voids can be mistaken for particles of shale. If the void is not completely filled with the clay lining, the resultant hollow particle will give the impression of an object that is dissolving from the center to the outside. Cutans can be distinguished in thin section with a petrographic microscope (Fitzpatrick, 1984: 207-222).

Samples and Methods

A set of Pomona sherds were examined with the SEM in order to clarify the nature of the aplastic matter and voids in them and to compare them to an undisturbed subsoil. All but one of the sherds are from the Pomona component at site 14MM213, at Hillsdale Lake, in eastern Kansas. This is one of the sites for which inter-observer error is known to have been a severe problem when optical analysis was performed.

The single exception is a granite-tempered sherd of Woodland period affiliation from site 14MM551, which is also located at

Hillsdale Lake. It was included to provide a comparison between the Pomona sherds and a sherd known to have been purposefully tempered. Since granite is not locally available in the vicinity of Hillsdale Lake, it was also likely that the vessel was made elsewhere and that the paste of the sherd would differ somewhat from the Pomona pastes.

The soil sample chosen for examination is a clear example of the undisturbed formation of peds in a subsoil context. It consists of a plug of soil taken from the marrow cavity of the tibia of a white-tail deer (Odocoileus virginianus). The bone is from a Nebo Hill phase component of Late Archaic age at site 14MM1C dated to 3885 +/- 135 years B.P. (UGa-4084) and 4020 +/- 90 years B.P. (UGa-4085). The animal bone from this component had been broken, possibly for marrow extraction, allowing infiltration of soil into the marrow cavity. Inside the bone, the soil was exposed to approximately the same processes that produce peds in natural conditions (with the exception, of course, of the partial chemical buffering provided by the surrounding bone). Blocky peds, visible to the naked eye, were present both in the soil from the marrow cavity and in the matrix that surrounded the bone.

Two Pomona sherds were selected for comparison with the soil sample. One was identified as having grog temper by examination of a freshly broken edge with a 10X hand lens. The other was tentatively identified by similar examination as having a blocky paste that resulted from inadequate mixing.

Also chosen for comparison with the subsoil sample was a piece of daub from site 14MM213. The daub originated from a house in the Pomona component. The clay for the daub is thought to be derived from a shallow basin-shaped pit in which water was added and the resultant mud mixed by trampling. Such pits are frequently associated with Pomona houses. It was expected that the daub would lack any ped structure, both because of its near-surface origin and because of the mixing in the pit.

Two Pomona sherds were selected for examination of the flat cells. One, 14MM213-146, contains cells that are lined with a lighter-colored substance than the surrounding paste. The second sherd lacked any visible lining of the flat cells, but exhibited several ferrous-manganese oxide concretions and one readily visible angular void.

All of the specimens were broken to expose cross-sections that had not been subjected to soil abrasion or post-excavation modification. These were mounted on aluminum stubs and coated with approximately 200 Angstroms of gold in a sputter coater. They were then examined using an ETEC Omniscan scanning electron microscope (SEM) in the secondary electron mode.

Another 12 Pomona sherds, all from site 14MM213, were prepared as thin sections to determine the nature of the linings seen on the walls of some voids. These thin-sections were compared to thin sections of soil features illustrated in standard manuals (Brewer, 1964; Fitzpatrick, 1984). The preparation of thin sections of potsherds without the loss of original fabric, especially void spaces, is difficult and is unfamiliar to most archaeologists. For this reason, the process used is described here in detail.

Because low-fired sherds are quite soft, they must be impregnated with epoxy prior to grinding. Petropoxy 154 resin and curing agent were used. The following recipe allows for the preparation of about six sherds that are 25 mm in diameter and roughly 7 mm thick. Amounts can be increased proportionately for larger numbers of specimens. Processing large batches of specimens reduces the preparation time per specimen.

Ten ml of Petropoxy 154 is first heated to 100°C. Then 100 mg of Petropoxy 154 blue dye is added. The purpose of the dye is to allow the investigator to distinguish between the original voids in the specimen and any that might be created during cutting and grinding. The resin-dye mixture is then heated in an oven at 130°C for two hours, with frequent stirring. The heating and stirring should continue until the powdered dye is completely dissolved in the resin. Once the mixing is complete, one ml of Petropoxy 154 curing agent is added, using a syringe, and the resin is stirred for 30 seconds. A small amount (2-3 ml) of isopropyl alcohol can be added to the mixture to decrease its viscosity. This will reduce problems with air bubbles and will increase the degree of penetration of the specimens.

To saturate the specimens with the epoxy, a vacuum impregnator capable of drawing 500 mm of mercury is used. Each sherd is immersed in the resin (20 ml plastic cups are handy for this) and placed in the vacuum chamber. Once the vacuum reaches the maximum limit (550 mm of mercury in this case), the pump is turned off and the specimens are left in the sealed chamber for five minutes. Air bubbles will form at the surface of the resin. After the five minutes have passed, the chamber is opened and the liquid stirred gently to remove the bubbles. This procedure is repeated until air bubbles no longer form.

The sherds are then placed in an oven for one hour at 130°C. This heat treatment hardens the epoxy. Larger sherds, especially thick ones, may require longer heating times. Depending on the porosity of the sherds, some experimentation may be necessary to achieve complete impregnation of the sherds. Incomplete dosing may lead to the formation of air bubbles when the sherds are attached to the microscope slides, leading to problems during the grinding of the thin sections.

Upon cooling, the sherds are ground flat along one edge with a 600 grit grinding wheel. At this point it will be obvious whether the sherd interior has been completely impregnated. If not, epoxy can be applied to the ground surface and the sherd returned to the vacuum chamber to complete the process.

It is useful to examine the edge of the ground sherd under a binocular microscope even before more epoxy is added. Careful scanning of the ground surfaces during this study revealed details of the fabric that might have been missed had more epoxy been added and the edge re-ground. Once any needed epoxy has been added and the edge refinished, microscopic examination alternating with grinding will allow the investigator to choose a cross-section that clearly reveals the features of interest.

Once a surface with features of interest has been exposed, the sherd is mounted on a petrographic glass slide using the same ten-to-one mixture of resin to curing agent, but without the dye. Just enough epoxy to cover the ground surface should be applied to the sherd prior to placing its ground edge down on the slide. (Too much epoxy can crack the slide.) Next, the slide and specimen are heated to 100°C on a hotplate, and the junction between them inspected for air bubbles. It is critical that such bubbles be removed because they prevent grinding of the specimen to the appropriate thickness. Removal is accomplished by applying pressure on the sherd and by sliding it slightly across the slide.

When the air bubbles have been removed, the seal between specimen and slide is hardened by placing them in the oven at 130°C for three hours. A butterfly style paper clip can be used as a vise to help ensure a good bonding. The slides are then removed and allowed to cool. Following this, the sherd is cut with a saw or ground on a wheel to a thickness of approximately 100 microns. The specimen is then ground to the standard 30 micron thickness for thin section analysis with 600 faceting grit on a glass plate. The thickness of the section is checked using a petrographic microscope with a standard color chart.

Results

Examination of the undisturbed subsoil sample with the SEM (Fig. 2a) showed the presence of well-defined blocky peds. Some were separated from one another by crevice-like voids, and most were penetrated by small circular to oval voids 22 to 134 microns in maximum dimension.

Both the sherd identified optically as having a blocky paste and the one identified as having grog temper (Fig. 2b) were basically identical to the subsoil. Indeed, close examination of the latter led to reclassifying it as untempered, with a poorly mixed subsoil paste. The crevice-like breaks between peds were visible in both sherd specimens. Smaller voids in a variety of shapes and sizes were also recorded. Some were round to oval, while others were irregular in outline.

These specimens were also compared to the daub samples and the intentionally tempered sherd. The daub (Fig. 3a) is distinct in appearance from all of the pottery, even at low power. It lacks any suggestion of the presence of peds or concretions. Only a few small voids are present. It is granular in appearance, apparently as a result of the presence of numerous loess grains typically found in near-surface zones of the upland soils of the region.

The granite-tempered sherd (Fig. 3b) has a paste that is similar in general texture to the rest of the pottery. It lacks, however, any sign of ped structure, concretions, or flat voids. The temper tends to be large in comparison to the inclusions seen in the Pomona sherds, and the paste is free of all but loess-sized inclusions.

Most of the Pomona sherds exhibited a variety of features (Fig. 4a). A single broken edge might have concretions, peds, flat smooth

a

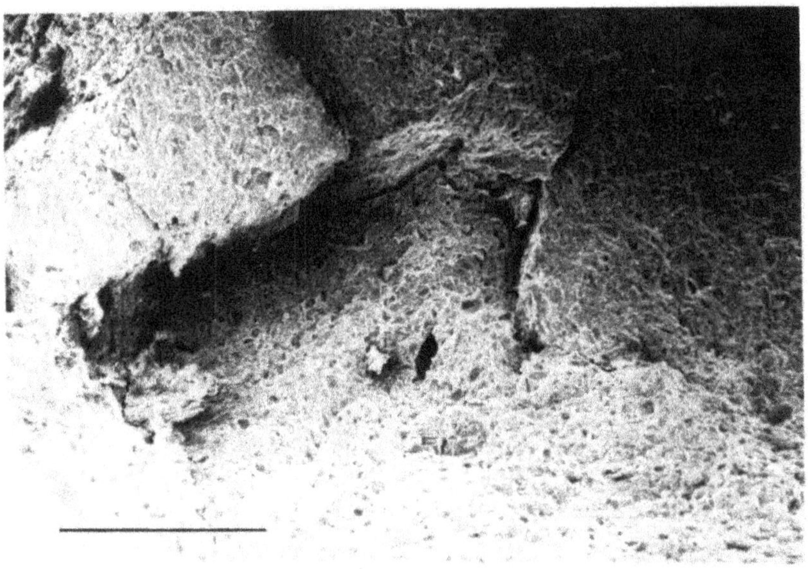

b

Fig. 2. a. Undisturbed subsoil peds, showing inter- and intra-ped voids. Scale bar = 1000 microns. b. Peds preserved in a Pomona sherd. Scale bar = 500 microns.

a

b

Fig. 3. a. Daub sample, showing a lack of large voids and other large subsoil features. Scale bar = 500 microns. b. Granite-tempered sherd. A quartz crystal is at bottom center in front of a feldspar crystal. Scale bar = 200 microns.

voids, flat irregular voids, smooth rounded voids, and smooth angular voids. Some voids were lined, while others were unlined. Close examination of these under a variety of magnifications clarified the origins of some of the features.

At least one angular void with smooth edges (Fig. 4a:1) was created by a solid object in the paste at the time the vessel was formed. The object was later removed, either by combustion during firing or by leaching after deposition. That the void was filled at the time the vessel was shaped is evident from the distortion of an adjacent flat, irregular void around the angular void (Fig. 4a:2). The walls of the void are flat and featureless, even under high magnification, rendering identification of the particle it contained an impossible task.

Flat, irregular voids (Fig. 4a:2, Fig. 4b) may have been formed by two processes. Some are spaces in the original subsoil that were flattened during the shaping of the vessel with a paddle and anvil. Others may have been created by the strains resulting from the shaping process. These voids tend to lie parallel to the surfaces of the sherds, except where they are adjacent to solid objects in the paste. At low magnifications, they (Fig. 4a:4) are readily mistaken for the flat, smooth-walled cells apparently left by leached shell temper (Fig. 4a:3). At higher magnifications with the SEM, however, it is possible to look down into the voids and distinguish these types by the smoothness of their walls. The voids presumably left by leached shell temper (Fig. 5a:2) have smooth, featureless sides, even at very high magnifications. The flat irregular voids, in contrast, have sides which retain the irregularities of the original subsoil voids (Fig. 4b).

Some other fairly smooth-walled voids appear to have been created by organic objects (Fig. 4a:5). They tend to have rounded edges and are circular to elliptical in outline. Some are likely to be molds of fine rootlets. A few (Fig. 5b) retain what appear to be opal phytolith structures. A copy of Figure 5 was examined by P. Twiss (pers. comm.), who was able to identify the structure in the void as a probable opal phytolith, but, because of the distortion presumably caused by firing of the vessel, could not determine whether it was derived from a monocotyledon or a dicotyledon.

Still other voids (Fig. 4a:6, Fig. 5a:1) are lined with a dense substance that tends to charge under the SEM beam. Examination under a petrographic microscope in thin section indicates that these are clay linings, which have low conductivity and hence charge readily. Given that the sherds are low-fired, it is conceivable that some linings were created after deposition by soil formation processes.

Other sherds (Figs. 4a and 5a) contain both lined and unlined voids. The lining in these cases appears to differ in nature from the above examples. This makes a post-depositional origin for these linings unlikely. It is difficult to conceive of soil conditions such that one void would become lined while another of similar shape and size only a millimeter away would not. It is more likely that the voids have become lined prior to collecting of the clay from the

a

b

Fig. 4. a. Typical Pomona sherd exhibiting subsoil features: large angular void (1), irregular voids (2, 4), smooth flat void (3), oval voids (5), cutan-lined void (6), and ferrous-manganese oxide concretion (7). Scale bar = 1000 microns. b. Irregular void at high magnification (1600X). Small vesicles in clay matrix are visible above the void. Scale bar = 10 microns.

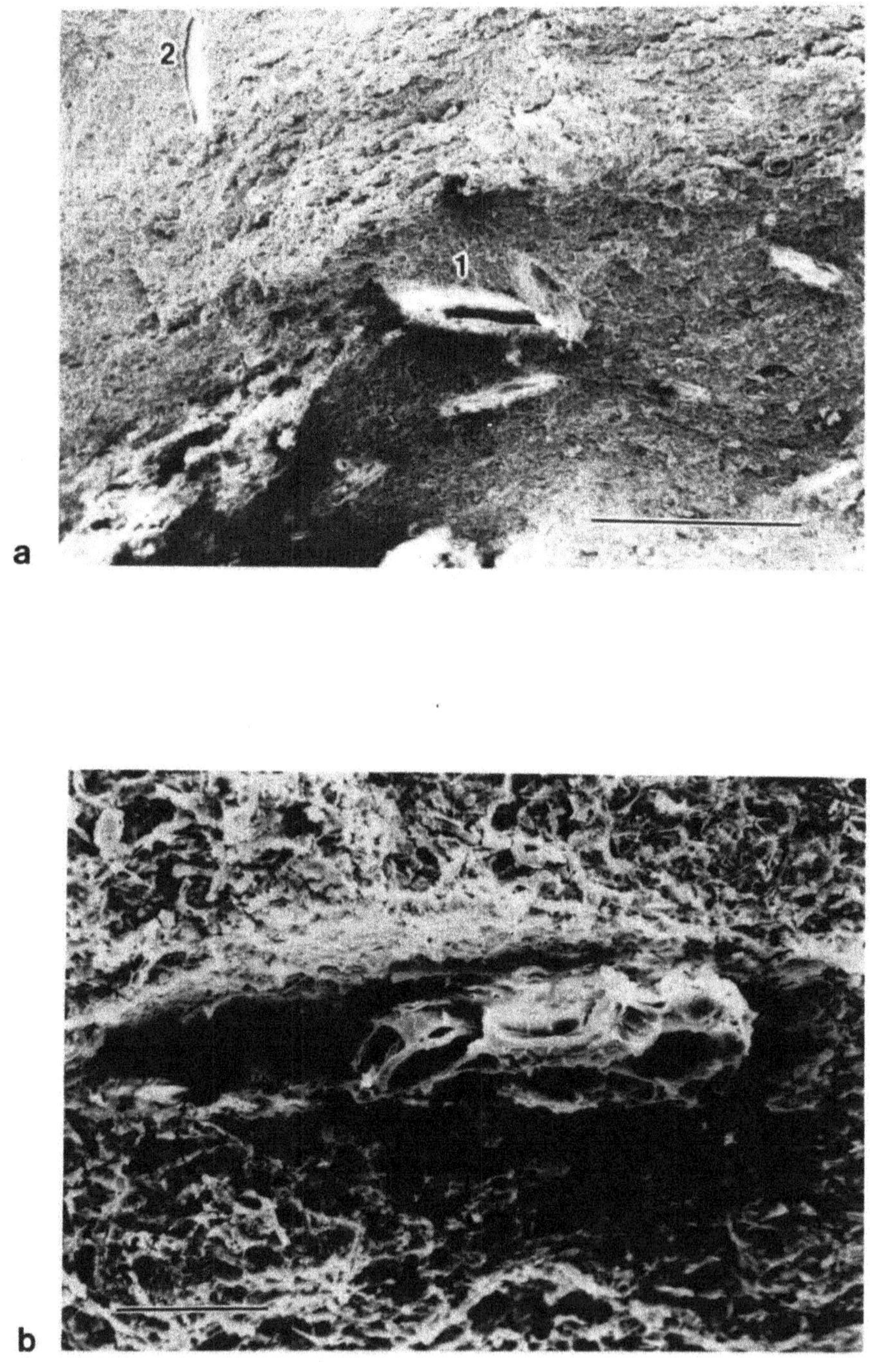

Fig. 5. a. Cutan-lined voids (1) and smooth void (2) from which the cutan has broken away. Scale bar = 500 microns. b. Void produced by organic material. Remnants of the organic structure are visible in the right-hand side of the void. Scale bar = 50 microns.

subsoil by the potter. The reason for this became clear when the sherds were examined in thin section (see below).

Concretions, when viewed in cross-section (Fig. 4a:7) are shown to have a texture very similar to that of the surrounding paste. This is not surprising, since they are composed of subsoil clay particles that have become cemented with iron and manganese oxides. All but the smallest concretions are readily seen with the naked eye or a low-power hand lens. Reddish to black in color, their outer surfaces sometimes are shiny when freshly exposed. On eroded sherd surfaces, however, the concretions mimic grog temper, and even their rounded shape cannot distinguish them from pieces of crushed pottery, the outlines of which are rounded by erosion. In thin section, the concretions tend to be marked with concentric reddish-brown iron oxide stains.

A fortuitous discovery made while examining the voids under high magnification (1600X) was evidence that some sherds had been heated to the point that they had begun to bloat (Fig. 4b). Small gas bubbles were visible on broken paste surfaces. These may be derived from the combustion of minute organic particles which would have been removed if the clay had been aged properly. If the vessels had been fired at higher temperatures, the bloating would have weakened the vessel walls to the point that they would have begun to flow, distorting the vessel shape.

The specimens that were thin-sectioned were examined with a binocular microscope after one face had been ground flat. Some showed a variety of types of concretions and voids. A few of the voids were surrounded with dark gray paste that was fired in a reduced state because combustion of organic material in the void used up the available oxygen. Reduced clay around some voids was visible on one specimen prior to coating with gold for SEM examination. These features, however, were invisible in the SEM, since they did not involve any alteration of the topography of the broken edge.

One specimen (Fig. 6a) contained features that might readily be mistaken for grog temper. The paste of the sherd is nearly free of sand, but a structure within the sherd contains abundant fine sand and a concretion (that also contains sand). The structure is probably a vermiform, i.e. a filled worm channel, containing material translocated from another soil horizon (cf. Fitzpatrick, 1984: Fig. 7.22).

In another specimen (Fig. 6b), several regular voids contained a white mineral. Dilute hydrochloric acid was applied to determine whether the material was effervescent. Since it was not, it cannot be mussel shell temper. Additional work will be needed to determine its composition. The same specimen contained voids, such as the large one in the center of the photograph (Fig. 6b), which were clearly lined with a cutan. The smaller, flat voids have similar linings, visible in the photograph as dark gray rims.

Examination of the sherds in thin section with a petrographic microscope helped to clarify some of the features in the paste. Identification of temper and inclusion minerals is straightforward with the petrographic microscope. Figure 7 shows a sherd that is tempered with fine quartz sand and that has smooth, flat voids as inclusions.

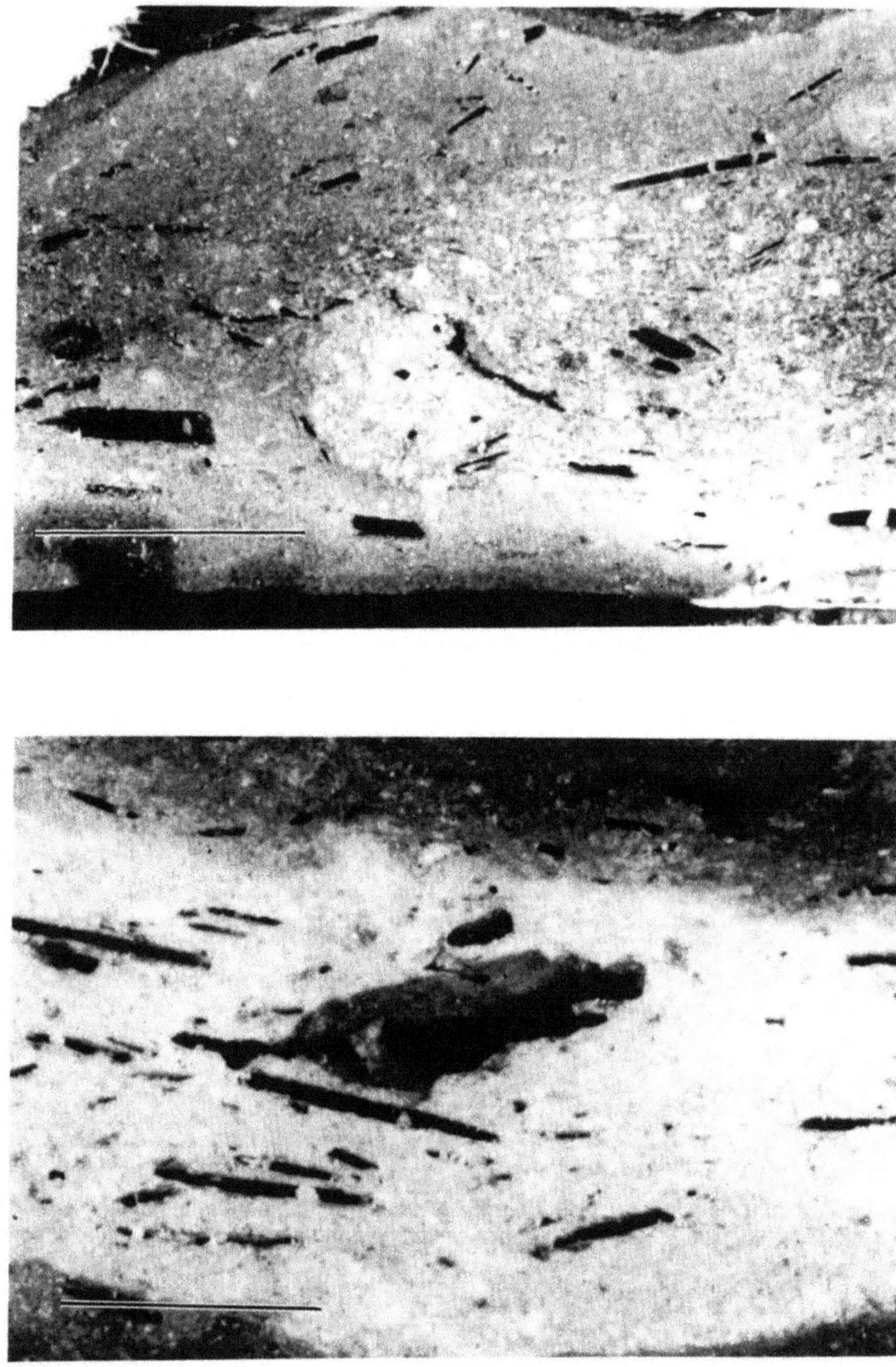

Fig. 6. a. Optical microscopic view of the cross-section of a sherd that contains a vermiform, the contents of which contain fine sand. The rounded edge of the vermiform distinguishes it from grog. Scale bar = 3 mm. b. Optical microscopic view of the cross-section of a sherd with a variety of voids lined with cutans. The lining of the large void in the center is readily visible by color differences between it and the surrounding paste. Scale bar = 3 mm.

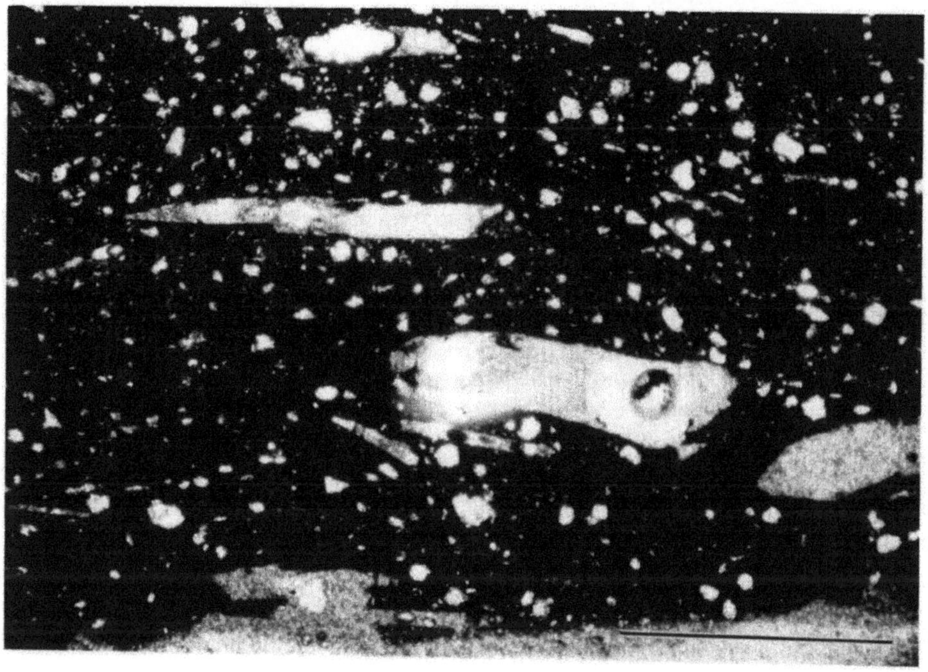

Fig. 7. Thin-section view of a sand-tempered sherd that contains smooth flat voids as inclusions. A thin, dark lining of organic material coats these voids. The lining is most apparent along the top of the lowest void in the photograph. Scale bar = 2 mm.

The smooth flat voids in this sherd have a thin, dark lining, apparently made up of an organic material. Irregular voids in the same sherd do not exhibit this lining. Since the irregular voids are either inter-ped voids or products of the process of manufacture, this is a clue as to the source of the lining material. When both unlined and lined voids are interspersed in a single sherd, they have both been exposed to the same environment of deposition. In this case, it is safe to conclude that the material forming the lining of some of the voids must be derived from the particles that formed the voids. Thus, the organic lining seen in the regular voids in Figure 7 must derive from an organic material that once filled them.

The process of forming the vessels was also clear in thin section. The clay particles in the Pomona sherds were moderately to highly aligned with the surfaces of the sherds. The orientation of clay particles can be observed when the thin sections are observed using crossed polarizers. If a quartz plate is added, the orientation is apparent in color changes that occur when the microscope stage is rotated. The moderate to high orientation of the clay apparently is the result of use of the paddle and anvil method of shaping the pots (J. Gundersen, pers. comm). Such strong alignment is not apparent in a series of Middle Woodland period vessels for which thin sections were available.

One sherd that contained highly aligned clay particles also contained a ped. Since the ped did not dissolve during the mixing of the paste, the clay particles in it did not have a chance to become aligned during manufacture of the vessel. When the microscope stage was rotated, the ped remained the same color, while the surrounding paste changed colors. This provides an additional means of distinguishing peds (which will not have aligned clay platelets) from grog temper (which will retain its original orientation, even though this should be different from that of the surrounding paste).

Conclusions

Examination of Pomona sherds with the SEM provides strong support for the hypothesis that they were made from subsoil clays. The bulk of Pomona ceramics appear to be untempered. The numerous inclusions in them are also present in the clay-rich subsoils of the Osage Cuestas physiographic province. Because the clays were not ground or were only poorly ground prior to use, subsoil peds, concretions and voids were frequently preserved in the pottery. Vessels made from clay that was mixed only minimally are often identified as having grog temper. Sherds that contain a large number of ferrous-manganese concretions are likewise often misidentified.

Some voids compressed during manufacture are similar in appearance to voids left by leached shell temper. Some voids appear to have been present in the subsoil; others may be produced during the manufacturing process. Argillic cutans in some voids mimic shale particles. Analysis of Pomona ceramics has been complicated by the confusion of these inclusions with the grog, shell and other true tempers that are occasionally found in sherds from some Pomona sites.

Because the subsoil clays were not aged prior to use in ceramics, the paste is rich in organics. Large particles burned, leaving reduced areas of clay surrounding voids or organic linings on the walls of the voids. More generally distributed minute organic particles imbedded in the clay caused the clays to bloat if they were fired at high temperatures. Pomona pottery, thus fired at low temperatures is softer than both earlier and later ceramics from the same region that contain true tempers.

Pomona ceramics may be seen as an adaptation to a particular environment. The clay rich subsoils of the Osage Cuestas province contain numerous insoluble to partly soluble particles that precluded the necessity of tempering the clay. Indeed, in order to retain the particles, processing the clay prior to use was not only unnecessary, but had to be abandoned. This, in turn, mandated a low firing temperature. The numerous, varied inclusions mimic various temper types, especially on the eroded surfaces of low-fired sherds. As a result, archaeologists accustomed to looking for temper have been led astray.

Future Work

While this study has shown that other voids can be mistaken for those left by leached shell temper, more work is needed. Experiments with shell temper must be done so that the voids created when shell temper is leached from pottery can be characterized precisely enough to determine which of the flat voids in Pomona pottery were produced in this way. Studies of samples from the argillic horizons of soils of the Osage Cuestas are also necessary. Both thin-section and SEM examination of these subsoils are needed to document the presence and composition of the features seen in Pomona pottery. SEM and X-ray mapping would be an appropriate strategy for identifying inclusions in both the subsoils and the pottery.

Acknowledgements

Dr. Page Twiss (Kansas State University) examined Figure 5b and identified the structure in the void as a possible phytolith. Dr. James Gundersen (Wichita State University) kindly provided assistance with the thin-section analysis and identification of various features seen in the thin sections.

References

Bell, E.H. and Gilmore, G.H. 1936. The Nehawka and Table Rock Foci of the Nebraska Aspect. In E.H. Bell, Chapters in Nebraska Archaeology, pp. 301-356. University of Nebraska, Lincoln, Nebraska.

Blakeslee, D.J. and Rohn, A.H. 1986. Man and Environment in Northeastern Kansas: The Hillsdale Lake Project. U.S. Army Corps of Engineers, Kansas City District.

Brewer, R. 1964. Fabric and Mineral Analysis of Soils. John Wiley and Sons, New York.

Brown, K.L. 1984. Pomona: a Plains Village variant in Eastern Kansas and Western Missouri. Unpublished Ph.D. dissertation, Department of Anthropology, University of Kansas.

Fenneman, N.M. 1931. Physiography of the Western United States. McGraw-Hill, New York.

Fitzpatrick, E.A. 1984. Micromorphology of Soils. Chapman and Hall, London.

Rohn, A.H. and Daniel, K.J. 1984. Rocky Ford Archaeology, Hillsdale Lake, Kansas. Wichita State University Publications in Anthropology, No. 2, Wichita, Kansas.

Shepard, A.O. 1954. Ceramics for the archaeologist. Publication 609, Carnegie Institution of Washington.

Soil Conservation Service. 1975. Soil Taxonomy: a Basic System of Soil Classification for Making and Interpreting Soil Surveys. U.S. Department of Agriculture Handbook, No. 436, Washington, D.C.

Soil Conservation Service. 1979. Soil Survey of Johnson County, Kansas. U.S. Department of Agriculture, Washington, D.C.

Wilmeth, R. 1970. Excavations in the Pomona Reservoir. Kansas State Historical Society, Anthropological Series, No. 5, Topeka, Kansas.

Witty, T. 1967. The Pomona focus. Kansas Anthropological Association Newsletter 12 (9): 1-5.

Witty, T. 1981. The Pomona focus, known and unknown. The Missouri Archaeologist 42: 77-84.

PART IV:

STONE ARTIFACTS AND MINERALOGY

QUARTZ GRAINS STUDIES: ENVIRONMENTAL RECONSTRUCTION FOR ARCHAEOLOGISTS BY SCANNING ELECTRON MICROSCOPY

P.A. Bull and A.W. Magee

Summary

Scanning electron microscope (SEM) analysis of quartz grains taken from archaeological sites can throw much light upon the problematic geologic layers barren of artefacts and other kinds of sedimentary evidence. SEM analysis can identify palaeoenvironmental histories of the material and indicate their provenance and palaeoenergy conditions of emplacement. A case study of Tabun Cave, Mount Carmel, Israel is provided as an example of SEM reconstruction techniques.

Introduction

In recent years archaeologists have begun to take very tentative steps towards quartz grains studies by SEM in order to reconstruct past environments in archaeological sites (e.g. Tankard, 1974; Tankard and Sweitzer, 1975; and Meie et al., 1981). These sites range from classic cave sequences such as those of the Peking Man site (Zhoukoudian) to essentially geologic sites which contain minimal archaeological material. The problems which exists for these archaeologically-trained SEM "beginners" is, on the face of it, quite daunting. There are over seven hundred published papers on the use of the SEM as a tool in environmental discrimination (see Bull, 1981; Krinsley and Trusty, 1986; Bull et al., 1987). With some of the papers, at least, there is testimony to inadequate training and grasp of SEM rationale. These published "red herrings" serve to confuse rather than help readers. As a result of this literary mine-field, this paper intends to set out as simply as possible some do's and don'ts of sample collection, preparation analysis and interpretation procedures with particular reference to archaeological problems.

It is perhaps to the advantage of archaeologists that many of the problems in SEM interpretation of quartz grains have already been sorted out. From the early days in the 1960's when it was possible to examine but a few grains, photograph them in the SEM and then publish the results, we have now moved into a more scientific method of enquiry. So long as the record is preserved on the surface of the grains to be examined, it is now possible to reconstruct in great detail the palaeoenvironmental history of a sample. This is of particular interest to archaeologists who, when examining sedimentary layers, often designate them "barren" of archaeological artefacts. If these layers contain quartz sand grains, it is probable that they are of more value in environmental reconstruction than the artefacts themselves. It is important to remember that we can often tell more from the bits of sediment which archaeologists throw away than they can from artefacts which they keep and study intently. It is not just a case of throwing the baby out with the bath water; it is rather that the bath water was never collected in the first place. Used in conjunction with other more standard archaeological investigation techniques, the SEM can provide a most powerful adjunct to the reconstruction of prehistory.

The basic rationale of the technique is quite straight forward. Quartz grains, which have been weathered and eroded from their parent rock, are often transported on their journey ultimately to the sea by a number of different modes of transportation. Such modes include transportation by glaciers, rivers, and wind. Each of these different modes of transportation move grains in a different manner. Glaciers, for example, move material by grinding and crushing; hence the grains tend to exhibit very sharp edges and high relief (Fig. 1). Associated with this grinding action, grains often exhibit surfaces which appear to have been scratched and crushed, and generally show much disruption. Quartz grains which have been transported by fluvial action, or indeed by sub-aqueous action such as movement in the sea, show surfaces which have been generally rounded; edges are very often abraded and the grain surface often exhibits ubiquitous disoriented V-pit structures (Fig. 2). Those grains moved by aeolian transportation show broadly similar features of reduced relief and dulled edges. This is, of course, because the mechanism of transportation by rolling and bouncing the grains is roughly the same as that of sub-aqueous transportation, although the medium of transport is different. The most characteristic features found on the surfaces of quartz grains which have been modified by aeolian action are: dish-shaped concavities (Fig. 3) and crescentic impact pits. It is perhaps important to note here that whilst particular types of sediment transport can impart features on the surface of quartz grains that appear unique to that environment, we are in fact dealing with "sets" of energy conditions which themselves may be characteristic of that environment. This is not a pedantic point since energy conditions need not always be unique to one particular type of transportation. It is crucial therefore to seek other lines of corroborative evidence such as sedimentary structures and grain-size distributions and not use the SEM as a one step indicator of any particular environment.

One of the great advantages of investigating quartz surface textures is that quartz is a very resistant mineral. It can often pass through a large number of different environments, even during one sedimentary cycle. It is possible therefore, to envisage a situation where it may be possible from the combination of the textures on the surface of grains, to identify a complex palaeo-history for the grain. A grain may start its life in a Millstone Grit outcrop, be weathered and eroded by a glacier downstream and when the glacier melts, be deposited in a soil sequence, where it is later eroded by a river, transported to the sea, and ultimately thrown up onto a dune ridge or storm beach. This ridge may then be submerged and later investigated via a borehole sunk on the shore. SEM techniques can be used to reconstruct this history (Bull and Culver, 1980).

It is also equally possible to reconstruct the environmental history of a sand grain through a number of different sedimentary cycles. Quartz grains laid down in a deposit that is ultimately lithified to a sandstone, will contain surface features which are representative of the transportation mechanisms prior to lithification of that rock. Whilst, of course, this can confuse present day analysts who investigate the quartz grains that have now been removed from that

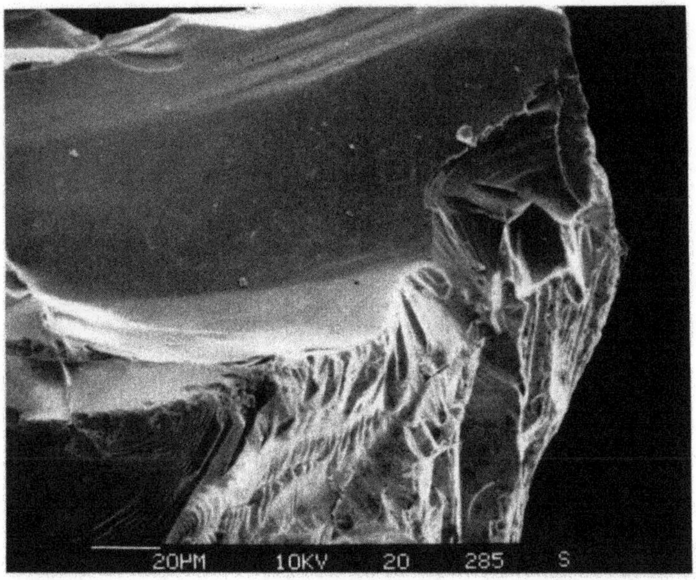

Fig. 1. SEM photomicrograph of quartz grain taken from a glacial snout in Kashmir, India. Note the sharp edges, high relief, conchoidal fractures, small adhering particles, arc-shaped steps and scratches.

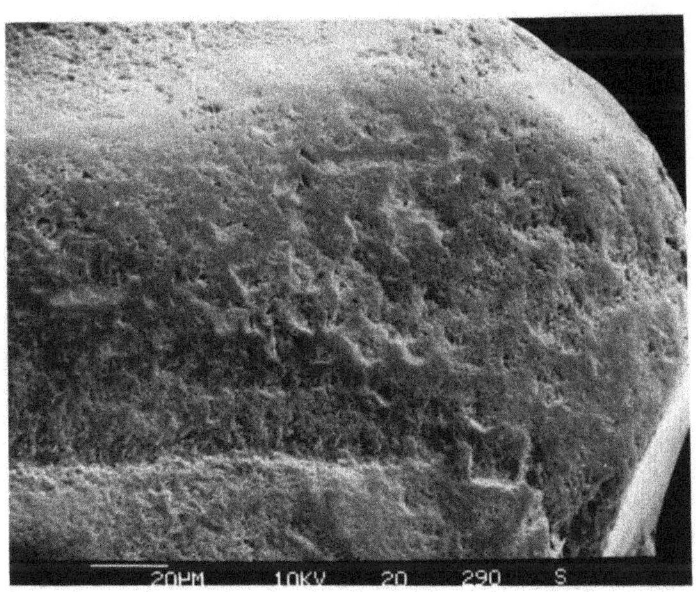

Fig. 2. SEM photomicrograph of a quartz grain taken from Caswell Bay, Swansea, Wales. Note the ubiquitous non-oriented V-pits upon a generally well-rounded, low relief grain. Compare with Figure 1.

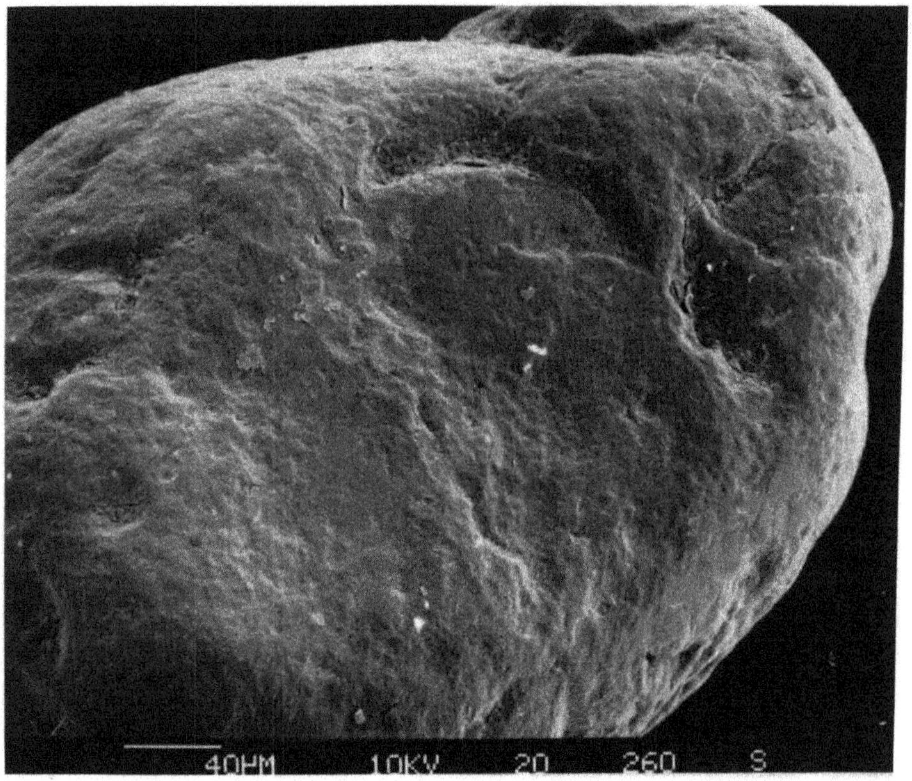

Fig. 3. SEM photomicrograph of a quartz grain from the Saudi Arabian desert. The grain is well-rounded, exhibits low relief and gentle dish-shaped concavities.

rock by weathering processes, geologists can use the palimpsest of surface textures to indicate from which rock body this particular material has been derived (Scholle and Hoyt, 1973; Coch, 1975). Not only, therefore, can palaeoenvironmental histories be reconstructed from the sand grain surfaces, but in certain instances we can also identify provenance. Examples of both palaeoenvironmental reconstruction and provenance studies are detailed later in this paper.

Sample Collection and Preparation

The most powerful usage of the SEM technique in environmental discrimination for archaeologists, is that of section analysis. This is where samples are taken sequentially through a vertical section, or through chronostratigraphically distinct layers. The advantage of sampling in this manner is that any gross differences in the layers can be attributed to climatic or environmental change. If this SEM analysis is allied to some geochronometric dating technique such as Uranium/Thorium dating, detailed interpretations of the environment (involving both climate and modes of sediment transport) can be ascribed an actual date.

It is most important to follow the correct laboratory preparation techniques for the sediment, since it is possible to introduce inadvertent operative bias in both the cleaning and selection of quartz grains for SEM analysis that will prejudice the final results. To insure meaningful interpretation of the quartz sediment, at least 30 to 50 grains must be analysed (Tovey and Wong, 1978). These grains must be picked from the residue of the treated sediment. Firstly, five grams of the raw sample should be boiled in 10% hydrochloric acid for 20 minutes and then thoroughly washed in distilled water. To the sediment is then added dilute sodium hexametaphosphate (Calgon solution). The mixture is then shaken and washed in more distilled water. If the grains appear particularly red in colour, it is likely they are coated with iron oxide. This can be removed by boiling the material in a stannous chloride solution for 20 minutes. Following further washing in distilled water, the samples are then air-dried or gently oven-dried at less than 40°C. From the residue, between 30 to 50 quartz grains should be selected (see Bull, 1978) and mounted on specimen stubs suitable for the particular SEM being used for the analysis. Before machine analysis, the stubs should be coded, preferably by some person other than the SEM researcher. This double blind test enables accurate and confident predictions to be made at the interpretive stage when all the samples have been analysed. The grains mounted on the stub are then sputter-coated with a thin layer of gold-palladium or carbon.

During SEM viewing it is most important to utilize the time on the machine to the fullest. This normally involves some form of check list of features. There are perhaps some 50 or more textures that have been recognized, although for any particular sample it is usual that only the primary 30 or so be investigated. Table 1 presents some surface feature categories that are commonly found on sand grains. They are divided into the mechanical features, morphological features, and chemical features. The mechanical features include modifications such as grain breakage, edge abrasion, conchoidal fractures, stepping, striations and grinding features. These alterations may have been formed by the last transport mechanism (e.g. water, wind or glacier), or may indeed by inherited from a previous cycle of sedimentary erosion. It is always important, however, to ascertain whether the textures derive from the first cycle quartz, that is quartz that has come from a nearby granitic source. There has been much confusion in the published literature concerning the differentiation between first cycle quartz textures and those derived from glacial grinding. The reader is urged to consult the following references for clarification (Krinsley and Takahashi, 1962; Krinsley and Donahue, 1968; Margolis and Kennet, 1971; Krinsley and Doornkamp, 1973; Whalley and Krinsley, 1974; Eyles, 1978).

The morphological surface feature categories listed in Table 1 give some indication of the roundness, angularity, and general relief of the surface of the grain. These categories can be altered as required. The chemical features are more varied and include both precipitation and solutional features. Many of these chemical features have rather exotic names, including scaling, carapace, euhedral crystals, chattermarks, dulled surface, and anastomosis. The reader is urged to consult Krinsley and Doornkamp's (1973) excellent atlas.

For laboratory analysis, Table 1 would include the textures as set out in the horizontal rows. However, the vertical columns comprise 30 to 50 boxes, each of which represents one grain to be analysed. Table 2, however, contains agglomerated surface feature percentages, portrayed in a visual manner. Using such a data presentation format, many thousands of observations can be provided in one diagram. Indeed the surface feature diagram presented in Goudie and Bull (1984) covers one page, yet succinctly summarizes some seventy-six thousand observations. Not only is the editor kept happy by such brief presentation but the publication of such data provides for later re-evaluation by other researchers. Together with the SEM micrographs and laboratory notes, even the most detailed and seemingly unimportant material can be provided for later analyses. This ensures, when a valuable archaeological site is sampled, that the data collected can be recorded fully and published.

Interpretive Procedures

Once the laboratory analysis has been completed and the check-list (Table 2) has been compiled, interpretation of the results can be undertaken. Table 2 provides detail of analyses carried out on a set of samples taken in section through Tabun Cave, Mount Carmel, Israel and will be used to illustrate the interpretive technique. The table provides fine detail of some 42,500 observations and measurements, yet allows up-section variation in environmental history to be plotted.

Tabun Cave was initially excavated by Garrod and Bates (1937) to a depth of 16 metres. They identified seven culturostratigraphic layers based primarily on archaeological evidence. The sampling used for this analysis was undertaken by Goldberg from Jelinek's (1981) excavations and is briefly described in Bull and Goldberg (1985). Table 2 provides 25 sample layers which correspond to Garrod and Bates' (1937) stratigraphy in the following manner (Bull and Goldberg first):

Table 1. Typically used SEM "score chart". The surface features and the categories are to the right, whilst the top row of numbers represents each grain in the sample.

Surface feature categories																						
34. Chattermarks																						
33. Euhedral silica																						
32. Amorphous ppt. (silica)																						
31. Carapace																						
30. Scaling																						
29. Solution crevasses																						
28. Solution pits																						
27. Dulled surface																						
26. Anastomosis																						
25. Oriented etch pits																						
24. High relief																						
23. Medium relief																						
22. Low relief																						
21. Angular																						
20. Subangular																						
19. Subrounded																						
18. Rounded																						
17. Dish-shaped concavities																						
16. Mechanical V-pits																						
15. Curved scratches																						
14. Straight scratches																						
13. Meandering ridges																						
12. Fracture plates																						
11. Adhering particles																						
10. Imbricate grinding																						
9. Parallel striations																						
8. Arcuate steps																						
7. Straight steps																						
6. Conchoidals (> 10 μ)																						
5. Conchoidals (< 10 μ)																						
4. Breakage blocks (>10 μ)																						
3. Breakage blocks (<10 μ)																						
2. Edge abrasion																						
1. Complete grain breakage																						

Sample number

Table 2. Surface feature summary chart for a section through deposits in Tabun Cave, Mount Carmel, Israel. Each box contains a converted percentage (summarized into classes by graphical representation) of the presence or absence of surface feature categories identified from 50 grains per sample.

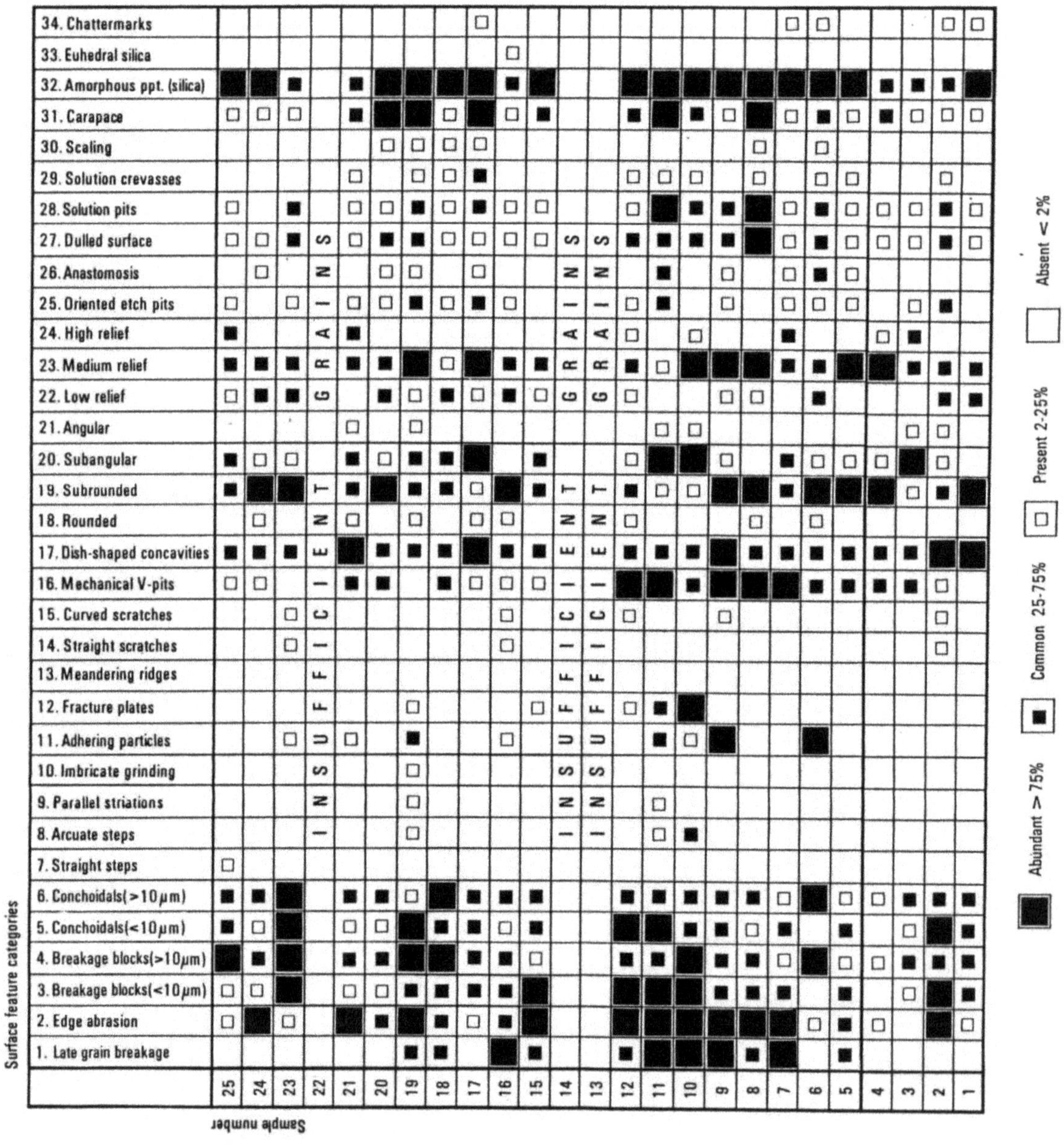

1,2(G); 3,4,5(F); 6,7(E/F); 8,9,10,11,12,13,14,15(E); 16,17,18(Ea); 19,20,21,22,23(D); 24,25(B/C). Considering that all the samples have been double-coded (not in sequence), the reconstruction of Table 2 is quite remarkable since the palaeoenvironmental trends portrayed are quite outstanding.

Table 2 shows, in the first instance, that there were insufficient quartz grains in the raw sample of 13, 14 and 22 to effect any sensible analysis and reconstruction. This is a most important point. If, say, only 5% of the material is quartz, it is quite meaningless to analyse the quartz and to ascribe the palaeoenvironmental reconstruction to the whole sample.

Generally, however, Table 2 shows that most grains contain conchoidal fractures and breakage blocks of different sizes, but there is no up-sectional variation in occurrence. Whilst this has no direct implication, it at least suggests that there are not materials now present in one or more layers that are distinctly varied in provenance from any others. This does not mean that all of the grains come from the <u>same</u> source, but that they do not seem to have come from markedly <u>different</u> sources. Few grains contain surface features in categories 7-15. These, if found in abundance on many grains often suggest glacial modification. Whilst the grains have not come from a local glacial source material, they may have had these small textural features eroded during subsequent water or wind rounding. Many grains exhibit mechanical V-pits on their surfaces; these are quite suggestive of fluvial or marine modification during traction and saltation processes. Predominant, however, are the V-pits found on the grains from samples 3 to 12; other samples show signs of V-pits, but it is this area of the section which has clearly been exposed to sub-aqueous modification.

Since all of the samples produced grains that have their angular edges abraded (category 2) and the grains are generally sub-rounded and of reduced surface relief (due to mechanical abrasion), it would seem likely that a large proportion of the grains have been modified by wind/water action. Furthermore, the very characteristic dish-shaped concavities (category 17; Table 2) found on the surfaces of most grains suggest that there has been extensive grain surface modification in aeolian conditions.

Once the grains were deposited, very little chemical or diagenetic alteration to the grains occurred, as testified by the general lack of grains exhibiting textures 25-34. There is, however, one very important additional feature present on Table 2 which can unscramble the trends mentioned above. Late grain breakage (category 1) is an unusual phenomenon not found on many naturally produced quartz grains. The grains, often rounded, can be found <u>in situ</u> broken in half. The palaeoenergy conditions necessary to break a free moving grain (in water or wind) are in excess of 10 m/s (water) or 35-40 m/s (wind) (Bull, unpub.). Clearly such conditions cannot prevail in nature very often, yet samples 7-19 show almost complete grain breakage in each sample analysed. It is interpreted here, that the grains were broken when they were thrown against an immovable object (a cave wall is an excellent target) during high storm conditions. Further, it would appear that these high storm events, reflected in samples 7-19 (Garrod and Bates' 1937 culturostratigraphic layers designated E and Ea) were marine conditions. Note the high intensity edge abrasion, high percentage of V-pits, and subrounded and medium relief grains in these layers. These subaqueous emplacements would have been similar in every way (except the marine modification) to those aeolian derived samples above and below. The source and immediately subsequent palaeoenvironmental histories of these grains may all have been similar. These events are portrayed diagrammatically in Table 3. Table 3 also shows that sample 16 is different from all other samples. It is only marginally different from other layers when compared on Table 2 (it has fewer breakage blocks and conchoidal fractures, more scratches, and less chemical activity than other samples taken as a whole) but distinguishes itself by its general appearance under SEM visual analysis (see Bull and Goldberg, 1985, for more detail). Care must be taken, then, if only information obtained from these summary charts is used. SEM laboratory notes and photomicrographs make important adjuncts to any full SEM analysis and this, as mentioned before, represents only one line of investigation in a study aimed at understanding a section or sequence of archaeological materials.

Conclusions

The importance of SEM studies in environmental reconstruction theory can be summarized by reference to the case study of Tabun Cave.

Firstly, pre-coded, non-sequential samples need to be provided in order that detailed and unbiased analysis of section trends can be made (one of the authors, P.A. Bull, had not even been to the site and knew nothing of its archaeological history).

Secondly, section trends need to be identified from tabulated semi-quantitative data. Once all of the different trends are identified they need then to be related to known chronostratigraphic or culturostratigraphic horizons. In the case of the Tabun Cave these trends and terrain boundaries matched almost exactly, adding conviction to the importance of these identified sequences of events.

Thirdly, analysis should be undertaken with caution. The semi-quantitative summary charts noting the presence or absence of surface features on quartz grains should not be used solely as a one-step indication of palaeohistory. In the case of Tabun Cave it is possible to infer but not identify a marine transgression into the area about the time of layers 7-17. "Absolute" identification can only come with reference to previous archaeological studies. In this case, as with many other research projects, the SEM analysis confirms and adds detail to previously supposed (but not proven) ideas. As an adjunct, or even a "last-resort" technique when all else has failed, SEM is a vital tool for the research archaeologist.

TABLE 3: Provenance and palaeoenvironmental history of Tabun Cave sediments.

SAMPLES		PALAEOHISTORY
This Paper	Garrod and Bates, 1937	
17-25	Ea, D, B/C	Source Rock Wind Transportation Dune deposition to cave (by wind)
16	Ea	Different to 1-15, 17-25
7-16	E, Ea	Source Rock Wind Transportation Dune Deposition Beach Storm Beach to cave
1-6	G, F, E/F	Source Rock Wind Transportation Dune Deposition Dune deposition to cave (by wind)

References

Bull, P.A. 1978. An SEM investigation of the deposits from the Minchin Hole and Bacon Hole, Gower. Abstract to the meeting, Deposits in Sea Caves, organized by the Willian Pengelly Cave Studies Trust Limited, March 11, 1978, London.

Bull, P.A. 1978. A statistical approach to scanning electron microscope analysis of cave sediments. In Whalley, W.B. (ed.), SEM Analysis in the Study of Sediments, pp. 212-226. Norwich: Geo Abstracts.

Bull, P.A. 1981. The scanning electron microscope as an adjunct to environmental reconstruction in archaeological sites. Proceedings of the 8th International Speleological Congress, Kentucky, U.S.A., July, pp. 340-342.

Bull, P.A. and Culver, S.J. 1980. The Quaternary deposits of Swansea Bay. In Collins, M.B. et al. (eds.), Industrial Embayments and their Environmental Problems, pp. 39-50. Oxford: Pergamon Press.

Bull, P.A. and Goldberg, P. 1985. Scanning electron microscope analysis of sediments from Tabun Cave, Mount Carmel, Israel. Journal of Archaeological Science 12: 177-185.

Bull, P.A., Whalley, W.B. and Magee, A.W. 1987. An annotated bibliography of environmental reconstruction by SEM: 1962-1985. British Geomorphological Research Group Technical Bulletin 35, 94 pp.

Coch, H. 1975. Scanning electron microscopy of quartz textures from granite grus and associated sediments, Black Hills, South Dakota. Unpublished M.Sci. thesis, Queen's College, City University of New York.

Eyles, N. 1978. Scanning electron microscopy and particle size analysis of debris from a British Columbian glacier: a comparative report. In Whalley, W.B. (ed.), Scanning Electron Microscopy in the Study of Sediments, pp. 227-242. Norwich: Geo Abstracts.

Garrod, D. and Bates, D. 1937. The Stone Age of Mount Carmel 1. Oxford: Clarendon Press.

Goudie, A.S. and Bull, P.A. 1984. Slope process change and colluvium deposition: an SEM analysis. Earth Surface Processes and Landforms 9: 289-299.

Jelinek, A.J. 1981. The Middle Palaeolithic in the South Levant from the perspective of the Tabun Cave. In Canlaville, P. and Cauvin, J. (eds.), Prehistoire du Levant, pp. 203-211. Colloque Internationaux du CNRS.

Krinsley, D.H. and Donahue, J. 1968. Diagenetic surface textures on quartz grains in limestone. Journal of Sedimentary Petrology 38: 859-862.

Krinsley, D.H. and Donahue, J. 1976. Composition of diagenetic and weathering microtextures on quartz sand from Meadowcroft Rock Shelter, Avella, Pennsylvania. American Association of Petroleum Geologists Bulletin 60: 4: 689-690.

Krinsley, D.H. and Doornkamp, J.C. 1973. Atlas of Quartz Sand Surface Textures, Cambridge: Cambridge University Press, 91pp.

Krinsley, D.H. and Takahashi, T. 1962. The surface textures of sand grains. An application of electron microscopy. Science 135: 923-929.

Krinsley, D.H. and Trusty, P. 1986. Sand grain surface textures. In Sieveking, G. de G. and Hart, M.B. (eds.), The Scientific Study of Flint and Chert. Proceedings of the Fourth International Flint Symposium held at Brighton Polytechnic 10-15 April 1983, pp. 201-208. Cambridge: Cambridge University Press.

Margolis, S.V. and Kennet, J.P. 1971. Cenozoic glacial history of Antarctica recorded in sub Antarctic deep-sea cores. American Journal of Science 271: 1-36.

Ren Meie, Liu Zechun, Jin Jinluo, Deng Xiyang, Wang Feiyan, Peng Buzhuo, Wang Xueyu and Wang Zonghan. 1981. Evolution of limestone caves in relation to the life of early man at Zhoukoudian, Beijing. Scientia Sinica 24: 5: 843-851.

Scholle, P.A. and Hoyt, D.E. 1973. Quartz grain surface textures from various source rocks. Geological Society of America Abstracts for 1973: 797-798.

Tankard, A.J. 1974. Surface textures of detrital grains: an application of electron microscopy to archaeology. South African Archaeological Society, Goodwin Series 2: 46-54.

Tankard, A.J. and Sweitzer, F.R. 1975. Textural analysis of cave sediments: Die Kelders, Cape Province, South Africa. In Davidson, D.A. and Shackley, M.L. (eds.), Geoarchaeology, pp. 289-316. London: Duckworth.

Tovey, N.K. and Wong, K.Y. 1978. Preparation, selection and interpretation problems in scanning electron microscope studies of sediments. In Whalley, W.B. (ed.), Scanning Electron Microscopy in the Study of Sediments. Norwich: Geo Abstracts.

Whalley, W.B. 1978. An SEM examination of quartz grains from sub-glacial and associated environments and some methods for their characterization. In Johari, O. (ed.), Scanning Electron Microscopy/1978/I: 353-360.

Whalley, W.B. and Krinsley, D.H. 1974. A scanning electron microscope study of surface textures of quartz grains from the glacial environments. Sedimentology 21(1): 87-105.

CHEMICAL ETCHING OF WEAR FEATURES ON EXPERIMENTAL QUARTZ TOOLS

Kjel Knutsson

Introduction

The nature of surface damage on the working edges of quartz tools which were used experimentally on different types of materials has previously been defined, using a scanning electron microscope (Knutsson 1985 and 1986). As a result it could be concluded that wear features like smoothly "polished" surfaces, which developed primarily on the outermost edges of tools are similar in appearance to wear on flint tools described by Keeley (1980). In addition to smooth areas many other wear features, mostly of mechanical origin, also develop on quartz during use (striations, pitting, etc.).

Based on results from a number of experiments with quartz, an analysis of 500 prehistoric artefacts from the late Middle Neolithic settlement of Bjurselet in Northern Sweden was performed to compare wear features. This analysis revealed that, in contrast to almost all experimental tools, the prehistoric specimens showed no smooth, material-specific "polish" zones. Mechanical wear, mostly in the form of striations, occurred more frequently on the prehistoric specimens. These features were also more enhanced; i.e. they looked broader and deeper than what could be seen on the fresh experimental tools. The surfaces of the archaeological tools, in areas where no culturally-induced wear could be found, showed different types of variable dissolution features not found on newly-struck flakes. These included crystallographically-oriented, V-shaped and amorphous looking etch pits (Fig. 1a, b). The density of etch pits on the tools varied considerably, with some surfaces still nearly intact to others which were totally eroded.

The difference between the prehistoric and experimental tools probably does not have a functional origin. The situation is more likely to be due to postdepositional weathering of the surfaces (see Krinsley and Doornkamp 1973). There were also marked differences among the prehistoric artefacts. My interpretation, following these observations, is that the postdepositional processes change the damaged surfaces in different ways depending on the type of wear that occurred during use. The obvious question following that assumption is, of course, how do the different wear traces change their appearance as a result of weathering processes? Only if the general principles behind wear-formation on quartz tools are understood can one hope to make reasonable predictions about how postdepositional weathering will alter this wear. In the following pages, I will address one aspect of this problem by looking at etched-out wear features on experimental quartz tools.

Chemical Etching of Wear Features

For many years, in the field of material science, chemical etching of worn surfaces has been used as a research tool to enhance surface damage and to make subsurface damage visible on ceramic machining

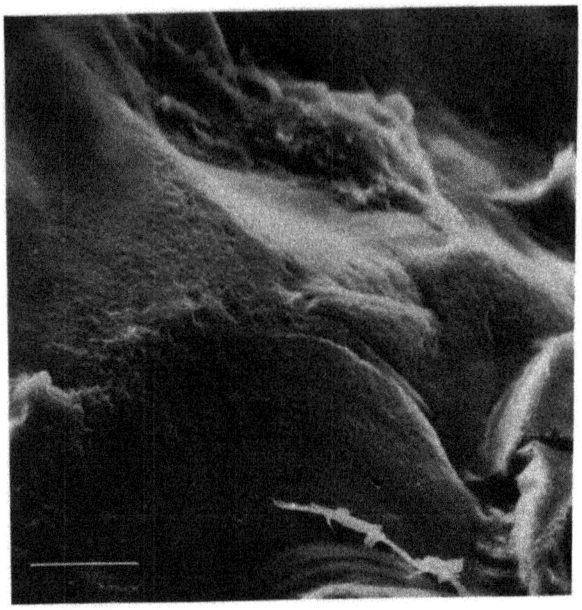

a

b

Fig.1. Evidence of weathering on the surfaces of flakes and tools from Bjurselet, seen as amorphous and crystallographically oriented etch pits. a, specimen number C 319; b, D 37. Scale bar: 10 microns.

damage and to make subsurface damage visible on ceramic machining materials (Koepke 1972, Hockey 1972, Rice 1972, Gielisse and Stanislao 1972, Stokes 1972, Kirchner, Gruver and Richard 1979). Within sedimentological research, chemical etching has also been used to study natural and artificially-induced defects in silicate crystals (Wegner and Christie 1983, 1985). Etching enhances and enlarges microcracks, slip planes, subgrain and grain boundaries, and dislocations in crystals.

It is reasonable, therefore, to assume that microstructural changes (wear) initiated on the edges of quartz tools during use will be highlighted and accentuated by chemical etching, thus contributing to the understanding of processes behind wear formation. At the same time it is hoped that this approach will perhaps give the first clues as to how different wear features and deformed surfaces alter due to dissolution, a process evidenced on the tools at the Bjurselet site. It must be stressed that the aim here is not to simulate conditions during natural weathering. This is primarily a controlled etching experiment, the aim being to study certain wear features enhanced by the etchant. The etchant chosen for the experiment is ammonium bifluoride (NH_4HF_2), generally thought to be a good etchant for quartz and especially suitable for etching out deformations and cracks (see Wegner and Christie 1983: 28).

Sample Selection, Preparation and Etching Procedures

Five tools were selected for the study from an experimental series manufactured earlier. They represent five different wear situations (Table 1) (see also Knutsson 1986). The type of quartz used has properties comparable to the prehistoric tools from the Bjurselet site. The experimental material consists of an ice-transported, homogeneous, monocrystalline variety of quartz, probably originating from pegmatite or migmatite veins where large crystals formed. However, the quartz is not rock crystal, that is, the outer crystal faces need not have been developed. Large numbers of milky streaks in the quartz indicate subparallel cracks developed by pressure when the crystal was still in the vein. In these cracks there is also evidence of second generation crystal growth (B. Collini, pers. comm.). Basically, all wear features observed on any individual tool in the experimental series developed in areas with the same crystallographic orientation. The orientation of cracks within the quartz symmetry might differ from tool to tool. No effort has so far been made to check crystal orientation on different experimental tools, although the resulting variation in mechanical strength might have some affect on wear formation. The selection of experimental tools for the study was based on a previous analysis of wear features on quartz (Knutsson 1985 and 1986) where experimental worked materials were categorized along a continuum, starting with wet and soft materials and ending with hard and dry ones. Variation in the appearance of the wear could be related to these variable material conditions. The materials chosen for this study represent a range of conditions, with wet and dry hide, fresh wood, soaked antler, and fresh bone being used.

Six scrapers from the Bjurselet site were also selected as a test sample for this study. The selection was based on an earlier analysis of 500 randomly selected tools from the site. The tools represent the

Table 1
ETCHED/ANALYZED SPECIMENS

Tool No.	Prehistoric Tool	Experimental Tool	Tool Type	Used on	Use Time	NH_4HF_2 Time	Replica
91	-	X	saw	dry wood	25 min	15 min	X
91	-	X	saw	dry wood	25 min	-	X
108	-	X	scraper	raw hide	30 min	15 min	X
108	-	X	scraper	raw hide	30 min	-	X
112	-	X	scraper	dry hide	30 min	15 min	X
112	-	X	scraper	dry hide	30 min	-	X
159	-	X	saw	soaked antler	15 min	15 min	X
159	-	X	saw	soaked antler	15 min	-	X
241	-	X	plane	dry wood	10 min	5 min	X
241	-	X	plane	dry wood	10 min	-	X
241	-	X	plane	dry wood	10 min	-	-
243	-	X	plane	fresh bone	10 min	10 min	X
243	-	X	plane	fresh bone	10 min	-	X
243	-	X	plane	fresh bone	10 min	-	-
247	-	X	plane	dry wood	10 min	10 min	-
A:52	X	-	scraper	-	-	-	-
B:25	X	-	scraper	-	-	-	X
C:10	X	-	scraper	-	-	-	X
C:13	X	-	scraper	-	-	-	X
C:82	X	-	scraper	-	-	-	-
D:243	X	-	scraper	-	-	-	X

full range of variation of wear among the site's "scrapers".

Methods

All experimental and prehistoric tools from the earlier studies have passed through a cleaning process involving a 24 hour immersion in detergent, followed by cleaning in 5% HCl for the same length of time. Before analysis, the tools selected for this study were then soaked in detergent for 24 hours and held in an ultrasonic cleaning tank filled with distilled water. After air-drying, the edges of the tools were replicated with 2x2 cm pieces of Triafol acetate sheets, dissolved in acetone (for full description of the technique see Knutsson 1983; Knutsson and Hope 1984).

The replicated tools were then put in a NH_4HF_2 solution for between 5 and 15 minutes (see Table 1). During the whole experimental period, the beaker holding the etchant was placed under an exhaust hood. Subsequent to etching, the tools were rinsed in running water for several minutes and then left to dry. After drying, the specimens were given a final rinse in the ultrasonic cleaning tank. After drying again a second acetate mould was made in the same area of each tool where the first was made.

The acetate peels thus produced were examined under an incident light microscope and areas with typical wear features where high friction had occurred were marked. The areas of the peels taken before and after etching were then punched out, producing circular pieces with a diameter of 5 mm. The pairs of replicas representing the same tool before and after etching were then fastened to SEM specimen stubs with double-sided adhesive tape. Care was taken to ensure that the peels always had the edge oriented in the same direction. The stubs were then sputter-coated with gold to a thickness of 600 A. In spite of this relatively "thick" conductive layer, some specimen distortion due to beam damage occurred during scanning (see Figs. 8a, 9c).

Two of the experimental tools (scrapers used on wood and on bone) were broken into small fragments, so that pieces of the used edge could be mounted on specimen stubs and coated with gold. These two tools were scanned directly, thus avoiding the loss of resolution seen in the peels.

Finally, two of the selected prehistoric tools from the Bjurselet site were coated and scanned directly in the SEM. These were viewed in a new JEOL scanning electron microscope with a large specimen chamber, at the Department of Technology in Uppsala. All other analyses were conducted at the Centre for Biomedical Research in Uppsala, using a 15 year old JEOL U-3 SEM microscope. The accelerating voltage was kept at 20 KV in the case of the new JEOL and 15 KV in the case of the older microscope. Charging on the specimen surfaces was a problem with the older instrument. All documentation was carried out with a Hasselblad 6x6 camera loaded with Kodak Verichrome black and white film.

Brittle Fracture and Plastic Flow Wear in Stone Tool Use

During stone tool use, varying conditions interact to form wear. The chemical and structural variation between the different raw materials used to make the tools and the worked substances is thought to be responsible for observed material-specific wear features (Keeley 1980). Within material science, analysis of wear involving two bodies (tool and worked materials) in non-static experiments has provided a substantial body of data on wear mechanisms for different types of raw materials (Vingsbo 1980). The closest analogies to prehistoric stone tools can be taken from wear experiments on ceramics where brittle and semibrittle materials such as magnesium oxide, aluminum oxide, silica glass, etc. have been studied.

The aim of these tribological experiments has been to study surface and subsurface defects on machined materials, primarily simulating secondary fabrication stages like grinding, lapping and polishing. The ultimate goal is to eliminate surface defects which reduce the strength and use lives of ceramic machining parts (see Stokes 1972 and Evans 1979).

Although the goals are radically different, the research methods based on experimental simulations of wear situations and thus, to some extent, also the analytical and documentational techniques needed, are roughly the same in material science and microwear analysis of prehistoric stone tools. Thus, besides the specific information on mechanisms behind the wear formation in brittle and semi-brittle materials, experimental designs could also, at least partly or in a modified form, be adopted from material science by lithic use wear analysts. Comparing the two disciplines today, it is evident that the theoretical and methodological level of microwear analysis is unnecessarily "old fashioned". The most important step for microwear analysis now is to advance from a more descriptive to a more analytical, cognitive approach. For this, greater reliance on the use of SEM is inevitable.

Abrasive damage, which probably constitutes the major part of wear during stone tool use (Knutsson 1985, 1986), is generally classified into two major types when brittle and semi-brittle materials are involved: 1) surface damage consisting of cracks and flaws (scratches, notches, etc.) and 2) subsurface damage consisting of lattice defects, such as dislocations, microtwinning and phase transformations (Hockey 1972: 333). Microtwinning refers to a set of two or more microscopic crystals grown together along a crystal plane or around a symmetry axis. Twinning normally occurs around a few axes or a few planes on which favorable interatomic spacing is found. Orientated differently, the crystals will react variably.

Both the surface and subsurface damage is closely associated with specific observable features such as burnishing, cracking and pitting. These features have frequently been in lithic use wear analysis. Surface wear features are common on both brittle and semi-brittle materials. Experimental replication of quartz tool use has produced numerous striations, patches of abrasion and burnished areas on the tool surfaces (Knutsson 1985 and 1986). Subsurface wear, on the other hand, has generally been thought to occur only in semi-brittle or ductile materials, where plastic deformation is known to occur at normal temperatures. Studies on completely brittle materials, like quartz, have produced only limited evidence of subsurface damage. The generation of isolated dislocation traces or microtwinning during experimental abrasion are the only features which have been observed (Hockey 1972: 333). However, more recent studies show that in a completely brittle material, like aluminum oxide, plastic deformation does occur at a more substantial scale than previously thought. Wheel-ground surfaces of large grain, polycrystal aluminum oxide revealed, after thinning with argon ion bombardment, a discrete, highly deformed layer immediately adjacent to the worked surface (Hockey 1972: 334). Selected area diffraction confirmed the existence of intense lattice strains. The depth of the strained layer was between 0.25 - 1.0 microns, the variation being dependent on raw material and size of the abrasive grain. Hockey (1972: 337) concludes after the analysis: "plastic deformation does play an important role in the removal by abrasion", in brittle material, that is. Comparable studies on semi-brittle materials (such as magnesium oxide) show a similar wear situation, but the subsurface damage (plastic flow wear) is much more extensive (Koepke and Stokes 1972: 242f). After grinding quartz particles in a stone mill, Liedstrom (1968) reports that the depth of the deformed layer of quartz, amounts to around 100 A, i.e., comparable to that in completely brittle aluminum oxide and much less than in semi-brittle magnesium oxide (Liedstrom 1968: 15).

The way different kinds of materials react to wear is actually of minor importance for lithic use wear analysis, since normally only brittle materials are analyzed. Of greater interest is how variables other than the physical properties of the worked material influence wear development. Variables affecting wear described in material science are numerous although not always directly comparable to variables created during simulations of prehistoric tool use. To use information on wear processes described in material science we have to "translate" these processes into comparable units of tool use. For example, "wheel feed rate" is comparable to stroke frequency, "depth of wheel cut" equals pressure exerted during use, and so forth. This transformation is by no means a simple one. The best solution would be, of course, to conduct our own specialized experiments, since we can hardly expect others to do them for us.

Striations on Experimental Quartz Tools

Striations are very common on all of the experimentally used quartz tools, and are, therefore, good indicators of tool use. Some of the variations observed in the form of striations may be attributed to differences in worked materials (Knutsson 1985 and 1986). The striations are primarily thought to be the visual representations of linear arrays of incipient cone cracks, due to high stresses developed between the tool surface and individual abrasive grains (Knutsson 1986).

Experimental studies of crack formation in brittle materials like silica glass, using both sharp and blunt indenters in static loading situations, has demonstrated that the form of the indenter is crucial to the type of crack that will develop (Hagan, Swain and Field 1979: 16). Sharp indenters form radial, lateral and central cracks (Fig. 2). Flatter or rounder indenters form smaller lateral and radial cracks, but, also Herzian cone and so-called bulbous cracks (Kirchner, Gruver and Richard 1979: 24). During dynamic loading situations like tool use, where the force is both tangential and vertical, successive cracks

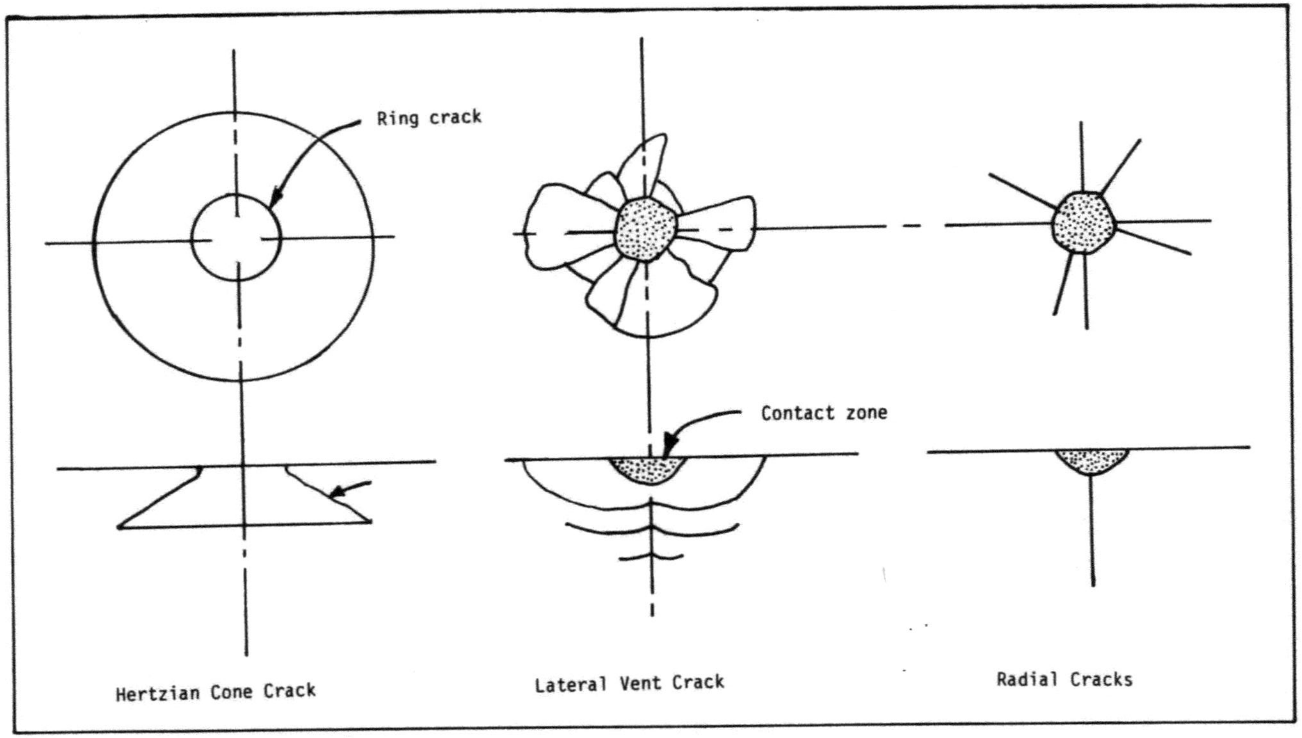

<u>Fig. 2</u>. Basic fracture types developed on brittle materials by localized impact (after Kirchner, Gruver and Richard 1979, Fig. 1).

of the above-mentioned types will form. Striation on used edges of quartz tools are thought to represent this situation.

The distinction between cracks caused by different types of indenters used on isotropic materials is not so evident when anisotropic materials such as quartz are used, because the crystal structure puts restrictions on the ways cracks develop and propagate. The angle of the worn surfaces in relation to the crystal symmetry of the material is thus of some importance here (van Groenau, Maan and Veldkamp 1978: 50). Crystalline material is more apt to produce cone fractures even with sharp indenters. In the case of quartz, this could probably be studied by doing a series of controlled experiments employing single point wear situations with abrasive grains (indenters) of controlled tip radius (for similar applications of this technique see Boese van Geroenau, Maan and Veldkamp 1979).

In the subsurface zone a plastic flow or deformed zone will often develop just beneath the abrasive grains during grinding operations (Fig. 3). Radial, median and lateral cracks are nucleated at these points due to the heterogeneous nature of this deformation. The nucleation and propagation of the median crack is therefore effected by the wedging action of the deformation zone beneath the indenter and not by the indenter itself (Hagan, Swain and Field 1979: 19). The driving force behind the propagation of the lateral crack consists of the unloading residual stresses which occur when the indenter (grain) has passed. The Hertzian cone crack created by spherical indenters or rounded grains is produced as a result of a dominating tensile stress which develops around the indenter.

Apart from the form of the abrasive grain, which is generally not controllable in primitive tool use situations, other factors also interact to form specific wear situations and features of brittle fracture connected with them. A possible difference between isotropic and anisotropic materials was mentioned above. Loading force on the individual grains is another important variable, where high loading is clearly related to the formation of central and radial cracks (Hagan, Swain and Field 1972: 178). In this connection, the size of the abrasive grains is of interest. Fragments of quartz dragged over the surface of the tool during simulations of prehistoric tool use are probably of varying sizes. Thus, different grains cause variable loading situations. Grain size cannot be controlled in experimental replication of quartz tool usage, but may range up to 20 microns. This variation in the size of abrasive grains which can be expected in tool use situations may explain the observed variation in size and depth of striations documented on experimental tools (Knutsson 1985: 18).

It was stated above that the striations formed on experimental tools during use are caused by quartz particles sliding over the tool surface under pressure, thereby producing mainly linear arrays of cracks. This wear situation, then, represents quartz on quartz wear. As Koepke and Stokes (1970) have demonstrated, the relationship between worked material and the chemical and physical structure of the abrasive grains could also be an important variable in the wear situation. A harder material in the abrasive grain is more apt to produce microcutting and, therefore, imitate brittle fracture wear, whereas grains which are similar in structure to the worked material are more

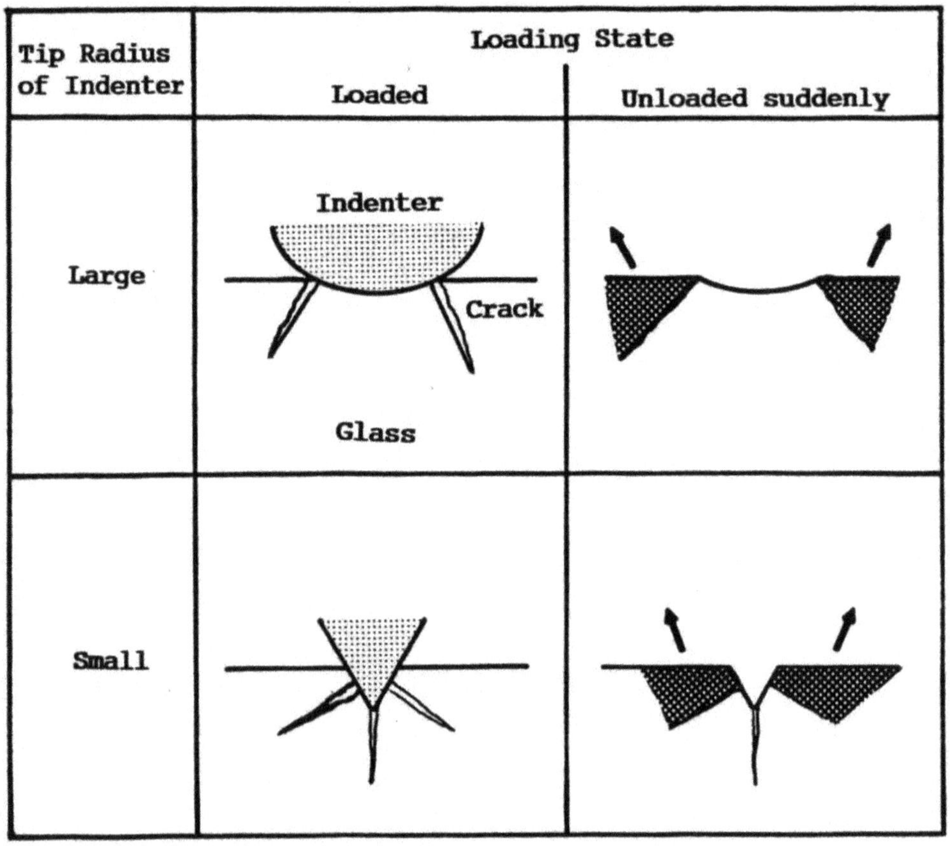

Fig. 3. A schematic diagram indicating typical conditions that apply to an abrading particle during abrasive wear. This is similar to wear situations during experimental tool use (after Evans 1979, Fig. 2a, b).

likely to cause plastic flow wear. The latter case will normally be the most important in tool use situations. In addition to mineral components of the worked material, the introduction of dust and grit particles from the environment may also be an important contributor to the abrasive medium. This is clearly the case with regards to the mineral tanning of hides, where materials like ochre or chalk are introduced between the tool and the hide.

The visible striations seen on experimental tools (compare Fig. 4 and 5), are, as mentioned, characterized by linear arrays of lenticular or crescent-shaped pits. These are most probably caused by linear arrangements of cracks, where the "shoulder" of each crack is torn away due to a combination of tangential and vertical forces in dynamic loading situations, such as tool use (compare Fig. 5 with Imanaka, Fujino and Mineta 1972: 54). Most cracks are thus hidden underneath the surface of the tool. The dimensions and the depth of these "shoulder" splinters depend primarily on the intensity of contact between the tool and the worked material and on the material's physical and chemical properties (Hagan, Swain and Field 1979: 19).

In low friction areas of tool edges, isolated striations may form, whereas high friction areas (eminences along the edge) will suffer severe loss of material due to the repeated chipping away of crack "shoulders". At least part of the rounding of edges and high points in the microtopography noted on experimental quartz tools could be explained by this repeated "shoulder" breakage.

This rather simplified discussion of attritional wear formation on quartz tool edges caused by brittle fracture events indicates that the appearance of the resulting features depends on many variables which may be expected to be present in prehistoric tool use situations. It is therefore reasonable to believe that the working of various raw materials will cause brittle fracture wear with a range of different dimensions. Like smooth polish zones, striations could, therefore, also be a rough indicator of the type of worked material and/or the intensity of frictional forces in specific wear situations (compare Knutsson 1985/1986 and Mansur Franhomme 1983). The variable situations in which wear is formed on stone tools clearly makes this an interesting possibility.

Plastic Deformation on Experimental Quartz Tools

Plastic deformation on quartz tool surfaces is thought to be an important aspect of wear formation. It is probably connected primarily with hard and/or dry raw materials (Knutsson 1986). Surface features such as burnishing, linear troughs, and sleeks (linear plastic deformation or smooth-walled striations), are the visible indications of this type of wear (Knutsson 1986: Figs. 4c, 18a, Hayden 1979: 147, 192).

The plastic deformation of single quartz crystals is caused by dislocation gliding. It is the permanent shear strain which results from the application of a shear stress greater than the material's elastic limit (Nicolas and Poirier 1976: 72). Physically, it can be categorized as translocations of parts of the crystal along slip planes (usually dense atomic planes). Plastic deformation in quartz requires

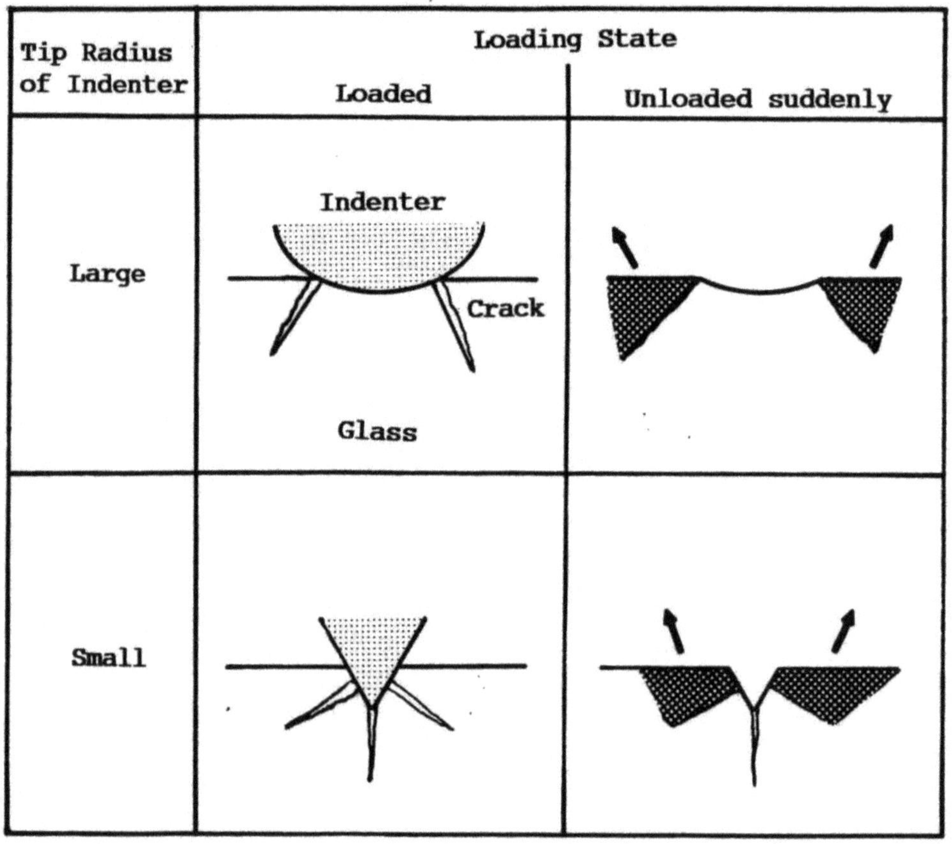

Fig. 4. Schematic diagram of splintering of glass scratched by hard indenter (from Imanaka et al. 1972).

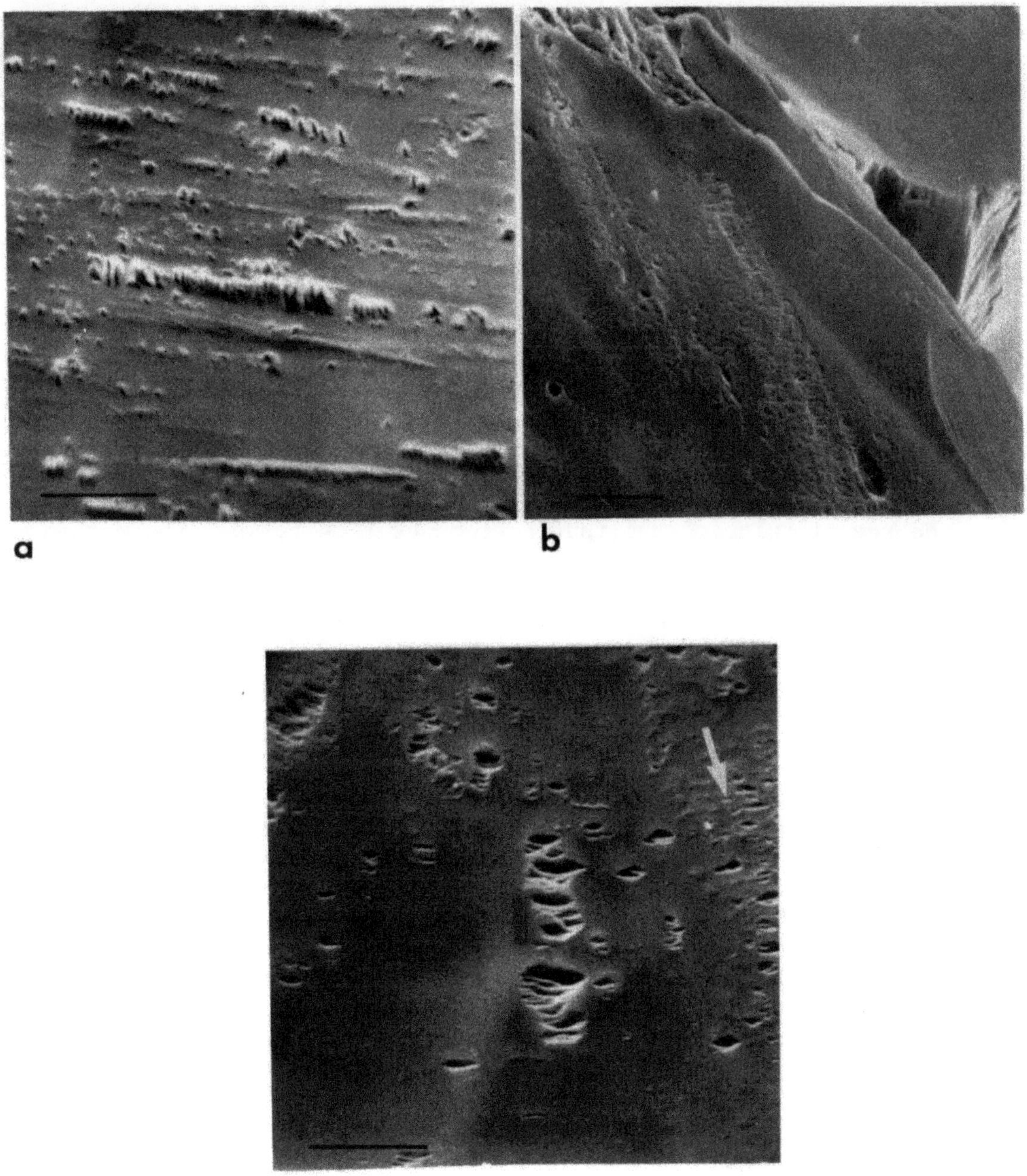

<u>Fig. 5.</u> Examples of striations formed by "shoulder" breakage of incipient Hertzian cone cracks in linear arrangements on experimental quartz tools. a, wood borer used for 30 minutes; b, scraper used on rawhide for 30 minutes; c, Burin used on antler for 15 minutes. Scale bar: a, 5 microns; b, 10 microns; c, 5 microns.

the implementation of very strong forces, but the presence of numerous dislocations in almost all natural quartz crystals (like the stone tool material used here) facilitates deformation (Hagg 1963: 221). Because it is easier to move an already existing deformation than to create new ones, much plastic flow wear in quartz could perhaps best be explained by the slip of already existing dislocations.

Brittle materials do not normally behave plastically, but plastic flow has been seen to occur on brittle and semi-brittle materials, especially when high temperatures are present in wear situations in which high local, short-range stresses occur (Nicolas and Poirier 1976: 72). Hockey (1972: 338) states: "The fact that penetrating arrays of dislocations and microtwins, typical of bulk deformation, are found in the near surface regions of ground specimens but not in polished specimens, is considered to be due in part to the differences in surface heating produced [by] the two operations". Very high pressure can also generate plastic behavior in quartz (Malkin and Huerta 1972: 103), because the pressure prevents cracking. In these stress-intense situations the material will either disintegrate or relax stresses through dislocation sliding (see also Koepke and Stokes 1970: 245).

The presence of a lubricant or coolant at the tool-workpiece interface can drastically alter the wear conditions, lowering the degree of friction and thus heat generation between the worked material and the tool (Koepke 1972: 317). In prehistoric situations, the "lubricant" often consists of water or fat, which also softens the worked material. The effects of other important variables in the wear formation process, such as the hardness of the worked material or the loadings on abrasive grains, will thus be reduced by the presence of a "lubricant" in prehistoric tool use situations.

A typical example of this situation will be presented in the analysis of etched tool edges below. The difference observed in wear formation between tools used on fatty, fresh hides and fat-tanned, dry hides is dramatic. In fact, the two tools used on these materials each represent one of the two basic wear types, plastic flow wear and brittle fracture wear.

The influence of the load on the presence of crack types in glass is demonstrated by the fact that increasing loads result in an increase in brittle fracture formation. This is true of single-point wear situations, where the abrasive grain cuts freely. Where the abrasive grain is prevented from cutting freely, high friction and energy density at the tool/workpiece interface will result. Under conditions of this type, the worked material will react plastically to exerted pressure (Boese van Groenay, Maan and Veldkamp 1979: 50). These situations will occur, for example, when the grinding wheel gets clogged and voids between grains become filled up with material detached from the worked piece. Variations of conditions of this kind can also be expected during prehistoric tool use, where different worked materials have variable structural hardness and "grain holding" capabilities.

The specific situations leading to high temperatures and high pressures in short-range stress situations are complicated and will not be discussed further here (see Vingsbo 1980 for a summary of different types of wear situations). Here it is enough to be able to conclude that plastic flow does occur in quartz, and that variations, primarily in heat and pressure, at the tool/workpiece interface are important variables in wear formation. It is also important to note that the way quartz reacts to these types of short-range stress consists of dislocation sliding, which creates permanent deformations of the surface. Besides producing subsurface dislocations, this type of wear could also involve microtwinning in the disturbed zone.

On the basis of these rather cursory observations on the formation of brittle fracture and plastic flow wear, a tentative model for wear formation in quartz can be put forward (Fig. 6). Wear formation in quartz involves, of course, other kinds of wear events besides brittle fracture and plastic flow wear (see Knutsson 1985 and 1986), but the etching experiments presented here were primarily designed to study these two types of wear.

In the model it is predicted that part of the wear caused by hard, dry materials such as bone and dry hide is the result of plastic deformation, because these materials produce intense local pressure and heat generation in the contact zone between the abrasive material and the tool surface, where the stress fields contain significant hydrostatic components. Under these conditions full relaxation of the load cannot occur by fracture alone (compare Hockey 1972: 338). Softer and/or wetter materials such as fresh wood and reed will primarily produce brittle fracture wear. The water acts as a lubricant/coolant, and the pressure is lowered due to the softness of the materials.

By etching the experimental tools from the different wear experiments, it is anticipated that the variation between these wear types (see Knutsson 1986) will be enhanced and made visible. If plastic flow wear is present, it is thus reasonable to believe that the disturbance of the surface layer caused by dislocation sliding or microtwinning will show up as etched out traces of this wear event. Similar results have been accomplished on deformed brittle materials by sedimentologists (Wegner and Christie 1983, 1985), and material scientists (e.g. Hockey 1972). If brittle fracture, on the other hand, is the predominant type of wear, the etched part of the used area of the tool will have enlarged cracks and also visible microcracks, not seen on the tool before etching (compare Wegner and Christie 1983, 1985 on this point). In short, the differences between the two types of wear will be accentuated through etching and the wear itself will be enhanced. Besides enhancing the wear features described above, it is also hoped that the etchant will remove surface deposits such as easily etched, hydrated surface layers, which mask wear.

Striations in High Friction Areas on Etched and Unetched Experimental Tools

To study the way etching or dissolution affects wear an area of each experimental tool was replicated with acetate peels (Knutsson 1983: 10) before and after etching. After being mounted in pairs on specimen stubs and sputter-coated with gold, these negative casts of worn surfaces were observed in the SEM using a magnification range of between 300x-5000x. The micrographs of striations presented in Figure 7-11 are thus negative impressions of the worn surfaces in high

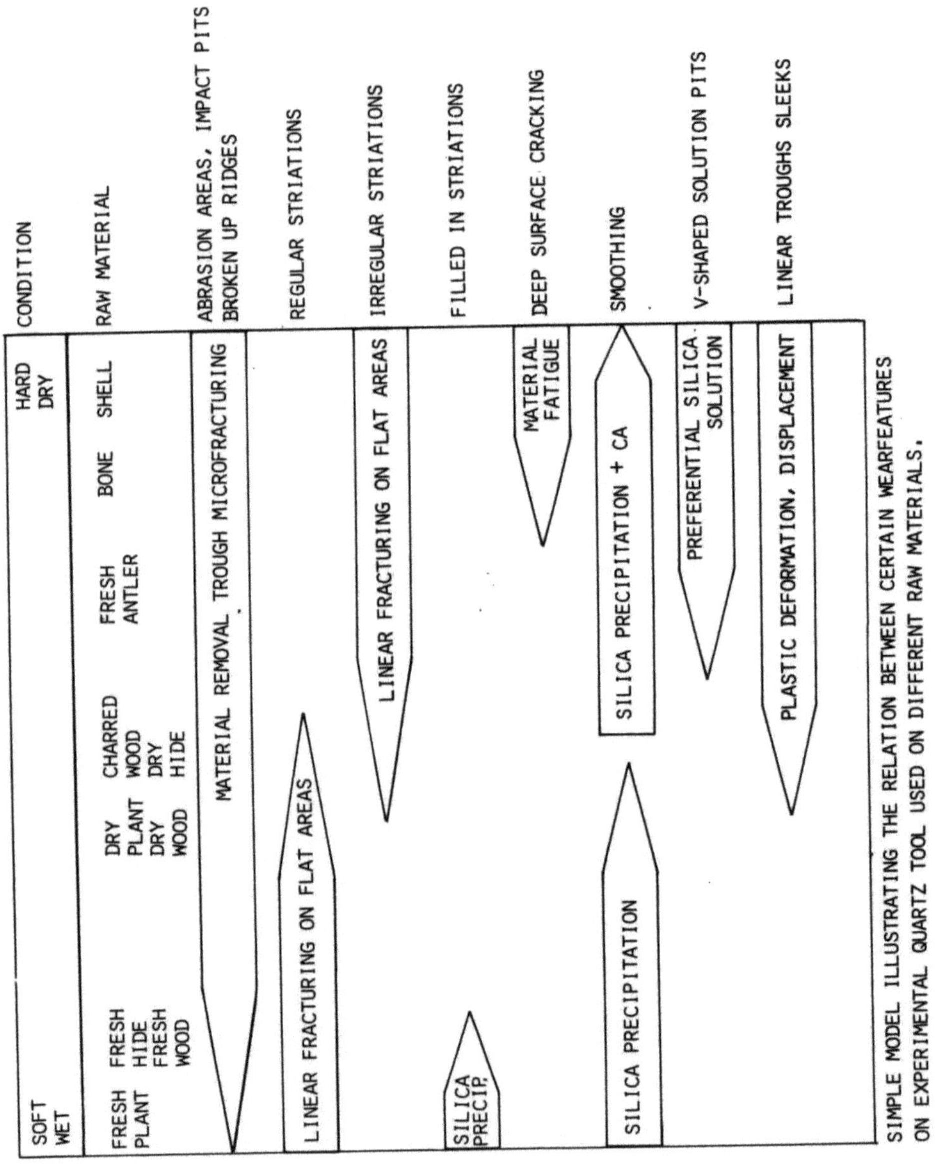

Fig. 6. Simplified composite diagram showing the systematic relationship between certain wear structures and worked material on experimental quartz tools (after Knutsson 1985, Fig. 7).

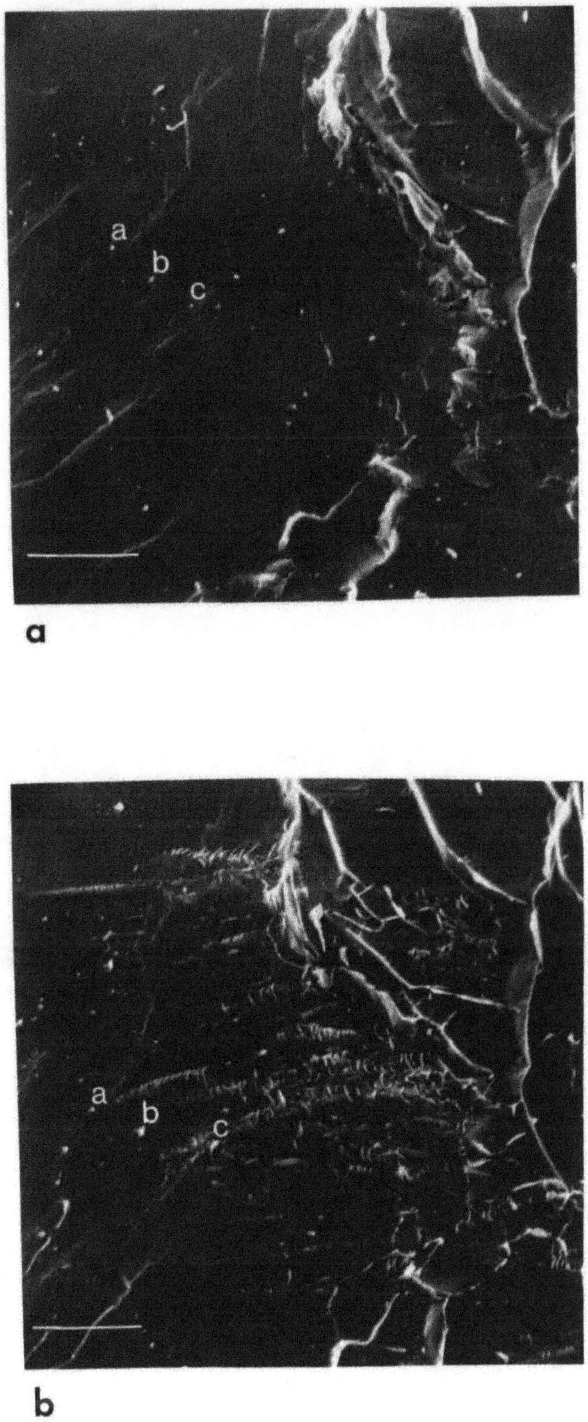

Fig. 7. Quartz flake used to saw antler for 30 minutes. a) unetched, b) etched in NH_4HF_2 for 15 minutes. After etching, vacuoles a-c (on Fig. 7b) and crevasses caused by fracture facets are slightly enhanced, but more importantly, striations consisting of linear arrays of incipient cone cracks are dramatically enhanced and made visible. Scale bar: a-b, 70 microns.

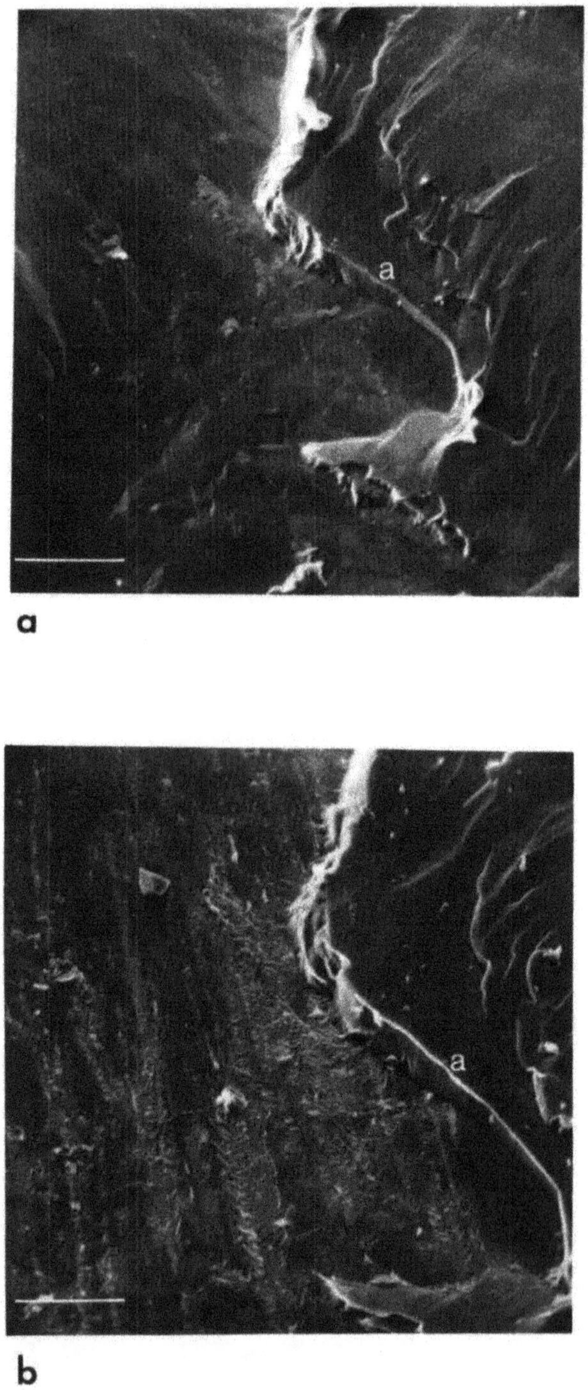

Fig. 8. Quartz flake used to scrape dry wood for 30 minutes. a) unetched, b) etched in NH_4HF_2 for 5 minutes. Crevasses caused by fracture facets (a) are enhanced after etching and numerous striations not visible before etching have been etched out. Striations consist of linear arrays of shallow conchoidal fractures. The undeformed surface does not seem to have been affected by the etchant. Scale bar: 70 microns.

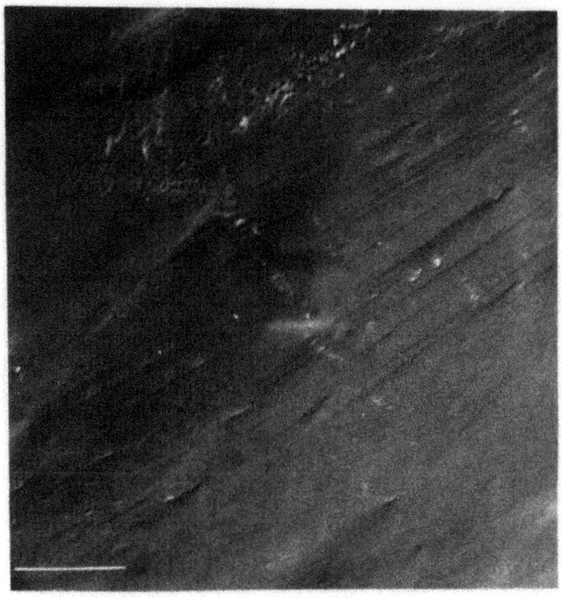

a

b

Fig. 9. Quartz scraper used to scrape raw deer hide for 30 minutes. a) before, b) after etching in NH_4HF_2 for 15 minutes. Evenly spaced and clearly defined striations have developed on a surface where, prior to etching, only faint shadows of a few striations could be seen. Between the "larger" striations on the etched surface are examples of very shallow types (arrows) and scattered impact pits or single incipient cone cracks. Scale bar: 20 microns.

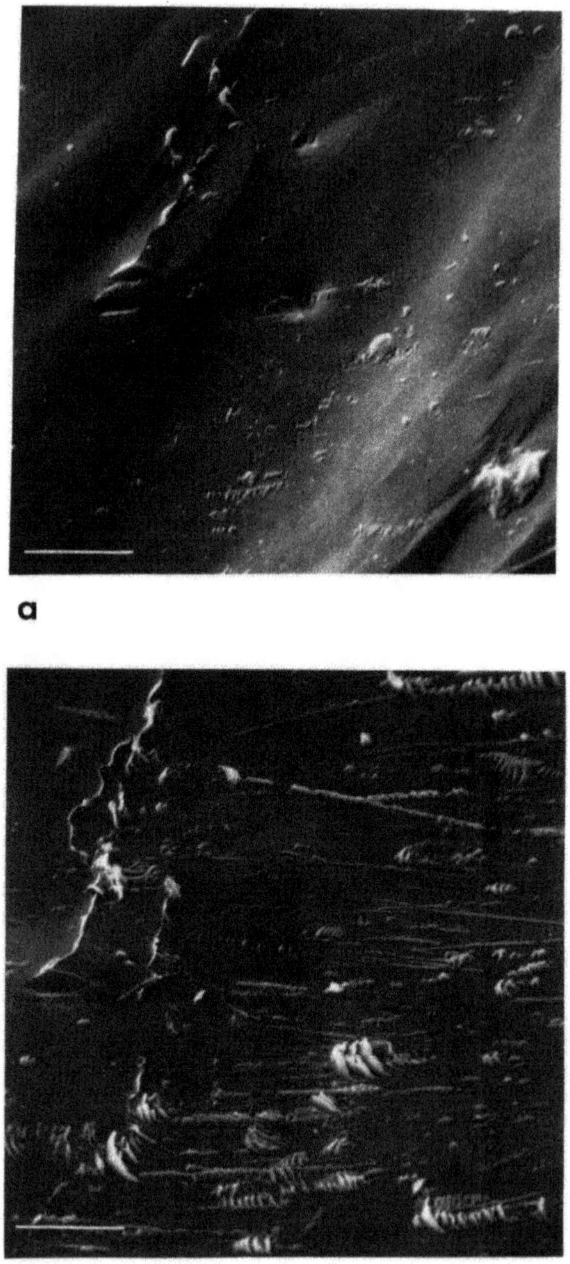

a

b

Fig. 10. Quartz flake used to scrape fresh elk bone for 15 minutes. a) before and b) after etching in NH_4HF_2 for 10 minutes. A slightly raised area of the tool surface shows wear in the form of striations caused by brittle fracture and plastic flow wear visible as sleeks. After etching, the surface is generally "streaky", with lines oriented in the direction of work and heavily highlighted conchoidal fractures. The sleeks form deep V-shaped grooves and/or arrays of weakly developed dislocations after etching. Scale bar: 20 microns.

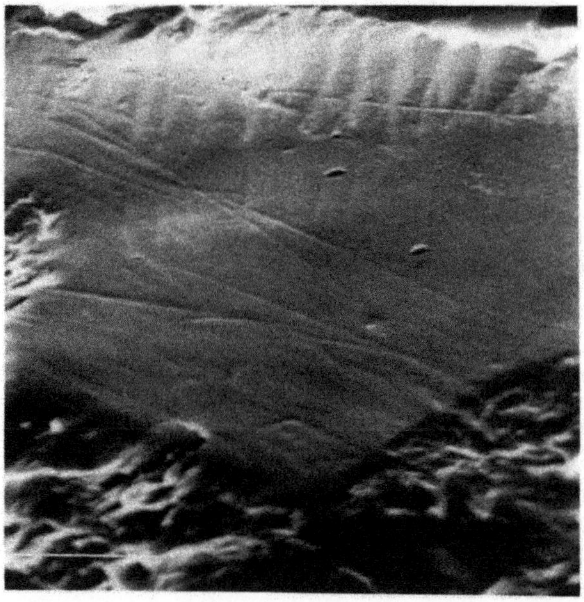

a

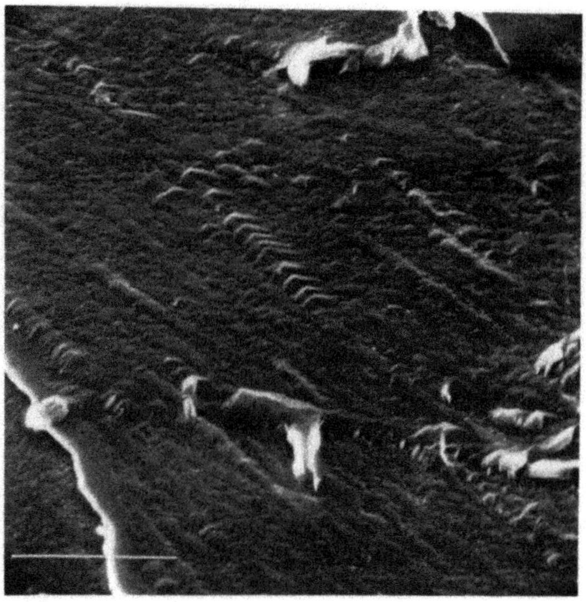

b

Fig. 11. Quartz scraper used to scrape dry, fat-tanned elk hide for 30 minutes. a) before, b) after etching in NH_4HF_2 for 15 minutes. Shallow dislocations are visible in an area which was covered by many sleeks before etching. The majority of dislocations are confined to narrow, linear regions, obviously defining traces of surface scratches. Scale bar: 10 microns.

friction areas of tool edges, before and after etching. An examination of these micrographs shows that etching has left the undeformed surfaces and the surfaces unaffected by wear almost intact, but fairly large cracks and vacuoles are slightly widened. On the other hand, surfaces disturbed by frictional wear undergo strong dissolution, which produces concentrations of etched-out linear arrays of cracks (Figs 7-11). Thus, the most obvious result of the etching experiment is that the number of visible wear features along the tool edges were enhanced dramatically by etching. This clearly indicates that surfaces on quartz tools subjected to wear are often much larger and more widespread than is indicated prior to etching. This observation may be comparable to that made by Mauger and Plisson (1986) in their etching experiments on flint in which they referred to "linear white patina", where etching has probably attacked disturbed areas of tool edges first. If the weathering of worn tools involves chemical dissolution of their surfaces, as clearly was the case in the material from Bjurselet, then it is reasonable to assume that the disturbed areas near the tool's edge which were subjected to frictional wear will be enhanced by weathering. This could at least partly explain the situation at Bjurselet, where highly visible wear features were observed on most of the analyzed tools.

The types of fracture which produce the striations discussed above and which were enlarged by the etchant do not vary in a set pattern according to the kind of stone involved, that is, the cracks forming the striations do not seem to be "material specific". The Hertzian cone fracture is the dominant or perhaps the only fracture type observed in striations. However, evaluation of the wear features is somewhat difficult, especially in heavily worn areas, where striations intersect. It was stated earlier that (as shown by Figs. 2 and 5), visible striations represent a series of chips broken off the tool surface at the shoulders of cracks. Analysis of the striations in this study has demonstrated that the basis for crack formation in different raw materials is the same, but that the size of chips and the way they are broken away from the shoulders of cracks may be more "material specific". This is also indicated by the variation in visible striation forms on unetched, experimental tools (Figs. 5a-c and Knutsson 1975:53). This hypothesis is clearly worth analyzing further, perhaps with the aid of milder etchants and with shorter use times for the experimental tools.

The etching experiment has shown that visible cracks on the surfaces of experimental tools are only indications of wear which is really much deeper. Incipient fractures beneath these features constitute the major part of the actual damage. In other words, only the damage that has been produced under the greatest stress causing "shoulder chip" formation will show up on experimental tools. Much damage in the form of incipient cone cracks will not be detectable on the experimental tool, and a "covering" layer of what is thought to be amorphous hydrated surface material also conceals striations which were visible at an earlier stage of the wear situation (compare Plisson and Mauger 1986). This is especially evident on the wood-working tool analyzed here (compare Figs. 12 and 13). Similarly, examples of latent scratches which appear after cleaning away a thin surface layer have also been reported on polished optical glass by material scientists (Izumitani, Miyade, Adachi and Harada 1979).

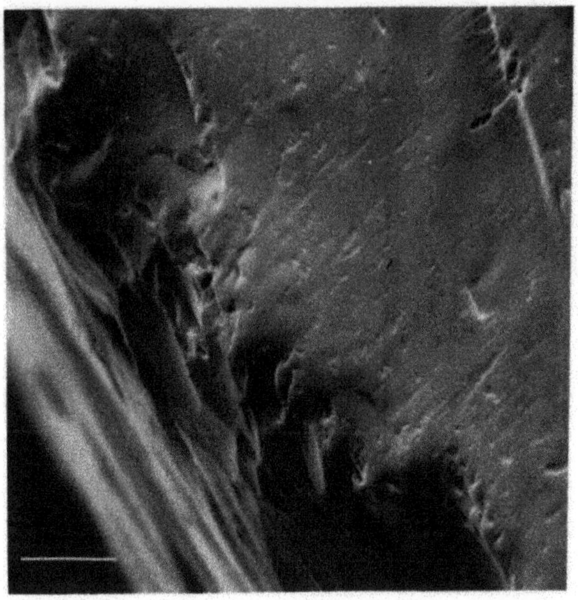

a

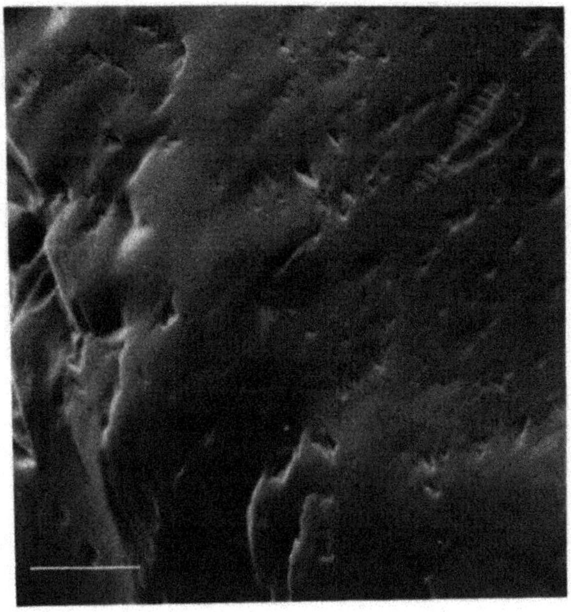

b

Fig. 12. a-b) Quartz flake used to scrape dry wood for 15 minutes. A smooth "polish" zone is characterized by linear troughs and features of brittle fracture origin such as striations and impact pits. Features seem to be "covered" by a smooth surface layer, possibly a hydrated layer of amorphous silica. Scale bar: a, 10 microns; b, 5 microns.

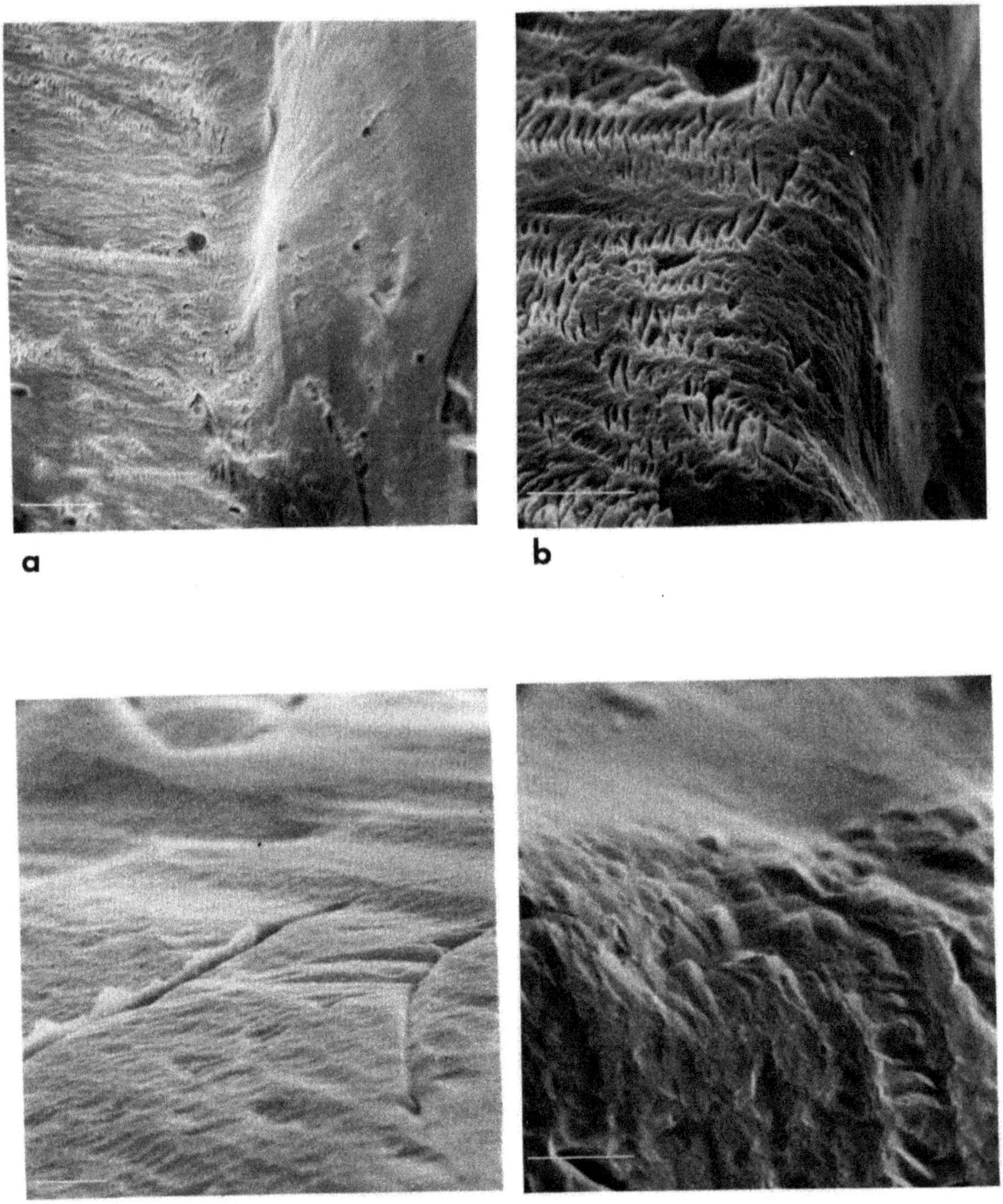

a b c d

Fig. 13. a-d. Quartz flake used to scrape dry wood for 15 minutes after etching in NH_4HF_2 for 15 minutes. a) and b) show how the area closest to the edge has suffered from severe brittle fracture wear. The outermost part, the apex of the edge, is worn down and rounded as a result of material loss; c) depicts linear trough-shaped structures indicating the direction of work. These features are probably caused by brittle fracture where "shoulders" of successively formed cone cracks have been broken away from the surface layer; d) pyramidal features on the edge apex of the wood plane indicate plastic flow wear in this region, although on a minor scale. Scale bar: a, 10 microns; b-d, 5 microns.

This experiment demonstrated that, apart from brittle fracture, there are also clear examples of plastic flow wear primarily on dry hide and bone-working tools, well in line with the hypothetical model presented earlier. However, soaked antler did not exhibit plastic flow wear of this magnitude either on the outer edges of the tools or on the tool surfaces which necessitates revision of the wear formation model. More experiments are certainly needed here.

In the highest friction areas (tool edges), experimental tools used on bone and dry hide (Figs. 10a, b and 11a, b) showed evidence of plastic flow wear before etching in the form of sleeks and linear troughs. Only the plane used to work bone displayed plastic flow wear on high points of the tool's surface (Knutsson 1985: 59). These areas exhibited highlighted dislocation traces after etching. This is especially evident in areas closest to the edges, where high loadings on the abrasive material can be expected, causing high frictional forces. The scraper used on dry hide is especially worn. Dislocations which take pyramidal forms (Fig. 11a, b) are clearly associated with surface structures such as sleeks, produced by individual abrasive particles.

If the outermost edges of the original, unreplicated surfaces of planes used on wood and bone are compared, it is evident that they represent two very different wear situations; in the former case brittle fracture wear dominates, while in the latter plastic flow wear is more significant (Fig. 15a-d). After etching, the wood plane displayed clusters of incipient cracks of the Hertzian type where linearity coincides with tool movement (Fig. 13a, b). The smooth, covering layer is completely etched away. Between the cracks, there are areas with shallow flaking or cracking (Fig. 13c). A closer look at these areas reveals a "grooved" surface, with the grooves indicating the direction of tool use. These grooves seem to indicate extensive material loss with a considerable part of the tool surface removed by mechanical splintering of the surface along the edge. Most of the above-mentioned features are invisible on the smooth surfaces of the experimental tools prior to etching (compare Fig. 12b and 13b). These shallow grooves are probably responsible for the undulating surface of the typical wood "polish". The fact that some striations are visible before etching and others are not could be explained by the fact that they represent different generations of striations, the latest generation being the only one visible (compare Plisson and Mauger 1986). The smooth, burnished surface covering the striations of earlier generations may be caused by mechanical polishing.

Subsurface damage due to plastic flow wear is also displayed on parts of the wood plane. Pyramidal forms, possibly etched-out traces of dislocations or microtwins (A. Wegner, pers. comm.), can be seen on the outermost part of the edge (Fig. 13d) in areas of greatest friction. The superficial similarity between the smooth wear surfaces on bone and wood planes, characterized by linear troughs and striations (Fig. 12a, b and 14a, b), is lost after treatment in the bifluoride etchant. Beneath the smooth surface of the bone plane, subsurface damage in the form of pyramidal dislocations or microtwins indicates plastic flow wear in the area closest to the edge, where frictional forces have been great. This is very distinct from the extensive

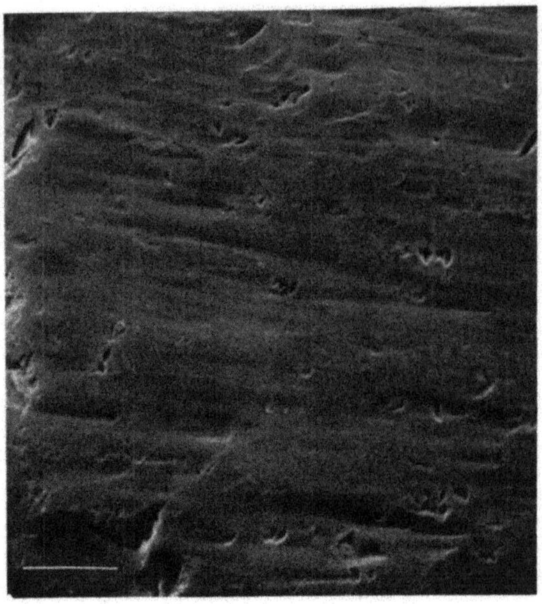

a

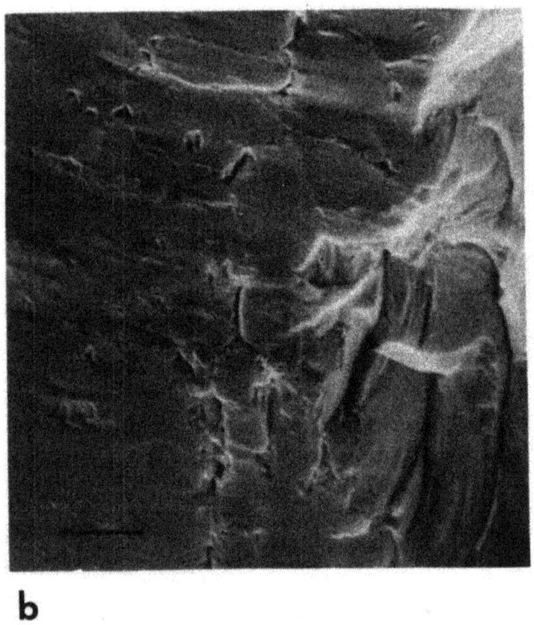

b

Fig. 14. a-b. Quartz flake used to scrape fresh elk bone for 15 minutes. Unetched surfaces. a. Linear trough-shaped structures indicating the direction of work. Among these features-believed to represent a plastic flow, burnished surface layer-there is also evidence of brittle fracture wear in the form of short striations and isolated impact pits. Scale bar: a, 10 microns; b, 5 microns. b. Smooth "polish" zone in the outermost part of the edge rim. The smooth surface exhibits slightly widened cracks perpendicular to tool movement.

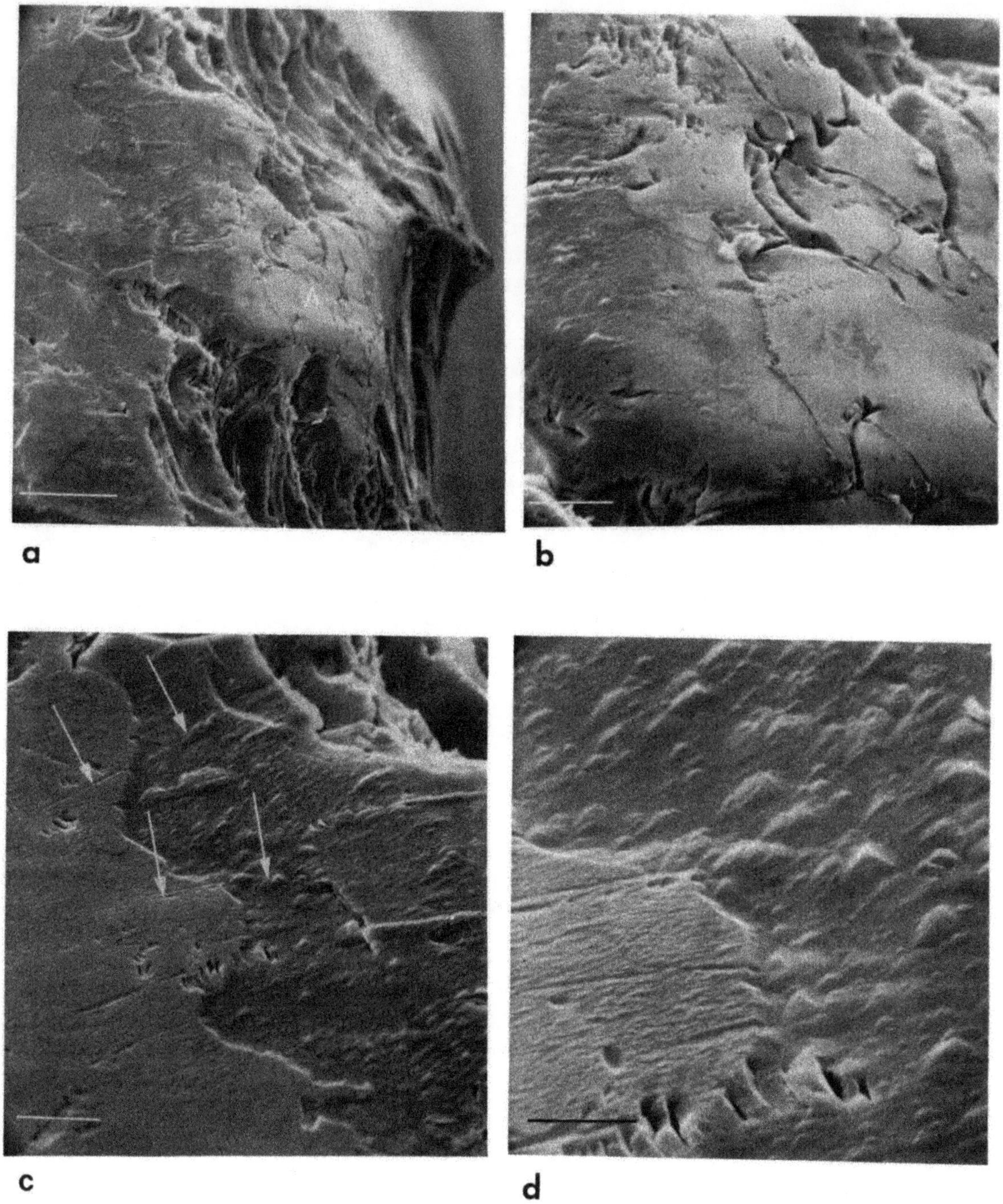

Fig. 15. a-d. Quartz flake used to scrape fresh elk bone for 15 minutes. Etched in NH_4HF_2 for 10 minutes. a. On the edge's apex, the smooth surface is still present after etching. In the area closest to the rim there is an area of pyramidal structures, A, which are traces of dislocations or microtwinning, probably indicating plastic flow wear in this region of the tool. Compare with Fig. 14a where the same area before etching is shown. b. The smooth polish zone has lost its streaky appearance after etching and the cracks seen in Fig. 14b are further widened. c-d. The density of dislocations is greater in a few linear regions which represent the distribution of sleeks on the surface before etching (arrows). Scale bar: a, 40 microns; b, 10 microns; c, 5 microns; d, 10 microns.

brittle fracture wear seen in the subsurface zone on the wood plane described above.

Broad sleeks in the surface layer are clearly connected with linear arrays of closely-spaced dislocation traces, strengthening the hypothesis that sleeks are formed under conditions of extreme pressure and high temperatures at the tool-workpiece interface. The dry hide scraper and the bone scraper thus show that the density of local dislocation traces on the tool edges, enhanced by etching, is dependent on the size and number of scratches produced during tool use on hard and/or dry materials (see Hockey 1972: 335).

In accordance with the findings of Plisson and Mauger (1986), my experiments show that the smooth "polish" zones on bone-working and woodworking tools seem to have different abilities to withstand attack from etchants. In Plisson and Mauger's experiment, wood polishes produced during different wear situations (wet and dry wood) also exhibited variable preservation during chemical attack. The dry wood polish was more resistant. This situation seems to indicate that "polishes" do have different origins and structures which cross the borders of actual worked materials. Evidently the amount of pressure and friction caused by specific material conditions is an important variable here. Dry wood theoretically causes more plastic flow wear, while wet wood gives rise to brittle fracture in combination with easily dissolved gel. Stokes (1972: 345) notes that "The steep temperature, pressure and plastic flow gradient at the tool-workpiece interface can cause phase transformation or recrystallization with preferred grain orientation in the finished surface layers." The occurrence of phase transformation of the outermost edge of bone-working tools could be the reason why bone "polish" shows a comparatively strong resistance to chemical attack. Dry varieties of worked materials cause not only plastic flow wear, but may also, under certain circumstances, cause phase transformation of the outermost edge of quartz tools.

In earlier observations of bone "polish" on quartz tools, thin cracks were seen in the polish zone. These were provisionally explained as material fatigue cracks caused by cyclic load fluctuations or etched-out grain boundaries underlying the "polish" surface (Knutsson 1986). Another explanation for these cracks is thermal quenching (A. Wegner, pers. comm.), i.e. surface cracks formed by the action of thermal stresses during tool use. A closer inspection of these cracks after etching reveals that they were more likely caused by stresses arising on the tool edge and relaxed by Hertzian crack formation during tool use (compare Fig. 15a, b arrows).

In addition to mechanical breakage and plastic deformation, part of the outermost edge of the wood plane displayed mechanical polishing, which left a rounded, smooth, featureless surface after etching (Fig. 16a). Clearly, this etching experiment has shown that wear is variable and that different wear surfaces respond differently to chemical attack. Is it then possible to see similar changes on archaeological material?

It is important not to overemphasize the role of abrasive grains in the form of quartz splinters as wear mechanisms. Although they are

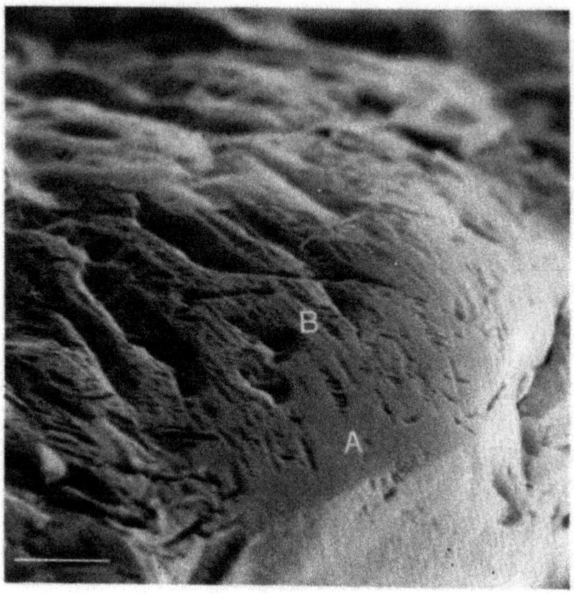

a

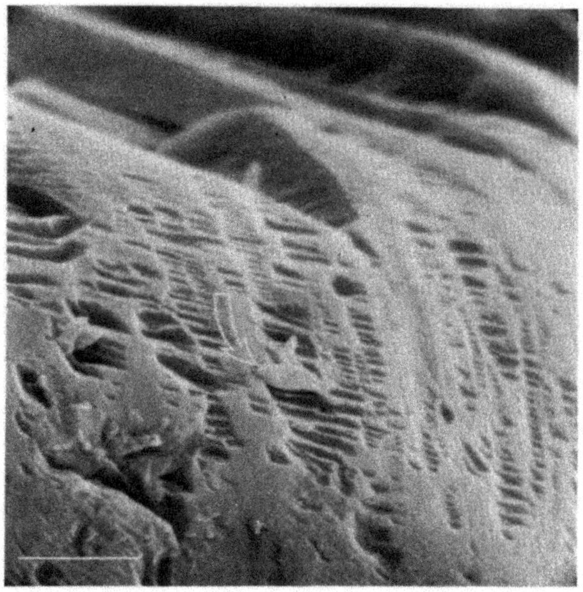

b

Fig. 16. a-b. Quartz flake used to scrape wood for 30 minutes, and etched in NH_4HF_2 for 5 minutes. After etching, the rounding of the outermost edge rim is still visible as a smooth, featureless area. A. Behind this area of lower friction are closely spaced striations (B). In one of the striations a "shoulder" chip is just about to loosen as a result of the etching treatment (arrow). This smooth surface could be the result of mechanical polishing, but its resistance to etching indicates that the surface could have undergone a phase transformation. Scale bar: a, 10 microns; b, 5 microns.

probably quite significant, the frictional forces between the tool edge and the worked material also play a part in creating wear.

Wear Features on Prehistoric Tools from the Bjurselet Site

Only a short comment is possible at this stage of research. A larger SEM analysis of the prehistoric tools is currently being carried out. Analysis of the 500 tools mentioned in the beginning of this paper was accomplished using a metallurgical microscope with incident light attachment.

What was defined as a typical dry hide scraping tool in the light microscope analysis, exhibited features of severe attrition (as shown in Fig. 17a, b). The heavily rounded edge and linear troughs perpendicular to the edge, are very characteristic of dry hide tools. The striations are irregular in outline, similar to those seen on experimental tools used on dry hide. Sleeks are also present in this area, but not in the quantities observed on the experimental tools. The experimental tools with myriads of sleeks along the edge were used to scrape fat-tanned elk hides. The experimental hides were completely dry and very stiff during the first stages of work, so wear was developed in an environment which caused extreme frictional force on the scraper's edge. Thus the scrapers from the Bjurselet site appear to have been used in an operation causing less friction. Perhaps they were employed during the final stage of dry scraping of fresh hides or in the actual currying process where semi-arid conditions prevail.

The Bjurselet scraper defined as a wood plane displays striations of a very different type (Fig. 17:C). These striations have a regular outline with a ground base consisting of closely-spaced, "shoulder" breaks. The lack of a smooth "polish" zone on the edge makes the attribution to wood wear tentative. Similar striations are found on experimental tools used on dry plants and rawhide. Although there is some evidence for the dissolution of striations, it is very minor. The striations stand out in very sharp contrast to the surrounding surface which lacks any sign of weathering.

Another prehistoric "wood plane" was analyzed directly under the SEM (Fig. 18b). The edge of this tool also does not show any smooth "polish" areas, but conchoidal breaks with step terminations expand onto the tool's ventral surface, indicating that the tool was used in a planing motion. Both regular and irregular striations and a large number of sleeks occur on this tool, a situation which is somewhat contradictory to the results obtained by experimentally working wood (Knutsson 1985, Fig. 6). The experimental scrapers were all used on fresh wood, so the presence of water and the pliability of the wood may have prohibited plastic flow wear. It is possible that the tool from Bjurselet was used as a plane on a medium soft material in semi-dry conditions.

One specimen identified as a "dry hide" scraper was also scanned directly in the SEM (Fig. 18a). The heavily rounded edge and broad linear troughs on the surface perpendicular to the edge are evidence of the severe conditions forming the wear on this tool. Closest to the edge is a highly disrupted area showing linear features and brittle fractures. A close look at this area reveals the presence of strongly

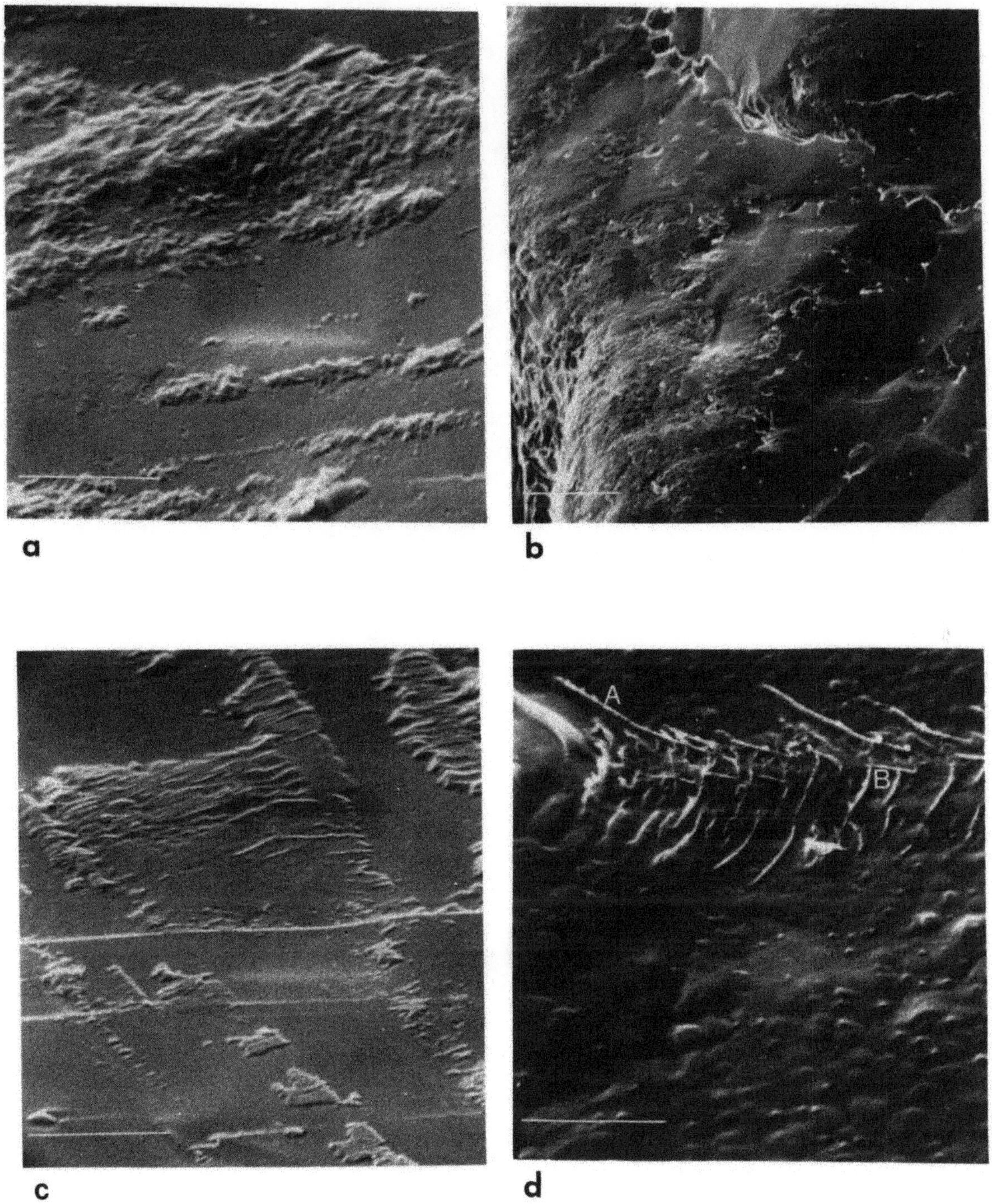

Fig. 17. a-d. Prehistoric quartz scrapers from the Bjurselet site. Negative triafol casts. a) Specimen number B:25. b) C:13. d) D:243. The easily detectable variation in form and structure of the striations on the different scrapers probably indicates that the striations originated in different wear situations. The irregular striations in a) probably have some connection with dry hide wear. The heavily rounded edge rim with mechanical, trough-shaped structures perpendicular to the edge in b) point in the same direction. c) has regular striations, each containing a series of "shoulder" breaks with straight fracture facets, suggesting work on wood. d) Unlike the experimental tools, this heavily weathered tool displays lateral and medial crack formation. Scale bar: a, 200 microns; b-d, 10 microns.

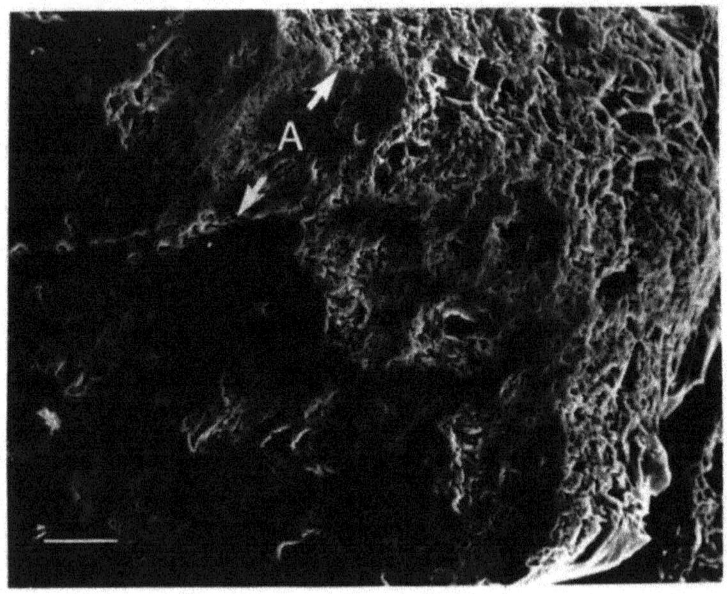

a

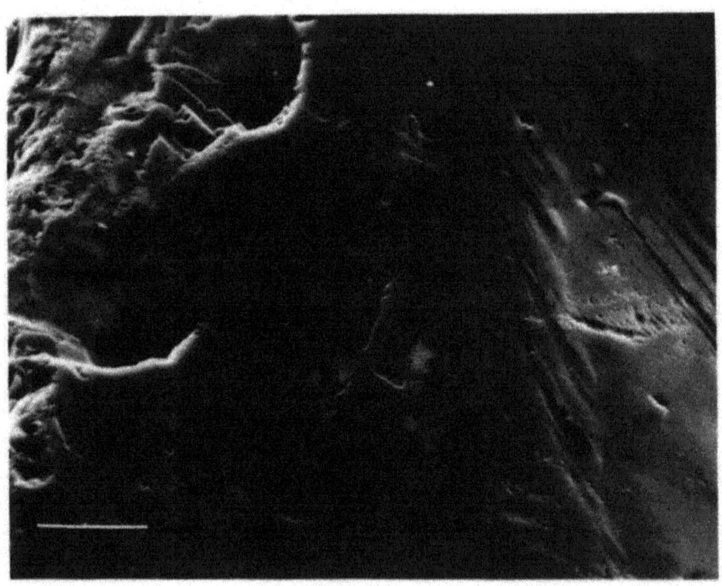

b

Fig. 18. a-b. Prehistoric quartz scrapers from the Bjurselet site. a) Specimen number D:35. b) D:52. a) shows a scraper edge that has suffered strong abrasive wear on the outermost edge, resulting in a heavily rounded rim. In the area closest to the edge, a highly disrupted surface has developed (A) with striations and linear troughs. The brittle fracture wear is etched out, but no signs of dislocation traces are present. b) The edge in this specimen shows only moderate abrasion on the outermost rim. Stepped flakes expand onto the flat ventral face, indicating this tool was used in a planing motion. Closest to the edge are striations and sleeks. There are no examples of dislocation traces. Scale bar: 20 microns.

dissolved abrasion areas containing short striations. Like the tool in Fig. 17a, b, this scraper lacks plastic flow wear, although a slightly undulating surface indicating the tool's direction of use is present. The lack of sleeks suggests that the wear was not formed under very dry conditions.

On one heavily weathered tool in the sample from Bjurselet, deep, natural etching of striations was found (Fig. 18b). Besides typical Hertzian cone cracks, this tool exhibited examples of lateral and medial crack formation. No comparable experimental evidence has been produced so far. Nevertheless, it is an interesting piece, because it shows that these types of fracture can occur in quartz.

These few examples of wear features on prehistoric specimens from the Bjurselet site clearly show that wear is present and that this wear varies from tool to tool. Until the whole assemblage has been analyzed in the SEM, no systematic grouping of wear types can be made.

Conclusions

Chemical etching has proven useful in the characterization of both natural and experimentally-induced microstructural features in silicate crystals within the fields of material science and mineralogy. In this study, chemical etching has been used as a research tool to analyze wear features on experimental quartz tools used in simulations of prehistoric tool use. The results indicate that substantial information can be gained from these relatively simple techniques, because wear features are enhanced and postdepositional surface layers are etched away. Wear formation processes are highlighted, because the etchant reacts variably with different types of wear.

The primary results of this study are:

1. Wear on crystalline materials is characterized by two basic types of damage: a) surface damage such as cracks and flaws; and b) subsurface damage such as dislocation slide traces, microtwinning and phase transformations.
2. Visible striations on quartz tool edges are the results of the "shoulder" breakage of linear arrays of cracks. This breakage is caused by a combination of tangential and vertical forces on individual abrasive grains during tool use.
3. The brittle fractures which form striations vary depending on the loading force on individual grains during use, the raw material from which the grains are made, their form and their size. The crystalline structure of a tool's material is also an important factor affecting variation in the types of fractures which develop.
4. Plastic deformation can occur on quartz even in relatively "low" force situations such as experimental tool use. This type of wear is strongly associated with frictional heat and short term stresses on tool edges generated during tool use.
5. Plastic flow wear in brittle materials like quartz is caused by dislocation sliding in the body of the material. Traces of sliding can be made visible by chemical etching.
6. On quartz tool edges experimentally used to work dry hide and bone, strong frictional forces with subsequent heat generation cause substantial plastic flow wear in high friction areas. This is revealed

by etched-out traces of dislocation sliding or microtwinning.

7. Wear on quartz tools used to work wood and rawhide is primarily due to brittle fracture, as demonstrated by numerous etched-out brittle fracture features close to the edges of tools.

8. The smooth surface layer typically seen on woodworking tools, is etched away easily, suggesting that it is composed of amorphous material.

9. On the outermost tips of woodworking tools there is evidence of plastic flow wear, indicated by a few dislocation traces. There is reason to believe that dry wood will cause more plastic flow wear than moist, fresh wood.

10. The smooth surface created on the outermost edge of tools used on bone is very resistant to the type of etchant employed in this experiment. This indicates that the tool area has undergone either a chemical or a structural change. It is hypothesized here that bone "polish" is the physical representation of a phase transformation of the areas of highest friction on tools used to work bone.

11. The chemical etching of worn areas on tool edges reveals many wear structures which are invisible before etching. This is especially true of the number of striations, which increases dramatically after etching. Thus, the disrupted surfaces caused by wear on quartz tools are more widespread and larger than the features on experimental tools revealed prior to etching.

12. Analysis of prehistoric tools from the Bjurselet site shows that weathering has enhanced and also changed wear features. Medial and lateral crack formation in connection with the formation of striations has been documented on some of these specimens.

13. So far, no examples of etched-out dislocations indicating plastic flow wear have been detected on prehistoric quartz tools.

One of the primary aims of this paper has been to test parts of a model for wear formation on quartz tools. The model has proven to be useful, since it was able to correctly predict the way wear on experimental tools would react to the etchant.

A secondary goal was to see if the variable wear features found on weathered quartz tools from the Bjurselet site indicated a difference in the original cause of these wear features. The prehistoric specimens analyzed to date do not allow any definite conclusions to be drawn as to the way wear traces are transformed as a result of weathering. Wear features directly comparable to the experimentally etched tools have not been found.

To conclude, there is, as suspected, no easily detectable relationship between weathering and wear types of different origin. To explain the variation in wear features observed on the Bjurselet material, a set of controlled etching experiments based on a thorough SEM analysis of the prehistoric tools must be conducted on modern quartz tools. Knowledge of wear formation and material science research methods will hopefully aid in designing this new set of experiments.

Acknowledgements

The experimental work was financed by the project "Nordarkeologi" under the head of Dr. Hans Christiansson, Uppsala University.

Important comments on the micrographs of the etched experimental tools have been made by Professor M.W. Wegner, the University of California (Department of Earth and Space Sciences). Professor B. Collin Department of Geology, Uppsala University, has provided me with literature on the mineralogy of quartz and commented on my ideas on wear processes during discussions. Dr. S. Hogmark at the Institute of Technology (Material Science Department), Uppsala, provided me with basic literature dealing with wear on ceramics. A grant from Marten Stenbergers Minnesfond, Institute of Archaeology, Uppsala, made it possible for me to attend the SEM in Archaeology Conference.

References

Evans, A.G. 1979. Abrasive wear in ceramics: an assessment. In B.J. Hockey (ed.), The Science of Ceramic Machining and Surface Finishing II. National Bureau of Standards, Gaithersburg, Maryland. pp 1-14.

Gielisse, P.J. and Stanislao, J. 1972. Mechanical methods of ceramic finishing. In S.J. Schneider and R.W. Rice (eds.), The Science of Ceramic Machining and Surface Finishing I. National Bureau of Standards Special Publication 348. Gaithersburg, Maryland. pp. 5-34.

Hagan, J.T., Swain, M.W. and Field, J.E. 1979. Nucleation of median and lateral cracks around Vickers Indentations in soda-lime glass. In B.J. Hockey (ed.), The Science of Ceramic Machining and Surface Finishing II. National Bureau of Standards, Gaithersburg, Maryland. pp. 15-22.

Hagg, G. 1963. Allmän och oorganisk kemi. Almquist and Wiksell, Stockholm.

Hayden, B. 1979. (Ed.) Lithic Use-Wear Analysis I. Studies in Archaeology, Academic Press, New York.

Hockey, B.J. 1972. Observations on mechanically abraded aluminum oxide crystals by transmission electron microscopy. In S.J. Schneider and R.W. Rice (Eds.), The Science of Ceramic Machining and Surface Finishing I. National Bureau of Standards Special Publication 348. Gaithersburg, Maryland. pp. 333-338.

Imanaka, O., Fujino, S. and Mineta, S. 1972. Direct observation of material removal process during grinding of ceramics by micro-flash technique. In S.J. Schneider and R.W. Rice (Eds.), The Science of Ceramic Machining and Surface Finishing I. National Bureau of Standards Special Publication 348. Gaithersburg, Maryland. pp. 37-44.

Izumitani, T., Miyade, E., Adachi, S. and Harada, S. 1979. Latent scratches on optical glasses that appear after cleaning. In B.J. Hockey (Ed.), The Science of Ceramic Machining and Surface Finishing II. National Bureau of Standards, Gaithersburg, Maryland. pp. 417-428.

Keeley, L.H. 1980. Experimental Determination of Stone Tool Uses: a Microwear Analysis. University of Chicago Press, Chicago, Illinois.

Kirchner, H.P., Gruver, R.M. and Richard, D.M. 1979. Fragmentation and damage penetration during abrasive machining of ceramics. In B.J. Hockey (Ed.), The Science of Ceramic Machining and Surface Finishing II. National Bureau of Standards, Gaithersburg, Maryland. pp. 23-42.

Knutsson, K. 1985. Yttopografiska studier av förhistoriska stenredskap 3. Svepelektronmikroskopi av experimentella kvartsredskap. Exempel på ur-lakning, utfallning, deformering och mekanisk notning. Tor, Vol XX. Uppsala, Sweden. pp. 45-73.

Knutsson, K. 1986. SEM-analysis of wear features on experimental quartz tools. Early Man News 35-46: Tubingen, West Germany.

Knutsson, K. (unpublished manuscript). SEM-analysis of acetate peel casts of quartz and flint surfaces. Uppsala University, Uppsala, Sweden.

Knutsson, K. and Hope, R. 1983. The application of acetate peels in lithic use wear analysis. Archaeometry 26(1): 49-61.

Koepke, B.G. and Stokes, R.J. 1970. A study of grinding damage in magnesium oxide single crystals. Journal of Material Science 5: 240-247.

Koepke, B.G. 1972. An assessment of surface and subsurface damage introduced in ceramics by semifinishing grinding operations. In S.J. Schneider and R.W. Rice (Eds.), The Science of Ceramic Machining and Surface Finishing I. National Bureau of Standards Special Publication 348, Gaithersburg, Maryland. pp. 317-331.

Krinsley, D. and Doornkamp, J.C. 1973. Atlas of Quartz Sand Surface Textures. Cambridge University Press, Cambridge.

Lidstrom, L. 1968. Surface and bond-forming properties of quartz and silicate minerals. Acta Polytechnica Scandinavia, 75a.

Nicolas, A. and Poirier, J.P. 1976. Crystalline Plasticity and Solid State Flow in Metamorphic Rocks. Selected Topics in Geological Sciences. Dept. of Geology, University of Durham, Durham.

Plisson, H. and Mauger, M. 1986. Chemical and mechanical alteration of microwear polishes: an experimental approach. In M.H. Newcomer and G. de G. Sieveking (Eds.), Proceedings of the Fourth International Flint Symposium, Brighton. Cambridge University Press, Cambridge.

Rice, R.W. 1972. The effect of grinding directions on the strength of ceramics. In S.J. Schneider and R.W. Rice (Eds.), The Science of Ceramic Machining and Surface Finishing I. National Bureau of Standards Special Publication 348, Gaithersburg, Maryland. pp. 365-375.

Stockes, R.J. 1972. Effects on surface finishing on mechanical and other physical properties of ceramics. In S.J. Schneider and R.W. Rice (Eds.), The Science of Ceramic Machining and Surface Finishing I. National Bureau of Standards Special Publication 348, Gaithersburg, Maryland. pp. 343-351.

Tixier, J., Initzian, M.L. and Roche, H. 1980. Préhistoire de la pierre taillèe. Terminologie et Technologie. Cercle de Recherches et d'Etudes Prehistoriques, Valbonne.

Wegner, M.W. and Christie, J.M. 1983. Chemical etching of deformation substructures in quartz. Physics and Chemistry of Minerals 9:67-78. Berlin, East Germany.

Wegner, M.W. and Christie, J.M. 1985. General chemical etchants for microstructures and defects in silicates. Physics and Chemistry of Minerals 12: 90-92. Berlin, East Germany.

Vingsbo, O. and Hogmark, S. 1979. Wear of Steels. I: Wear of Materials. American Society of Mechanical Engineers, New York. pp. 620-635.

THE USE OF RESIN REPLICAS FOR THE STUDY OF LITHIC USE-WEAR

Francesco d'Errico

Summary

A presentation of the techniques which have been used up to the present to detect and study traces of wear on prehistoric lithic material demonstrates the advantages and disadvantages of each technique. The difficulties that arise from the use of SEM are also the subject of specific discussion in order to remedy certain difficulties incurred with this equipment. A technique of observation of resin replicas obtained from silicon rubber moulds is described. Several kinds of silicon rubbers are compared. Examples are given to illustrate the use of this method on tools from the French Lower Paleolithic.

Introduction

Since the first observations were made in the study and identification of traces of use-wear on lithic implements (Semenov, 1964), researchers have been trying to perfect techniques and to make use of equipment which permits better identification and recording of these traces. The use of SEM has shown, since it was first applied to this field (Hayden and Kamminga, 1973; Brose, 1975), a remarkable improvement in the quality of observations. The mechanics of the formation of use-wear traces (Dauvois, 1977; Kamminga, 1979; Anderson, 1980, 1981, 1982; Masson et al., 1981; Meeks et al., 1982; Unger-Hamilton, 1984; d'Errico and Espinet, in press) as well as other natural or cultural alterations to lithic artifacts has been studied using the SEM (Inizan et al., 1976; Masson, 1981; Texier, 1981; Bradley and Clayton, in press).

On the other hand, however, the growing popularity of observation of use-wear traces by means of high magnification reflected-light microscopes (the high power approach) (Keeley, 1980; Moss and Newcomer, 1982; Vaughan, 1985, etc.) would be inexplicable if it were not for the fact that, together with its advantages, the use of SEM also poses some problems such as high operating costs, more laborious specimen preparation, the necessity of metallizing non-conductive surfaces, and size limitations for the specimen. The last two problems mentioned do not concern the examination of use-wear traces produced experimentally since the application of a conductive coating and the reduction of specimen size by means of breaking or sawing usually poses no problem on experimental tools. Original artifacts must be considered in a rather different light, however. With regard to these two difficulties, the study of use-wear traces on stone tools is not different from other branches of archaeological and palaeontological research for which the preservation of the integrity of an object acquires, because of the object's uniqueness and museographical value, an importance considered greater than the information which could be obtained from the microscopic examination of its surfaces. If the dimensions of the object are compatible with those of the chamber of the available SEM, good observations and photographs can be made without coating at low magnification (x10-200) and low voltage (e.g. 10 kv). Sometimes it can be advantageous to wrap the uncoated sample in a conductive film (such as a sheet of aluminum foil), leaving a hole over the area of the sample which you wish to observe. For greater magnification where metal coating is necessary, there are demetallization techniques available (Sela and Boyde, 1977; Bromage, 1984; d'Errico and Varetto, 1985). For larger stone implements, or those whose morphology makes handling difficult, it has been recommended that moulds made of silicon rubber (Ilkjaer, 1979), or varnish replicas (Knutsson, 1983) by used. These techniques have also been proposed for the observation of surfaces by means of reflected-light or transmitted-light microscopes (Beyries, 1981; d'Errico et al., 1982/84; Plisson, 1983; Knutsson and Hope, 1984). Apart from the obvious technical problems of viewing a negative replica, there can be some difficulty in identifying on the original the exact point examined by the SEM on the mould. Also, when a varnish peel is used, the result is only partially three-dimensional (Unrath and Lindemann, 1982/83). Moreover, the varnish, having little elasticity, tends to tear where it covers acute angles, points, and concavities (Unrath and Lindemann, 1982/83). Positive resin replicas made from silicon rubber moulds have been used in anthropology (Pfefferkorn and Boyde, 1974; Pameijer, 1979; Rose, 1983), taphonomy (Potts and Shipman, 1981; Bromage and Boyde, 1984; Bromage, 1984), and for the examination of use-wear traces on lithic implements (Pant, 1980; Walker, 1981; Unrath and Lindemann, 1982/83; d'Errico, 1985a).

The intention of this article is to compare some results obtained using different silicon rubbers for making moulds of original lithic implements. In particular, I wish to compare the use of traditional silicon rubbers and those used in dentistry, to investigate variations in the results, and to make some suggestions for avoiding problems that could arise during their use.

Materials and Methods

Using a variety of silicon rubbers, moulds were made of archaeological implements of flint, jasper, flinty limestone and quartz crystal, taken from sites of Vallonnet Cave, about 950,000 years ago (Rochebrune-Cap-Martin, the Maritime Alps, southern France), Lazaret Cave, about 100,000 y.a. (Nice, the Maritime Alps) and Arago Cave, about 400,000 y.a. (eastern Pyrenees) (Lumley, 1969, 1971, 1976), modern flint implements which were experimentally manufactured and used. Before each mould was made, the samples were washed in distilled water and a liquid cleanser, then kept in an ultra-sound tank filled with acetone for about 10 minutes. They were then rinsed in distilled water and left to dry at room temperature. Moulds were made of either one or both edges of the implements using Rhodorsil R.T.V. 573A silicon rubber (Rhone-Poulenc, France), using both a "standard" catalyst and the fast setting 60R catalyst; with Wacker RTC M 521 silicon rubber (Wacker, D-8000, Munchen), and with the fast setting silicon rubbers used in dentistry Xantopren Blue, Xantopren Plus, and Provil L (Bayer, Dental D-5090, Leverkusen). The dental rubbers were used according to established procedures (Rose, 1983). The moulds in Rhodorsil R.T.V. 573A and Wacker silicon rubbers were made using a technique similar to that generally applied to the production of casts (d'Errico, 1985).

The stone implement is set into a base of modelling clay with the area to be moulded exposed (Lefranc et Bourgeois, France). This base is then surrounded by a wall of modelling clay. With a soft brush, a fine solution of vaseline diluted with purified petroleum is applied over the base. After mixing the silicon rubber with the appropriate catalyst either one of three procedures can be followed in order to avoid producing bubbles in the rubber. The rubber can be placed under a vacuum bell for about 10 seconds, it can be dropped into the cavity formed by the modelling clay from a height of about 1 or 1.5 metres, or it can be applied with a syringe (Rose, 1983). Six to twelve hours are necessary for the rubber to solidify, depending on the catalyst used. After being separated from the original, the negative mould is left to rest for another 12 hours, to allow the rubber to cure before being used to make a resin replica.

All the positive replicas were made of epoxy resin (RBS, T2L Chimie S.A., 11230 Chalabre, France). The resin casts were mounted on metal stubs with a conductive adhesive and coated with gold-palladium. For observation, a Jeol J.S.M. 255 III scanning electron microscope was used at a voltage of 15 Kv. The photographs were taken with Ilford PanF 26 film.

Results

Comparison of detail. No great differences in detail were noticed in the replicas made from the various silicon rubbers when they were observed at a magnification of 300-400x. At higher magnifications the surface morphology of some of the replicas begins to differ from that of the original surfaces. Marked changes begin to appear at magnifications of about 1,000x. At 450x, for example, the three replicas of the edge of a flint flake from Vallonnet Cave made of silicon rubber show a similar general morphology (Fig. 1a, b, c). Even at this magnification, however, consistent differences can be seen, emphasized here by the fact that a particularly uneven surface was chosen. The results rendered by the two types of Xantopren (Fig. 1a, b and 2a, b) appear similar at a magnification of 450x. The replica in Provil L (Fig. 1c) gives greater detail and permits reproduction of microporosity of the flint texture. Comparing it with the original object (uncoated), we can state that at this magnification, except for loss of clarity on a detail in the internal area of a cavity, the replica in Provil L is exactly like the surface of the original. At an even greater magnification (1000x), comparisons of the two Xantoprens (Fig. 1d and e; Fig. 2c and d), shows that the "Blue" type gives slightly less detail than the "Plus" type. In fact, with the Xantopren Plus the image is sharper and the granular appearance of the texture of the flint is partially reproduced. Even at a magnification of 1000x the replica in Provil L appears to conform perfectly to the original (Figs. 1f and 2e). When magnified 2,000 times, the differences between the original surfaces and those of the replicas are much greater. The surface of the Xantopren Blue replica appears rounded and poorly defined. The same, with some small differences, is also true for the "Plus" type. The Provil L replica appears to conform to the original up to a magnification of 3,000-3,500x (Fig. 2f), showing a lessening of detail which is gradual, and differs according to the zone observed.

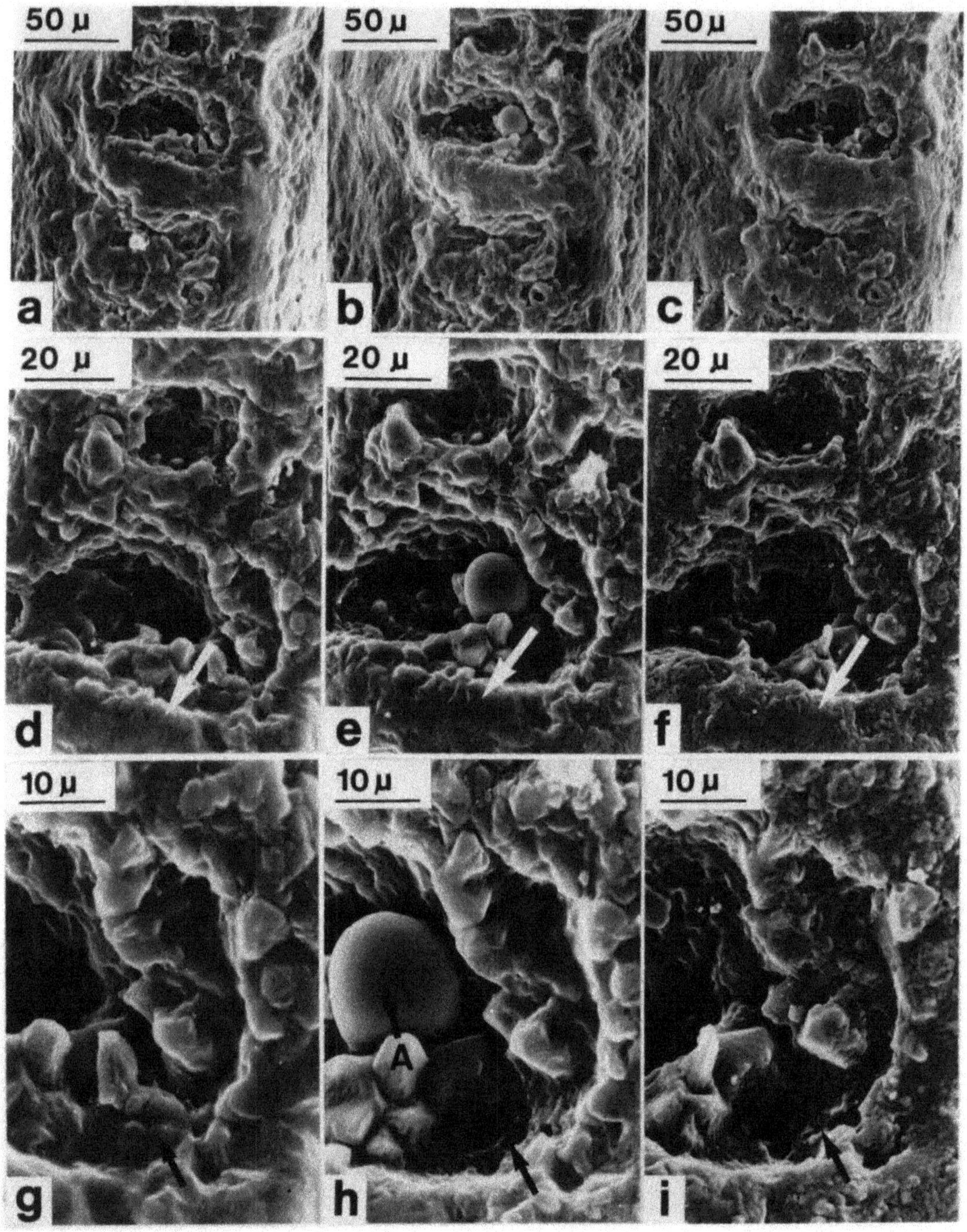

Fig. 1: Positive resin replica of the edge of a flint flake (Vallonnet Cave, Lower Paleolithic) obtained by using different silicon rubbers and photographed at different magnifications with SEM. The replicas were produced from moulds made of, in order from left to right: a,d,g: Xantopren Blue; b,e,h: Xantopren Plus; and c,f,i: Provil L. They were photographed at a magnification of: a,b,c=450x; d,e,f=1000x; and g,h,i=2000x. Notice the greater detail obtained with Provil L. That Xantopren Plus gives slightly better results than Xantopren Blue and that Provil L is better than Xantopren Plus can be seen from the area indicated by the arrow in d,e,f. (A) (in Fig. 1h): air bubble which remained between the original surface and the silicon rubber, later reproduced in the resin. The arrow in g,h,i indicates a particle of the surface removed between the first and the second time moulds were taken.

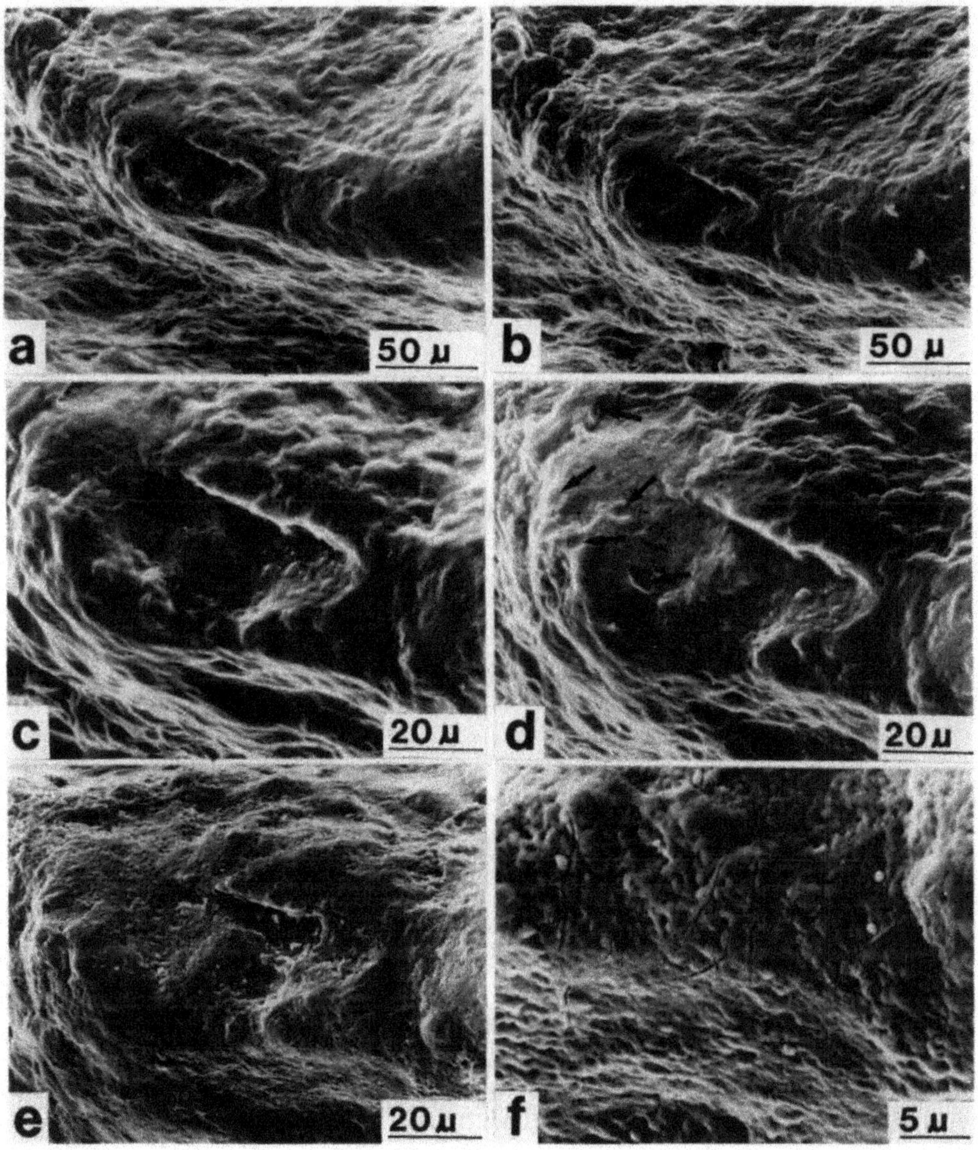

Fig. 2: Resin replicas of a flint flake (Vallonnet Cave, Lower Paleolithic) made from different silicon rubber moulds and photographed with SEM at different magnifications. a, c: Xantopren Blue; b,d: Xantopren Plus; and e,f: Provil L. Magnifications are: a, b=450x; c, d, e=1000x; f=3000x. The arrows in d. indicate areas not reproduced in the resin. Note the good detail offered by Provil L on microscopic cracks photographed at a high magnification.

Slow setting silicon rubbers, like Rhodorsil R.T.V. 573A and Wacker R.T.V. M 521, give results slightly inferior to those obtained with Xantopren Blue. It is important to note, however, that these results are more variable than those rendered by silicon rubber designed for dentistry. The use of a fast-setting catalyst increases the variability of the possible results with the Rhodorsil and Wacker rubbers. Once the Rhodorsil RTV 573A has become hardened it is usually easier to separate it from the surface of the replica than the Wacker RTV M 521. Both are, in fact, sold as products for use on surfaces previously covered with vaseline. The slow setting silicon rubbers tend to tear readily and pieces break off more easily than with the dental product. Minute lacerations can occur at two different moments: when the mould is separated from the original and when the resin replica is separated from the mould. According to our experience, microscopic fragmentation appears to be more common when the surfaces reproduced are highly porous (e.g. stone cortex) or vitreous (e.g. quartz crystal, obsidian, etc.).

Examples of the use of these two products can be seen on a Tayac point made of quartz crystal taken from Arago Cave (Fig. 3a and b) and on a flint dejete scraper from Lazaret Cave (Fig. 4b, c and d). On the Tayac point, abrasion is present on the edge itself and on the lower face of the implement near the edge (Fig. 3b). On the upper face micro-breakage is visible (Fig. 3a). For our aims it is of interest to note that the replica shows the surface morphology in great detail. In fact, negative bulbs, ripples, and striations, typical of the surfaces of fractured isotropic siliceous material, are present (Fig. 3a and b, feature A). This morphology can be observed with the same detail on a replica made from a slow setting silicon rubber mould, at a magnification of 2,000x.

The flint dejete scraper represented in Figure 4, shows particularly intensive wear on its edges after 10 minutes of experimental meat cutting. Like many other implements with converging retouch from the Lazaret Cave, it has been reduced in size by frequent resharpening of the edge through successive retouching (d'Errico, 1985). The use wear could be described as considerable rounding of the edge. The removal of material caused by wear has almost completely eliminated the concavities along thé edge of the implement and smoothed the granular texture of the flint (Fig. 4b). The rounded zone has characteristics of a surface which is both abraded and subjected to superficial compression of its texture, as previously recorded by the observation of others with SEM (Dauvois, 1979; Anderson, 1980; Meeks *et al.*, 1982).

Modification of the original surfaces. At a macroscopic level, the use of silicon rubber, regardless of type, has caused no serious modification in the morphology of the edges reproduced. There was a change in the colour of the object in one case, consequent to the use of Xantopren Blue. On the relatively porous surface a dark stain, similar in colour to that of the rubber used, remained after removal of the mould. Particles or lumps of silicon rubber which remain stuck to the original object could also be considered as modification of the surface. Microscopic modifications have occasionally been noted. In most cases, some particles of the surface have been removed (Fig. 1g, h

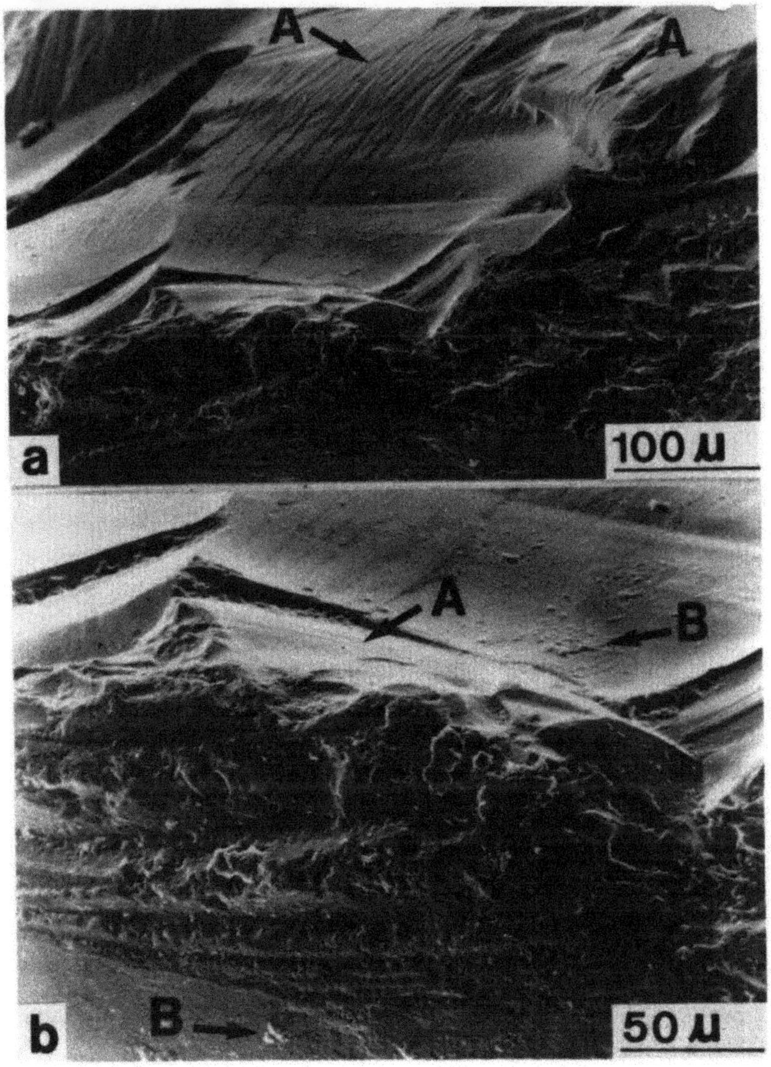

Fig. 3: Resin replica of the edge of a Tayac point in quartz crystal (Arago Cave) made from a Rhodorsil RTV 573A silicon rubber mould: a=200x; b=400x. Top: upper face of the implement. Bottom: lower face. (A) Superficial characteristics typical of the way in which siliceous isotropic material breaks (ripples, striations, etc.). (B) microscopic residues of silicon rubber.

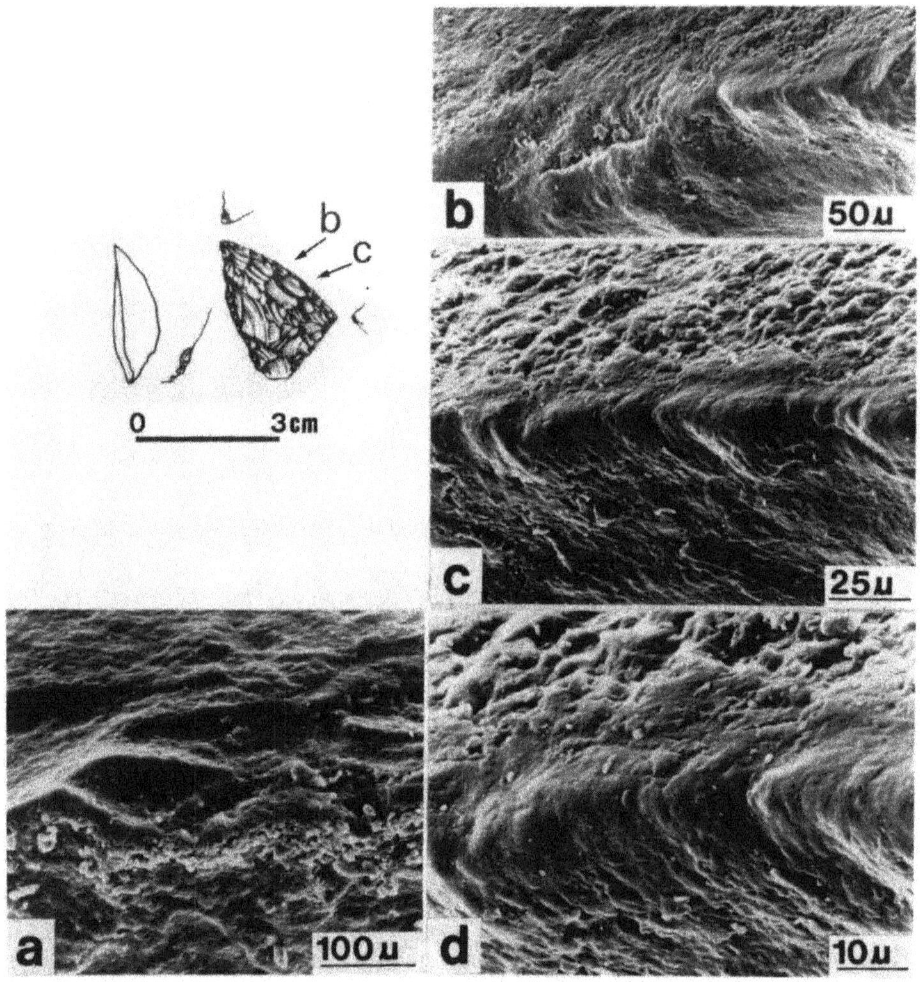

Fig. 4: a. edge of implement used experimentally to cut fresh meat for 10 minutes, showing marginal rounding of the edge. b., c., d.) resin cast from Rhodorsil RTV 573A mould of a dejete scraper from Lazaret Cave. Notice that the edge is considerably rounded and that the worn area has lost the granular texture of the surrounding surfaces. Magnification: a=200x, b=300x, c=700x, d=1500x (detail of c.).

and i). It is impossible to state whether this microscopic removal occurred during the cleaning process or during removal of the mould.

Discussion

Since the examination of wear traces with SEM should not damage archaeological stone artifacts, it is of interest to compare different materials for making replicas in order to perfect techniques which can protect the integrity of the original. Several researchers (Unrath and Lindemann, 1982/83; d'Errico, 1985) are working on this while pursuing their studies of use-wear traces on lithic implements with SEM.

It should be noted that all silicon rubbers we used give good detail up to the maximum degree of magnification usually used in the analysis of use-wear traces with reflected-light microscope. However, considerable differences can be noticed, as was pointed out, under examination by SEM at high magnification. According to our experience, the use of replicating materials that give greater detail makes it possible to observe samples at magnifications up to 2,000-3,000x without any impairment with regard to the interpretation of morphologic characteristics.

Taking the imprint does not, in itself, significantly alter the surface morphology of the original. The occasional removal of particles of the original does not constitute a very great disadvantage with moulding. Any pieces of silicon rubber, adhering to the resin positives can be easily recognized when observed under the SEM, and can be eliminated from the original surface with a soft brush soaked in purified gasoline. Air bubbles on the surface of the replica are also very easily recognized (Figs. 1b, c and h). A more complex problem is that of possible colour changes of the stone artifact, probably due to the absorption of silicon oils and dyes by its micro-porous surface. When the surface is very porous (e.g. flinty limestone, or slightly altered flint) it is inadvisable to make a mould with a slow setting silicon rubber (e.g. Rhodorsil RTV 573A, or Wacker 573 M 21). If quick-setting silicon rubbers used in dentistry are preferred, care must be taken to use a sufficient quantity of catalyst. This can reduce the initial penetration of the product. When stains occur they can be completely or partially removed using products like Flugene 113 (Trichlorine 1,11,2,trifluorine-122 ethane).

With silicon rubbers used in dentistry there is the advantage of very short preparation time. This facilitates the taking of imprints on the spot, when the archaeological objects are in a museum or distant locality. Because of the high cost of the dental rubbers, it is more economical to use slow setting rubbers for making moulds of large surfaces. They are also better, because of their softness, for making moulds of very thin implements, which could break if a stiffer quick-setting rubber were used. In any case, it is always advisable to take care when removing the moulds to avoid damaging the original or tearing the mould. As has already been pointed out several times, making replicas encourages exchange between researchers, allowing a greater circulation of information with minimum movement of originals.

Conclusions

If carried out with due care and attention, the technique of making resin replicas from silicon rubber moulds is well adapted to the study of use-wear traces on lithic implements with SEM and constitutes a valid alternative to coating the originals with metal. When original uncoated surfaces are analysed with an energy dispersive (EDS) microprobe, the successive observations of resin replicas of the parts analysed by the microprobe can provide excellent photographic documentation of the precise topographic location of the area where the analyses were carried out. A technique used in order to prevent modification of the original surface, often necessary with direct observation, clearly must not cause damage to the surfaces of the artifacts itself. For this reason, in addition to the investigation of the range of resolution in detail obtainable, that the comparison of different replicating materials was made. The consideration of the suitability of replicating artifacts in the face of particular problems posed by the study of use-wear is particularly important.

Acknowledgements

I wish to thank Professor H. de Lumley for making this research possible and giving me access to the lithic implements of Lazaret, Arago and Vallonnet Caves; Professor G. Giacobini and R. David for providing me with information and for participating in technical discussions; and C. Weber for her help during the observations with SEM. I also wish to thank S. Olsen, J. Espinet-Mouncadel, L. Comoglio, C. Lefebre, H. Stalens and the technical staff of the Rhone-Poulenc, France and Italy, and Bayer, France for their help. This paper was translated by Anne Topley.

References

Anderson, P.C. 1980. A testimony of prehistoric tasks: diagnostic residues on stone tool working edges. World Archaeology 12: 181-194.

Anderson-Gerfaud, P.C. 1981. Contribution Méthodologique à l'Analyse des Microtraces d'Utilisation sur les Outils Préhistorique. Thèse de 3e cycle, Université de Bordeaux I, no 1607.

Anderson-Gerfaud, P.C. 1982. Comment préciser l'utilisation agricole des outils préhistoriques? Cahiers de L'Euphrate 3: 149-164.

Beyries, S. 1981. Etudes des traces d'utilisation sur des empreintes en latex. Bulletin de la Societé Préhistorique Française 78: 198-199.

Bradley, R. and Clayton, C. 1987. The influence of flint microstructure on the formation of microwear polishes. In Proceedings Fourth International Flint Symposium. Cambridge: The University Press.

Bromage, T.G. and Boyde, A. 1984. Microscopic criteria for the determination of directionality of cutmarks on bone. American Journal of Physical Anthropology 65: 359-366.

Bromage, T.G. 1984. Interpretation of scanning electron microscopic images of abraded forming bone surface. American Journal of Physical Anthropology 64: 161-178.

Dauvois, M. 1977. Stigmates d'usure présentés par des outils de silex ayant travaillés l'os. Premiers résultats. In Méthodologie Appliquée à l'Industrie de l'Os Préhistorique. Paris: CNRS., 275-292.

Dauvois, M. 1979. Burins uses de la grotte du Crest à Viry (Saône-et-Loire). Congrés Préhistorique de France, XXI Session 2: 97-110.

d'Errico, F. 1985a. Traces d'usure sur l'industrie lithique: approche méthodologique et proposition d'une technique. L'Anthropologie 89: (5) 20-37.

d'Errico, F. 1985b. Étude fonctionnelle des outils à bords convergents de la Grotte du Lazaret (Nice, Alpes-Maritimes). Mémoire de D.E.A., Univ. Paris VI, 114 pp.

d'Errico, F., Giacobini, G. and Puech, P.-F. 1982/1983. Varnish replicas: a new method for the study of worked bone surfaces. Ossa 9-11: 29-51.

d'Errico, F. and Espinet-Mouncadel, J. 1986. L'emploi du microscope électronique à balayage pour l'étude expérimentale de traces d'usure. Bulletin de la Société Préhistorique Française 83(3): 91-96.

d'Errico, F. and Varetto, L. 1985. Scanning electron microscope study of microwear on original stone implements. Antropologia Contemporenea 8 (4): 17-24.

Hayden, B. and Kamminga, J. 1973. Gould, Koster and Sontz on "microwear": a critical review. Newsletter of Lithic Technology 2: 3-8.

Ilkjaer, J. 1979. A new method for observation and recording of use-wear. In Lithic Use-wear Analysis. New York: Academic Press, pp. 345-348.

Inizan, M.-L., Roche, H. and Tixier, J. 1976. Avantage d'un traitement thermique pour la taille des roches siliceuses. Quaternaria 19: 1-18.

Kamminga, J. 1979. The nature of use-polish and abrasive smoothing on stone tools. In Lithic Use-wear Analysis. New York: Academic Press, pp. 143-155.

Keeley, L.H. 1980. Experimental Determination of Stone Tool Uses: a Microwear Analysis. Chicago: University Press.

Knutsson, K. 1983. Yttopografiska studier av forhistoriska stenredskap. 1. Plastavtryck for dokumentation och analys av notningsspar. Tor 19: 5-32.

Knutsson, K. and Hope, R. 1984. The application of acetate peels in lithic usewear analysis. Archaeometry 26 (1): 49-61.

Lumley, H. de 1969. Une cabane acheuléenne dans le grotte du Lazaret (Nice). Paris: Mémoires de la Société Préhistorique Française.

Lumley-Woodyear, H. de 1971. Le Paléolithique inferieur et moyen du Midi méditerraneen dan son cadre géologique. Paris: V° Suppl. of Gallia-Préhistoire.

Lumley-Woodyear, H. de 1976. Les premières industries humaines en Provence. In La Préhistoire Française. Paris: CNRS, pp. 765-776.

Masson, A. 1981. Altération des silex préhistoriques: dissolution, néogénèses siliceuses, implications sédimentologiques et chronologiques. Comptes Rendus de L'académie des Sciences de Paris D292: 1533-1537.

Masson, A., Coquegniot, E. and Roy, S. 1981. Silice et traces d'usage: le lustré des faucilles. Nouvelles Archéologiques du Musée d'Histoire Naturelle de Lyon 19: 43-51.

Meeks, N.D., Sieveking, G. de G., Tite, M.S. and Cook, J. 1982. Glass and use-wear traces on flint sickles and similar phenomena. Journal of Archaeological Science 9: 317-340.

Moss, E.H. and Newcomer, M.H. 1982. Reconstruction of tool use at Pincevent: microwear and experiments. Studia Praehistorica Belgica 2" 298-312.

Pameijer, C.H. 1979. Replication techniques with new dental impression materials in combination with different negative impression materials. Scanning Electron Microscopy 2: 571-580.

Pant, K.R. 1979. Les traces d'utilisation sur les outils de pierre. Dossier de l'Archeologie 36: 86-89.

Pfefferkorn, G. and Boyde, A. 1974. Review of replica techniques for scanning electron microscopy. Scanning Electron Microscopy 1: 75-82.

Plisson, H. 1983. An application of casting techniques for observing and recording of microwear. Lithic Technology 12: 17-20.

Potts, R. and Shipman, P. 1981. Cutmarks made by stone tools on bones from Olduvai Gorge, Tanzania. Nature 291: 577-580.

Rose, J.J. 1983. A replication technique for scanning electron microscopy: application for anthropologists. American Journal of Physical Anthropology 62: 225-261.

Sela, J. and Boyde, A. 1977. Cyanide removal of gold from SEM preparation. Journal of Microscopy 3: 229-231.

Semenov, S.A. 1964. Prehistoric Technology. London: Cory, Adams and Mackay.

Shipman, P. 1981. Applications of scanning electron microscopy to taphonomic problems. Annals of the New York Academy of Science 376: 335-386.

Straus, L.D. and Walker, P.L. 1978. A technique for studying microscopic wear on artifacts surfaces. In Views of the past: essays on Old World Prehistory and Paleoanthropology. Mouton, The Hague. pp. 53-56.

Texier, P.J. 1981. Desilification des silex taillés. Quaternaria 23: 159-169.

Unger-Hamilton, R. 1984. The formation of use-wear polish on flint: beyond the "Deposit versus Abrasion" controversy. Journal of Archaeological Science 11: 91-98.

Unrath, G. and Lindemann, W. (1982/1983) Reproduktionsstoffe in der mikro-gebrauchsspurenforschung. Early Man News 7-8: 61-72.

Vaughan, P. 1981. Lithic microwear experimentation and the functional analysis of a lower magdalenien stone tool assemblage. Ph. D. Dissertation, Dept. of Anthropology, University of Pennsylvania, Philadelphia. (University Microfilms International, Ann Arbor, Michigan, no 82-08, 050).

STUDY OF UPPER PALEOLITHIC AND EPIPALEOLITHIC ENGRAVED PEBBLES

Francesco d'Errico

Summary

Until now, pebble engravings of Upper Paleolithic and Epipaleolithic have been studied with a stylistic approach relying on the counting of signs. Experimental work has been carried out by the author to look for important microscopic features capable of yielding a technological interpretation of pebble engravings. The results thus obtained are compared with replicas of engraved Azilian pebbles from the Rochedane site, in France.

Introduction

Complete understanding of wall art and art mobilier of the Upper Paleolithic and Epipaleolithic periods certainly depends to some extent on the technological interpretation of the manufacture of this art. As a result, some authors have pointed out, in engraved art mobilier, the importance of elements external to the design itself, the subject depicted, or the general artistic style, emphasizing instead the technology of production. So, the natural properties of the materials used and their influence on the artistic results, have been examined (Leroi-Gourhan, 1943; Delporte, 1973, 1976; Delporte and Mons, 1973).

In the case of pebble or bone artifacts with patterns of repeated notches or lines (either engraved or painted) like many of the European Epipaleolithic artistic objects, particular attention has been paid to the enumeration of the marks (Thévenin, 1983; Couraud, 1985). From the end of the 1960's on, because of ever increasing interest, the morphological characteristics of the engraved signs have been studied, in order to identify manufacturing techniques and the order in which they were executed (Pales and Tassin de Saint-Pereuse, 1968, 1976; Pales, 1969; Delporte and Mons, 1975).

Marshack (1969, 1970, 1972, 1976, 1979, 1984) demonstrated the potential of a systematic microscopic analysis of the surfaces of decorated objects and treated this as a separate subject of research, rather than an offshoot of the stylistic examination. Marshack generally carried out his observations with low magnification using a reflected-light microscope.

With recently developed techniques, we can observe art mobilier with SEM in their original state without conductive coatings or, if the object is in a well-preserved condition, resin casts made from silicone rubber moulds (Rose, 1983; Claugher, this volume; d'Errico, 1985 and this volume) may be preferred. This paper proposes the investigation of possible technological interpretations of engravings on pebbles through the use of SEM.

The morphology of engravings produced by a lithic implement on the surface of a pebble have been examined in order to recreate the mechanics of their formation. Research has also been conducted to determine macro- and micromorphological criteria indicative of the direction of movement of the engraving tool across the pebble's surface. When one engraving is superimposed upon another, it is often possible to establish the order in which they were made by microscopic examination.

Materials and Methods

Silicone rubber moulds were made of areas of engraved pebbles from the Azilian levels of the Rochedane site, France-Compte (Thévenin, 1982, 1983). The pebbles were chosen according to good macroscopic preservation of their surfaces. The surfaces to be reproduced were first cleaned with a soft brush soaked in acetone. The moulds were made by means of RTV 573A (Rhone Poulenc, France) or by means of Xantopren Plus, Bayer, Dental-D-5090 (Leverkusen, Germany) (Rose, 1983; d'Errico, 1985), depending on the size of the surfaces to be reproduced. Xantopren Plus was used for small areas because of its low cost and problems caused by penetration of silicon oil into porous surfaces of the pebble with other brands of rubber. The positive replicas were made in RBS resin (T2L Chimie S.A., Chalabre, France).

Experimentation

Twenty flint and jasper burins were experimentally made (Fig. 1). During experimental stone knapping, an attempt was made to reproduce the morphology, technology and dimensions of burins typical of the end of the French Upper Paleolithic and Epipaleolithic periods. These include small dihederal burins and small burins on truncation, often on thick flakes or fragments of blades. The implements thus obtained were used to make engravings on 30 pebbles made of limestone and flinty limestone with a fine or medium-fine compact texture. The pebbles were of different shapes and sizes varying from between 3 to 12 cm in length. Straight and curved lines of varying lengths were made, produced by a single stroke of the tool, or by several uni- or bidirectional strokes. The incised lines were initiated either at the centre of the pebble or at one edge. Engravings were also made with several lines crossing, forming a number of angles at the points of intersection. The pressure applied against the surface and the angle in which the burin was held were varied on different examples. Particular care was taken to experimentally replicate parallel series of lines, like those found in Azilian engraved pebbles (Thévenin, 1983). For each experimental pebble a brief description, including all the data regarding the above mentioned variables, was recorded on a file card.

In addition to the engraving work, another batch of pebbles was experimentally used as various tools such as hammerstones, different kinds of retouchers, etc. Preliminary examination of the surfaces of pebbles was carried out with a WILD type 308700 binocular reflected-light microscope. The pebbles were then put into an ultra-sound tank filled with acetone for 10 minutes to clean their surfaces.

Mounted on stubs with conductive glue, the experimentally engraved pebbles and the resin replicas of the surfaces of Azilian pebbles from

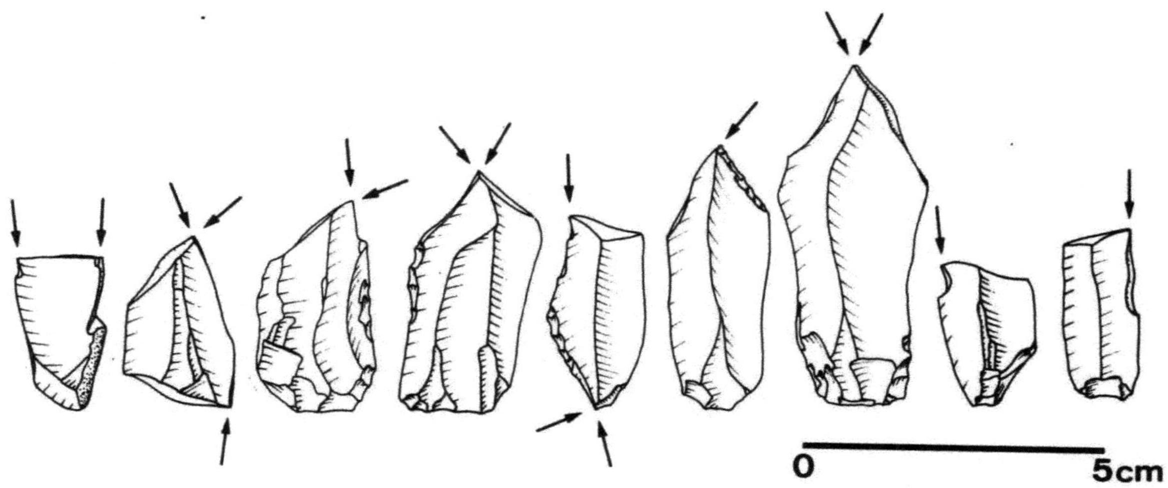

Fig. 1. Examples of burins experimentally used for making engravings on pebbles.

Rochedane, were coated with gold-palladium. A JEOL J.S.M. 255 III scanning electron microscope was used to observe the engravings, with a voltage of 15 Kv. Photographs were taken using an Ilford PanF 26 film. Nondestructive microanalysis with an energy dispersion x-ray spectrometer (E.D.S.) was carried out on some of experimental pebbles. For this, a Stereoscan 250 microscope together with a Link microprobe EDS was used.

Results

Contact either under pressure or by percussion between a flint tool and the surface of a limestone pebble causes removal of material from the pebble and deformation of its natural texture (Fig. 2a). The morphology of the visible surface after this contact depends, obviously, on the morphology of the part of the tool which came into contact with the pebble.

Retouchers. On the surface of pebbles used as retouchers it is possible to see, either with the naked eye or with a reflected-light microscope, percussion marks that are short, deep, subparallel cuts. Leading from these and perpendicular to them there are shallow striations of varying lengths depending on which part of the pebble was used to retouch. Observing the use area with the SEM, an abraded surface can be seen, on which the passage of the flint has produced areas where the natural granular texture of the pebble has become more compact (Fig. 2b). In these compact areas, which appear elongated and discontinuous, are parallel longitudinal striations. The direction followed by the retouched edge of the flint across the pebble is apparent from a series of oblique faults in the striations on the pebble which point outwardly in the direction opposite to the movement of the flint tool's edge (Fig. 2c).

Burin Engravings. An engraved line produced on a pebble with a burin point shows a varying morphology throughout its length. At the point where contact with the implement is initiated there is often a "staving in" of the pebble's surface (Fig. 3a, b). This forms a cavity which gives the beginning of the groove a "comet-like" appearance. Further along the striation takes an aspect of a longitudinal zone of abrasion. Inside this zone it is possible to distinguish areas with a granular and porous aspect (the internal texture of the pebble exposed by the removal of the surface) and sometimes, within these areas are one or more parallel broken bands of highly compact material (Fig. 4a, b). The bands of compact material generally have parallel striations running through them. An analysis by an E.D.S. microprobe showed the same composition both on the natural surfaces and on the compact areas. When the abrasion is on an engraving made from left to right by a right-handed person, the compact part occurs along the side of the groove closest to the experimenter (Fig. 3, 6). When there is a vertical engraving going downwards produced by a right-handed person, the band of compact material is visible both running along the centre of the engraving and along the left part. The presence of compact surfaces appears greater if the implement used as moved along the surface more slowly and under a greater pressure.

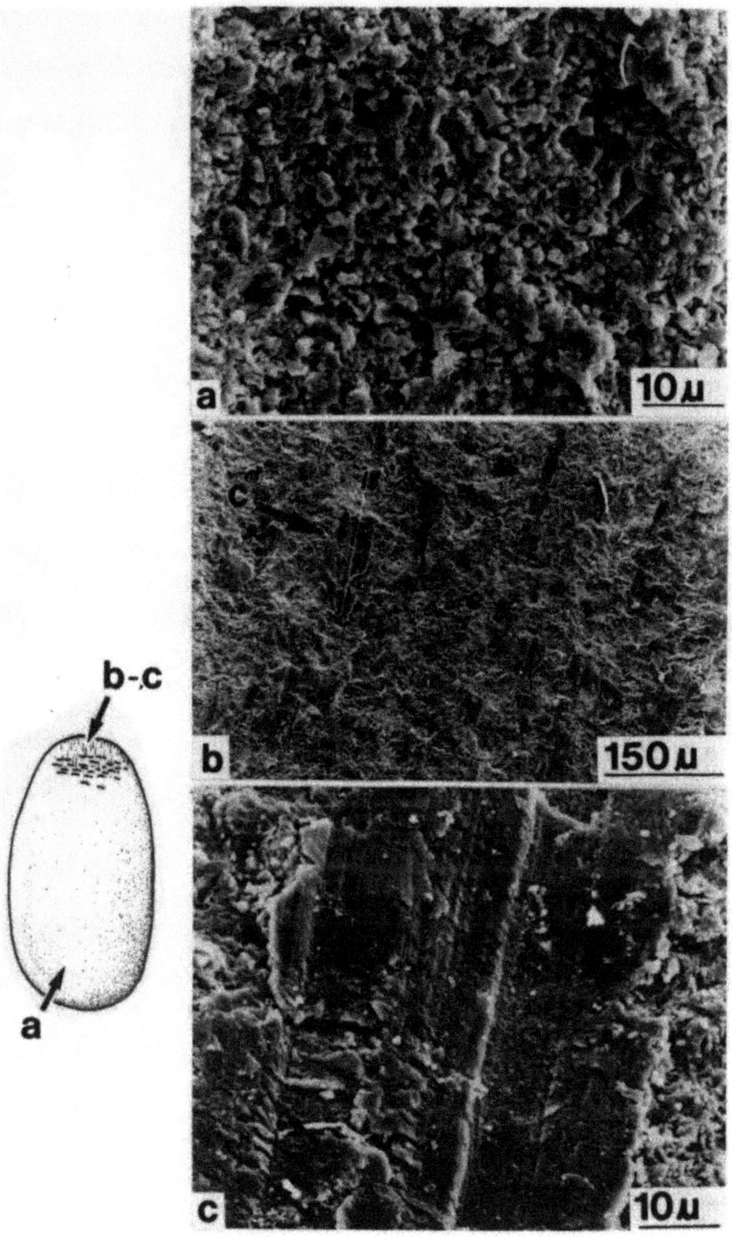

Fig. 2. (a) Natural surface of limestone pebble, x1500. (b) Abraded surface of the extremity of a pebble experimentally used as a retoucher, crossed by elongated compacted areas. (c) Detail of the preceding figure showing a compact area on which the oblique faults indicate the direction of the movement of the retoucher.

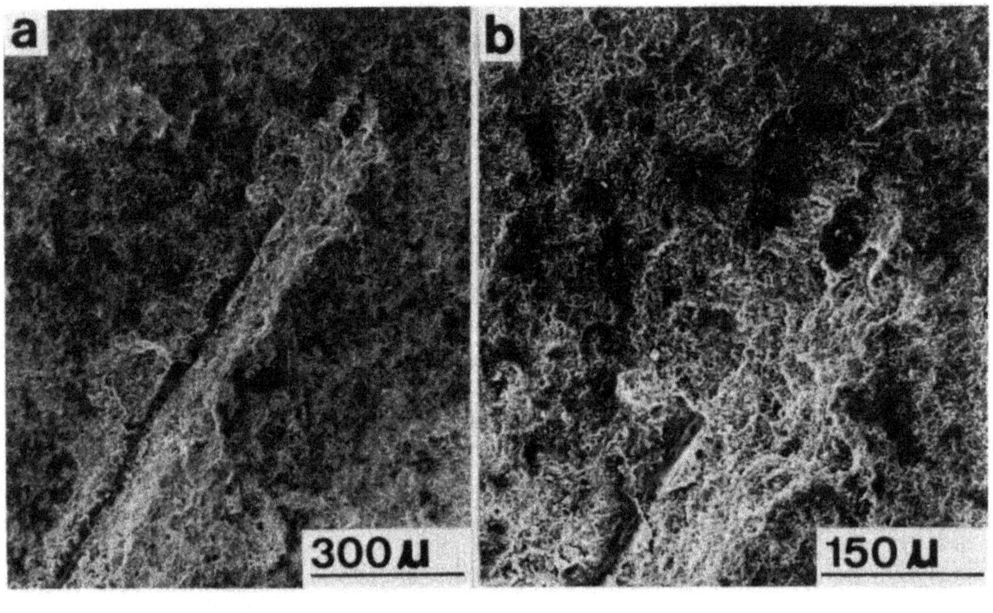

Fig. 3. Starting point of an engraving made experimentally moving from left to right (a, x70; b, x150). Note the cavity which gives the beginning of the engraved line a "comet-like" form.

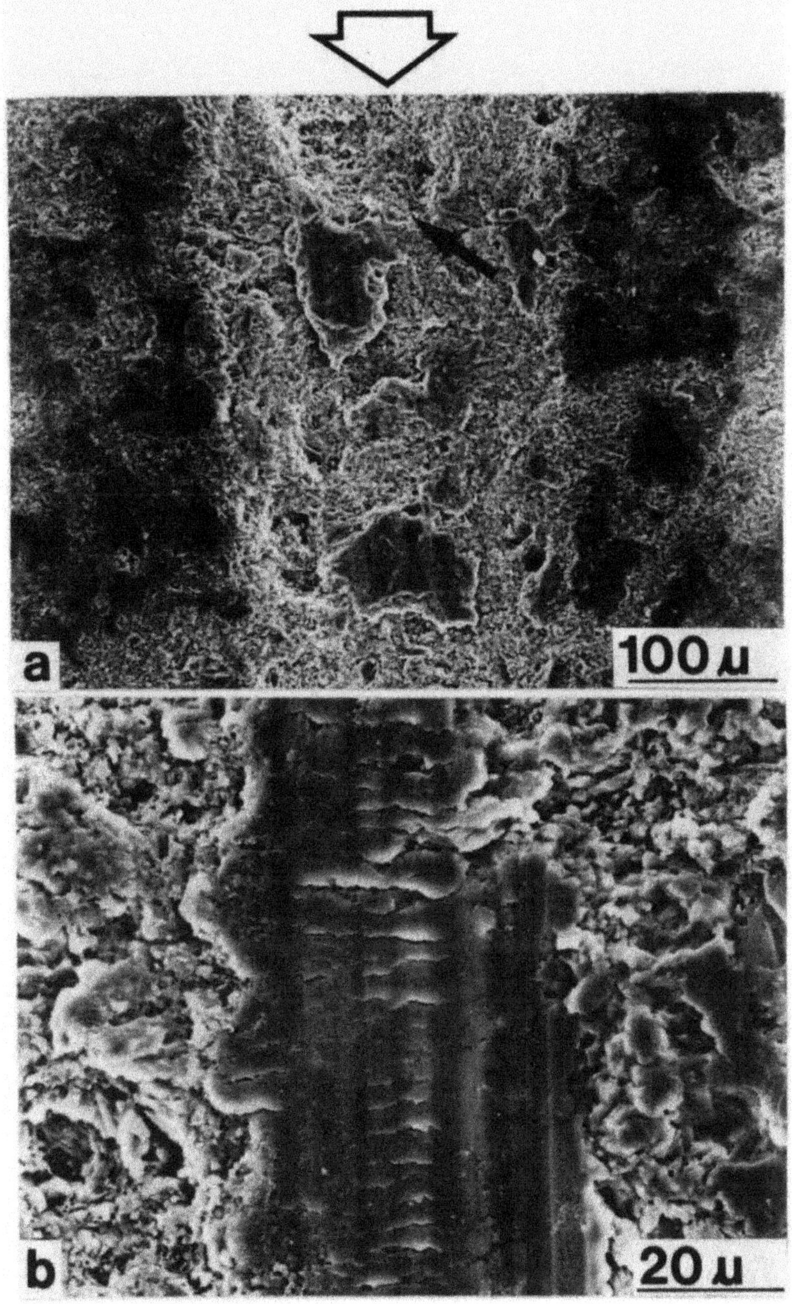

Fig. 4. Experimentally-made engraving produced from the direction indicated by the arrow above (a, x200; b, x1000). Note at the centre of the engraving the presence of compact and granular areas. The compact areas have parallel striations running along them with microscopic cracks. The black arrow in (a) locates a semicircular cavity, useful as an indication of direction of movement.

In thorough examination of experimentally-made engravings, two valid criteria for determining the direction of movement of the engraving tool have emerged. The first is to be found inside the compact areas. These often appear scaly at the edges or in the middle because of minute superficial lacerations. The scales are raised toward the direction opposite to that of the movement of the implement (Fig. 5). If the engravings have little or no compact areas, a useful criterion for judging the direction is the morphology of the porous areas, which may be rippled or have a succession of semi-circular cavities, with their convex surface facing the direction of movement (Fig. 4).

Intersections. In the case of engravings superimposed one over another and produced by a burin point, the order of their execution can be easily recognized when the striations contain compact areas (Fig. 6). The compact areas of the striation produced first are generally removed by the passage of the second at the point where they cross, and the compact areas of the more recently made engraving appear unbroken. The last engraving to be made often produces, when it removes part of a preceding one, an irregular cavity. In this cavity, the corners formed by the more recent engraving exiting the intersection are removed or crumbled (Fig. 6a). The presence of this cavity is also an indication of the direction of the more recently produced engraving. The contact between the two engravings can also cause lateral slipping of the more recent engraving at the intersection (Fig. 7). For a right-handed person this slipping will be downwards for an engraving going from left to right, and towards the left in an engraving running vertically downwards.

Azilian Material

While examining resin replicas of parts of the engraved pebbles taken from Azilian levels of the Rochedane Cave, some of the surface characteristics previously mentioned with regard to experimental material were recognized. The extremities of several engravings had the same morphology as that produced at the starting point of an experimental engraving. The indication of direction of movement of the tool, consisting of a succession of semi-circular cavities along the inside of an engraving, was found in several cases (Fig. 8c, d; Fig. 9) and was sometimes particularly evident. Similarly, the distinctive morphology of some intersections with chipping caused as the last engraving exited the intersection (Fig. 8b) was also visible on prehistoric material. Nevertheless, differences were noticed between experimental and archaeological engraving. Compact areas were rare in the case of the authentic engravings (Fig. 8d) and when they were visible they had an altered surface. The same can also be said for the unengraved surfaces of the original pebbles, which showed a much greater degree of post-depositional alteration than those of the experimental ones.

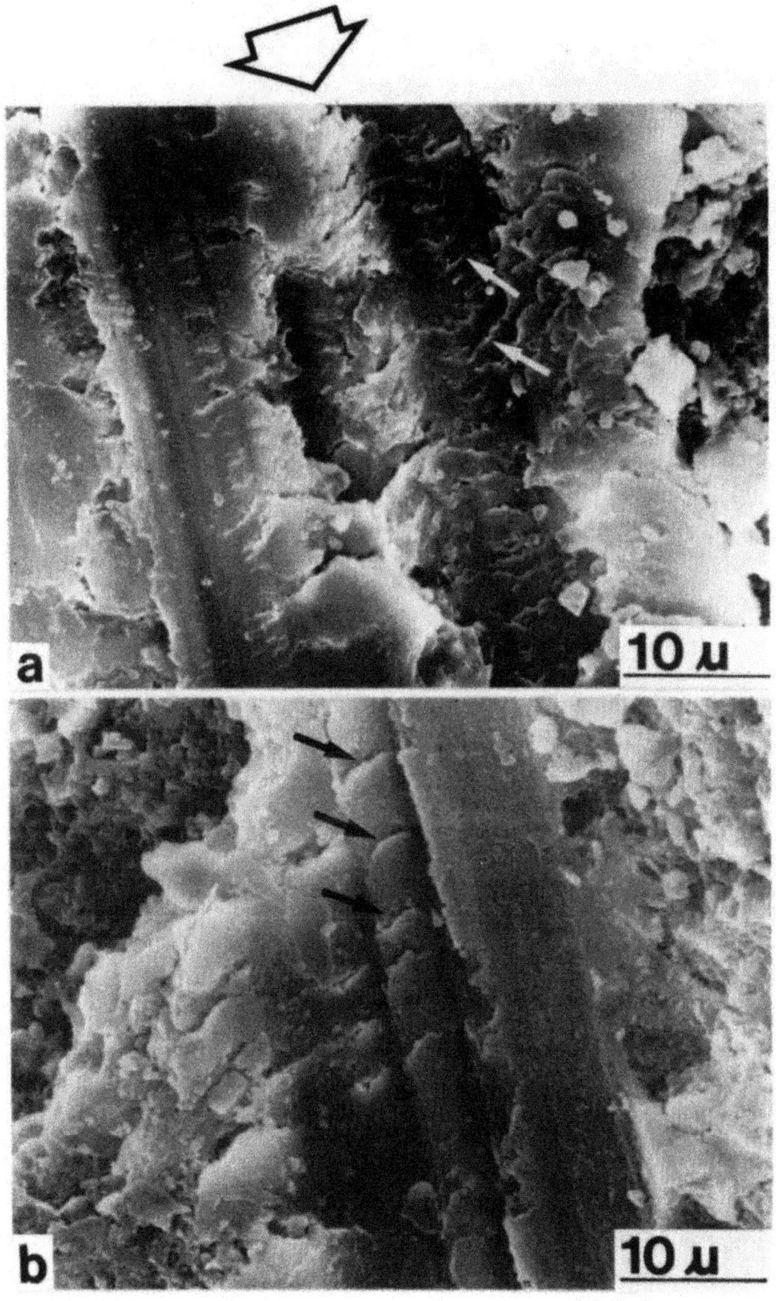

Fig. 5. Compact areas in the centre of an experimental engraving (a, x2000; b, x2000). The arrows indicate the edges of scales raised in the direction opposite to the movement of the engraving tool.

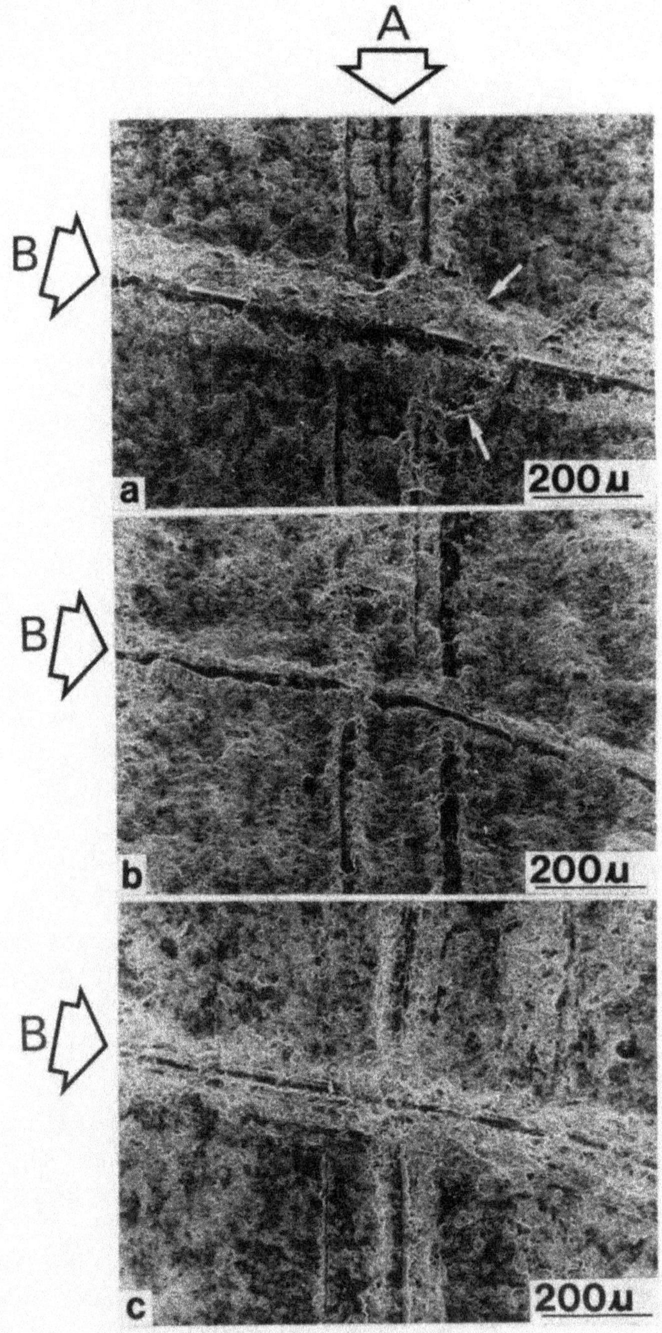

Fig. 6. Intersection of engravings. (A). Vertically downwards; (B) From left to right (a, b, c, x100). The more recent engraving (B) has removed part of the preceding one (A). The arrows in (a) indicate a cavity produced by the second engraving as it exited from the intersection with the first.

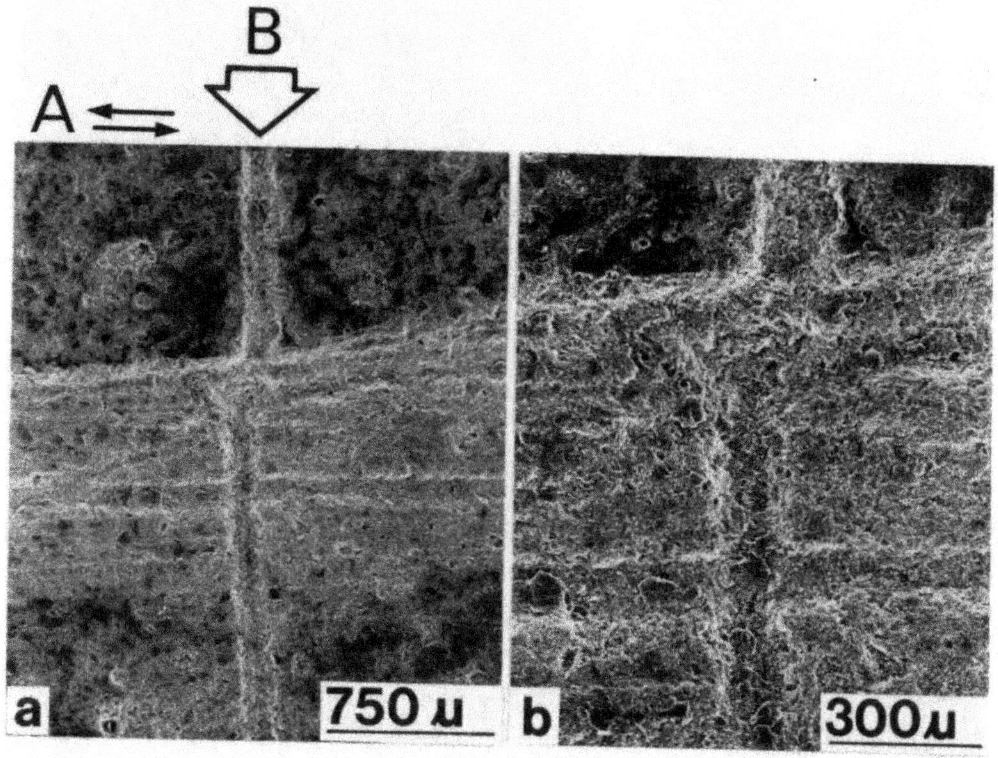

Fig. 7. Lateral slipping of a vertical downward engraving at the intersection with a series of engraved lines made with a bidirectional movement (a, x30; b, x70).

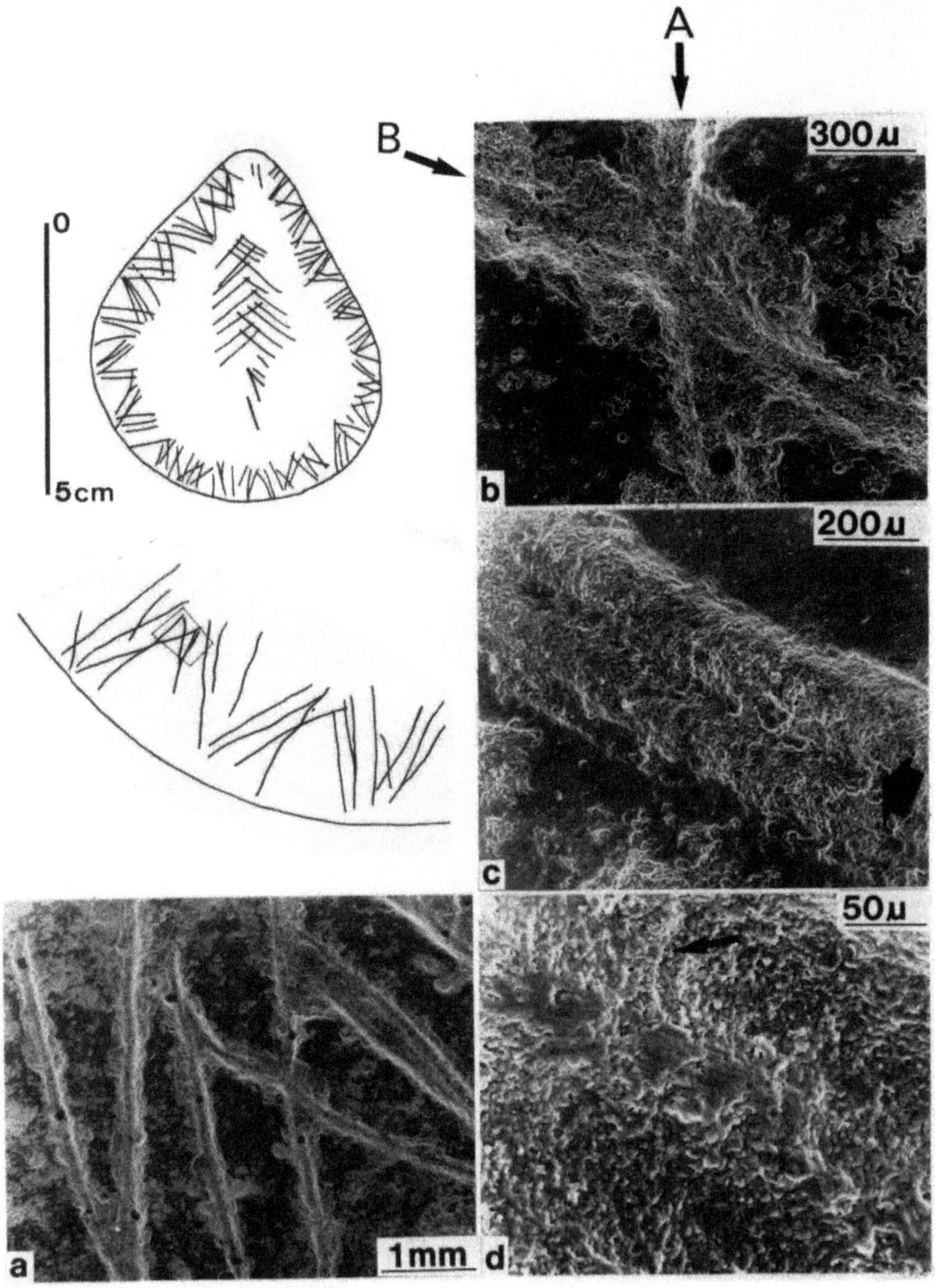

Fig. 8. Resin replica of engraved pebble from Rochedane; (a) Area covered by the square on the sketch, x20; (b) Detail of (a), x70. Notice the cavity produced at the exit of the intersection by the oblique engraving (the more recently made); (c) Engraving with a modified compact area at the centre with a succession of ripples (x100) indicated by the arrow in (d) (x300).

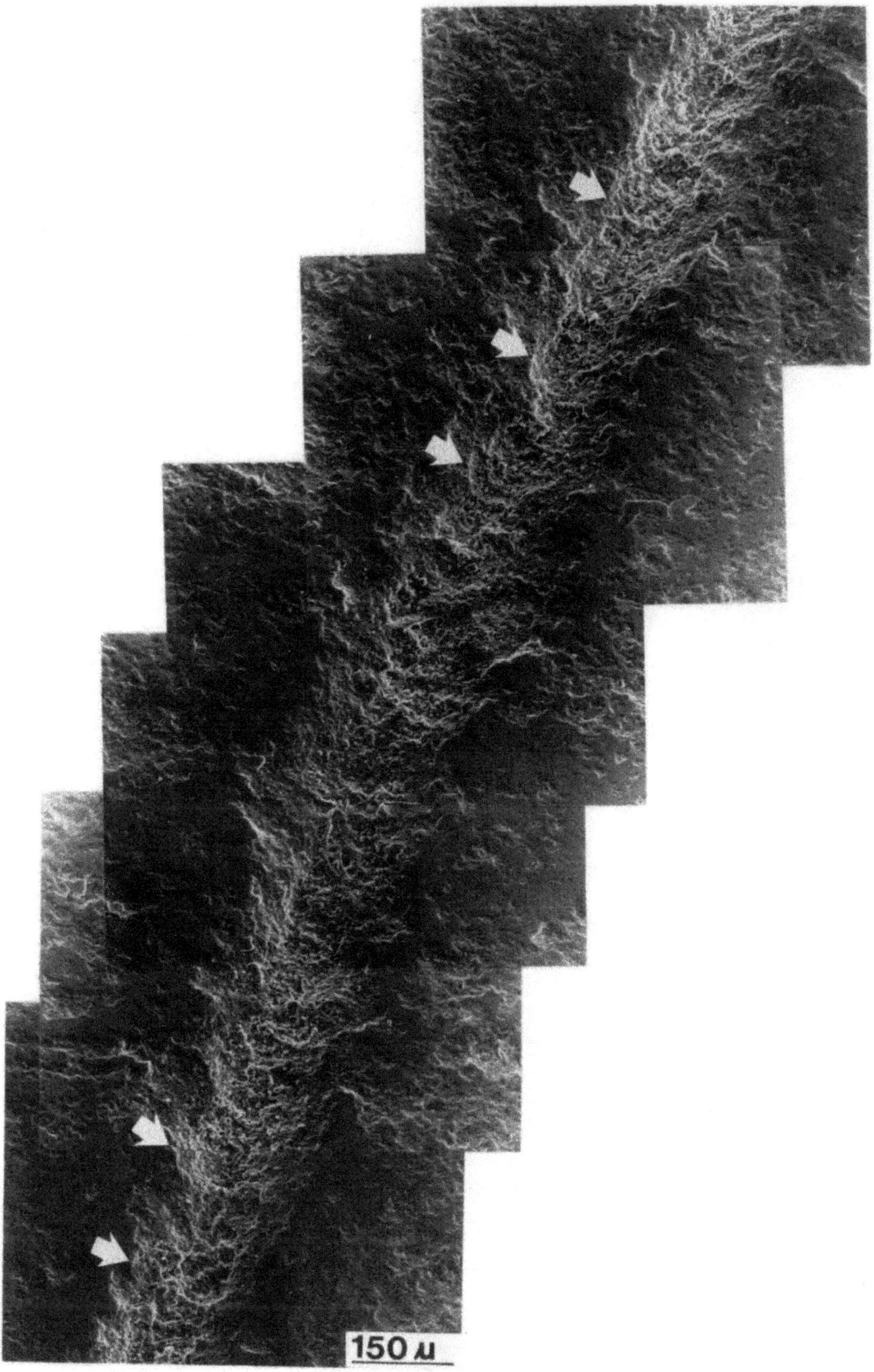

Fig. 9. Resin replica of engraving on a pebble from Rochedane. The white arrows point out a succession of ripples which can be used as a criterion for determining the direction of movement of the engraving tool.

Discussion

Previous researchers have already employed low power optical microscopy to the study of archaeological engraved pebbles (Marshack, 1970, 1972; Mons, 1972), and have often presented excellent photographs of the data (Marshack, 1969, 1970). This is, however, the first use of SEM on experimental and authentic engraved pebbles, and it is evident from this research that SEM provides a number of answers about the technology of engraving stone.

Engraving a pebble with a sharp flint implement not only produces a localized abrasion, but also subjects the material to elastic deformation which creates compact areas. The discontinuity recorded in the compact areas can probably be interpreted as microbreakages produced because the force exceeded the capacity of deformation of the limestone (Dauvois, 1979; Lieurade, 1978). It is beneficial to look at the "compact" areas for evidence regarding the direction of movement of the engraving tool. It is also worth noting that the directional characteristics encountered on engraved pebbles have many features in common with the criteria for directionality recently pointed out in butchering marks on bones (Bromage and Boyde, 1984). The rare compact areas found in original engravings may be due to the limited size of the sample examined, but may also be a problem of preservation. Future studies will have to deal with the retention on archaeological specimens of technological characteristics identified experimentally by SEM. Their preservation may, furthermore, be an important indicator of post-depositional modification of engraved art on pebbles.

After the presentation of this paper (April 1986), a research conducted with S.E.M. nitrocellulosic varnish replica and reflected light microscope, demonstrated experimentally (d'Errico 1987a; 1987b) the existence of other microtechnological features which indicate engravings made by the same tool. These new criteria, and those described below, were used on a large sample of French Azilian engraved pebbles (d'Errico in press) evincing interesting archaeological results.

Conclusions

As a result of the examination of experimentally engraved pebbles and replicas of Azilian originals with SEM, certain hypotheses regarding the mechanics of formation of the engravings can be put forward. Microscopic characteristics already identified through low power optical microscopy can be defined in greater detail by the use of SEM; and the existence of a whole new range of technological information was obtained from the interpretation of well preserved engraved surfaces.

Acknowledgements

I wish to thank Professor H. de Lumley for providing me the possibility of carrying out this research and A. Thevenin for allowing me to study the Rochedane engraved pebbles. This work is the first result of a cooperative research project with him. I also wish to thank Professor G. Giacobini and R. David for their useful advice and L. Varetto and C. Weber for their help with the SEM. S. Olsen, D. Vialou, M. Perpére, J. Espinet-Mouncadel, L. Comoglio and C. Lefebre also gave their kind assistance. This work was translated by Anne Topley.

References

Bromage, T.G. and Boyde, A. 1984. Microscopic criteria for the determination of directionality of cutmarks on bone. American Journal of Physical Anthropology 65: 359-366.

Couraud, C. 1985. L'Art Azilien. Origine-survivance. XX° suppl. Gallia Préhistoire. Paris: CNRS.

Dauvois, M. 1979. Burins usés de la grotte du Crest à Viry (Saone-et-Loire). Congrés Préhistorique de France XXI Session 2: 97-110.

Delporte, H. 1973. Les techniques de la gravure paléolithique. In Estudios dedicados al Professor Dr. Luis Pericot, Instituto de Arqueologia y prehistoria. Barcellona, Universitad de Barcelona, pp. 119-129.

Delporte, H. 1976. Typologie et technologie de l'art paléolithique mobilier: IX° Congrès U.I.S.P.P., Nice. Colloque XIV. Les courants stylistiques dans l'art mobilier du Paléolithique superieur. Preprint, pp. 37-53.

Delporte, H. and Mons, L. 1973. Notes de technologie et de morphologie de l'art paléolithique mobilier. Bulletin des Antiquités Nationales 5: 20-32.

Delporte, H. and Mons, L. 1975. Omoplate décoré du Mas d'Azil (Ariège). Bulletin des Antiquités Nationales 7: 15-23.

d'Errico, F. 1985. Traces d'usure sur l'industries lithique: approche méthodologique et proposition d'une technique. L'Anthropologie 89(5): 439-456.

d'Errico, F. 1987a. Nouveaux indices et nouvelles techniques microscopiques pour la lecture de l'art gravé mobilier. Comptes Rendus de l'Academie des Sciences de Paris 304(2-13): 761-764.

d'Errico, F. 1987b. Lecture technologique de l'art mobilier gravé. Nouvelles méthodes et premiers résultats sur les galets gravés de Rochedane. L'Anthropologie 91(4): 95-115.

d'Errico, F. in press. Memoire et rythmes au paléolithique: le mythe des calendriers lunaires. Actes du 2° Congrés International de Paléontologie Humaine, Turin.

Leroi-Gourhan, A. 1943. L'homme et la matière. Paris: Albin Michel.

Lieurade, H.P. 1978. Estimation des caractéristiques de résistence et d'endomagement en fatigue. In *Ecole d'été "La fatigue des matèriaux et des structures", Université de Sherbrooke, Québec*. Montreal: University Press, pp. 31-66.

Marshack, A. 1969. Polesini. A reexamination of the engraved Upper Paleolithic mobiliary materials of Italy by a new methodology. *Rivista di Scienze Preistoriche* 24 (2): 219-281.

Marshack, A. 1970. *Notation dans les gravures du Paléolithique superieur. Nouvelles méthodes d'analyse.* Bordeaux: Publications de l'Institut de Préhistoire de l'Université de Bordeaux.

Marshack, A. 1972. Cognitive aspects of Upper Paleolithic engraving. *Current Anthropology* 13 (3-4): 445-477.

Marshack, A. 1976. Some implications of paleolithic symbolic evidence for the origin of language. *Current Anthropology* 17 (2): 276-282.

Marshack, A. 1979. Upper Paleolithic symbol systems of the Russian plain. *Current Anthropology* 20 (2): 271-311.

Mons, L. 1972. Notes de technologie de l'art paléolithique mobilier (I-II). *Bulletin des Antiquités Nationales* 4: 14-21.

Pales, L. and Tassin de Saint-Pereuse, M. 1969. *Les gravures de la Marche I: Félins et ours.* Bordeaux: Publication de l'Institut de Préhistoire de l'Universitè.

Pales, L. and Tassin de Saint-Pereuse, M. 1968. Humains superposès de La Marche (document no. 1). In *La Prèhistoire, problèmes et tendances.* Paris: CNRS, pp. 327-336.

Pales, L. and Tassin de Saint-Pereuse, M. 1976. *Les gravures de La Marche. II: les humains.* Paris: Ophrys.

Rose, J.J. 1983. A replication technique for scanning electron microscopy: application for anthropologists. *American Journal of Physical Anthropology* 62: 225-261.

Thévenin, A. 1982. *Rochedane. l'Azilien, l'Épipaléolithique de l'Est de la France et les civilisations épipaléolithiques de l'Europe Occidentale.* Strasbourg: Mémoires de la Faculté des Sciences Sociales et Ethnologiques.

Thévenin, A. 1983. Les galets gravés et peints de l'abri de Rochedane (Doubs) et le problème de l'art azilien. *Gallia Prèhistoire* 1: 139-188.

PART V:

PLANT MATERIALS

METHODS USED IN CHARCOAL ANALYSIS AND THE RELATIONSHIP BETWEEN WOODS USED IN ARCHAEOLOGICAL TIMES AND THE PRESENT FUELWOOD CRISIS

Juliet Prior

Abstract

Swaziland, southern Africa, is rich in archaeological charcoals dating from the Middle Stone Age (200,000 to 20,000 bp) to Recent times. Charcoals from a Holocene river terrace required hardening before fracture, whereas pre-treatment under vacuum facilitated the examination of fragments from a surface scatter round an Iron Age smelting site. 97% of the Holocene and Recent fragments excavated from a rock shelter in the Lubombo mountains were identified using SEM. Their preservation was such that no cleaning or hardening treatment was required. The well-stratified rock shelter site provides evidence of those woods used as fuels throughout the climatic fluctuations of the Holocene. Past fuelwoods selected belong to only two families, the Combretaceae and the Leguminosae. Afforestation, using a range of such woods, may help to alleviate the present fuelwood crisis in areas where population growth and hence dependency upon renewable fuel resources is greatest.

Introduction

In 1940, Salisbury and Jane emphasised the potential of charcoals as indicators of vegetational change, when they published the results of analyses from Maiden Castle. Today, when the range and availability of techniques involving the use of scanning electron microscopy (SEM) is increasing rapidly, far more evidence could be obtained from a detailed examination of charcoal fragments, commonly occurring in both geological and archaeological deposits.

Southern Africa is one area of the world where the late Quaternary vegetational sequence is not well known (van Zinderen Bakker, 1978). Yet it is within this sub-continent that some of the earliest stages in human evolution occurred and cultural change often took place in response to climatic shift. Southern Africa today possesses a particularly rich, distinctive flora with a high level of endemism, indicating the extent to which floral evolution has occurred in isolation over a long period of time (Goldblatt, 1978:385). Despite the wealth of taxa, it is, as Acocks (1975:1) stresses, "possible to select relatively few species which will serve as indicators of different kinds of vegetation and of changes in vegetation". The climate and topography of the region as a whole are particularly suitable for charcoal preservation. Rainfall is comparatively low and there are a large number of cave and rock shelter deposits, providing evidence of prolonged occupation. Conversely, such a dry climate does not favour pollen preservation (Scott, 1982), so it is upon charcoal and other macrobotanical remains that archaeobotanists must place particular reliance. Deacon has pioneered charcoal studies in southern Africa, using evidence from sites in the Cape (Deacon, 1979; Deacon et al., 1983). His work and that of Tusenius (1986), emphasises both the diversity of the sub-continent and the need for an expansion of complementary research into different areas.

Regional Setting and Charcoal Retrieval

Swaziland, one of the smallest African countries, lying in the sub-tropics between 25°40'S and 27°20'S, has a wide environmental diversity on account of its varied topography. This diversity also existed in the past, since many archaeological sites span both the glacial and inter-glacial periods, which occurred during the Early, Middle and Late Stone Ages (Goudie and Price Williams, 1983). Sites situated in the northern Middleveld (Balekane), the central Middleveld (Mashila) and the Lubombo mountains of eastern Swaziland (Siphiso) (see Map 1) demonstrate some of the problems encountered during the retrieval of charcoal from a range of provenances: two subjected to weathering and one a shelter, partially protected by a rock overhang.

The site at Balekane in the northern Middleveld is a Holocene river terrace. Here, an exposure of mixed alluvial deposits is some 5 m in height, alongside a small tributary of the Komati river. Charcoal fragments were individually extracted from a length of about 1.5 m within an organic layer, visible as a dark band, at a depth of 3.5-4 m from the surface. Weathering of the exposure, exacerbated by numerous birds' nests, had made the charcoal soft and crumbly. It was therefore extracted together with some matrix, so that its structure was kept as intact as possible.

Mashila lies in the Mkondvo valley of central Swaziland. The charcoal at this site formed part of a surface scatter representing the remains of an Iron Age settlement currently being undermined by gully erosion (Price Williams, 1980). Again, fragments were individually lifted, though in this case the material was very hard, probably for two reasons. First, since associated material included rough iron slag and dimple-faced hammer stones for working the ore, it was assumed that some, if not all, of the fragments would have been charred at a high temperature, consistent with the extraction of iron. Secondly, the sub-tropical climate at Mashila, where the rainfall of approximately 800 mm per annum falls mainly during the warm summer season (Murdoch, 1978), results in a migration of soluble salts from the soil surface to the charcoal. The salts originate from the plagioclases released during the weathering of the local grano-dioritic rocks. Long term exposure to the elements also increases the adsorption of moisture and gases within the porous charcoal structure. In the present study, a random sample of 29 fragments was analysed from this site.

The topographical setting, dimensions and stratigraphy of the Siphiso rock shelter have already been described (Prior, 1983; Prior and Price Williams, 1985). The charcoal, comparatively hard and well preserved, formed a surface scatter without any obvious hearth concentrations. As at Mashila, the hardness was due to post-depositional migration of soluble salts through the strata. In this instance, the salts were released by the action of rain water on the feldspars present within the porous rhyolite of the roof. Each fragment was individually lifted and plotted three dimensionally, so that some estimate of the degree of fragmentation could be attempted.

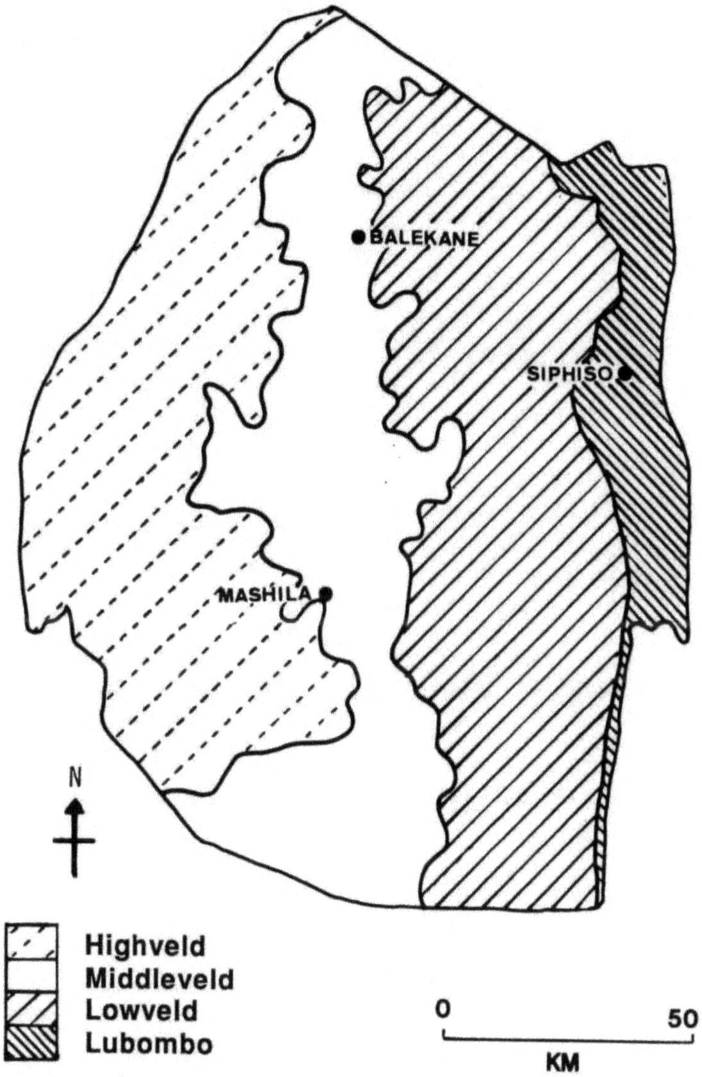

Map 1. Swaziland Environmental Diversity

It was felt that froth flotation or wet or dry sieving would preferentially damage the smaller or more fragile pieces. No sampling was attempted, as complete levels were analysed within each stratum, until a total of approximately 100 fragments per stratum had been examined. Excavation and analyses from this site are still in progress and will be published elsewhere.

Charcoals from all three sites were stored in airtight containers after collection, so that adsorption of atmospheric moisture was minimised.

Subsequent Treatment and SEM Technique. Pretreatment of Charcoal.

Balekane. The extracted fragments were freed from as much matrix as possible, then hardened overnight in a solution of polybutyl methacrylate (supplied as Vinalac by Mineralite Ltd.) dissolved in approximately twice its volume of ethyl acetate. After hardening, surplus resin was removed by washing the fragments several times in ethyl acetate. Finally, they were left in a fume cupboard until the solvent had evaporated, then thoroughly desiccated by pumping with a rotary vacuum pump over phosphorus pentoxide for at least six hours.

Mashila. The considerable quantity of gases and moisture were removed by placing prepared stubs under a vacuum of 10^{-6} torr. for several hours. The high vacuum, similar to that obtaining in the column of the scanning microscope, removed all adsorbed gases, thus facilitating subsequent sputter coating.

Siphiso. The comparatively dry conditions obtaining within the shelter enabled all the charcoal, including that from surface levels, to be examined without pretreatment.

SEM Technique

Each charcoal fragment was manually fractured after scoring and where possible, transverse, tangential longitudinal, and radial longitudinal planes were mounted together on aluminium stubs using Durofix and trails of silver dag, to increase conductivity. In the case of the smallest pieces, sometimes less than 1 mm in length, only transverse fractures were possible. All stubs were then coated with approximately 40 nm of gold in an atmosphere of argon, using a Polaron E5000 diode sputter coater. They were examined in a Phillips PSEM 500 at a kv of 25, a probe size of 32 nm and a range of magnifications from X50 to X6400. Identifications were sometimes possible using uncoated charcoal, but the gold surface greatly enhanced resolution.

Ancient charcoal can only be satisfactorily examined by comparison with charred reference material. The size of the modern samples collected and the routine charring procedure have already been described (Prior and Price Williams, 1985). The time and temperature of charring and the optimal initial moisture content of the samples were previously determined experimentally, using two species of highly divergent anatomical character (Prior and Alvin, 1983; Prior and Alvin, 1986). As Beall et al. (1974) stress, a thorough understanding of the physical changes that occur during carbonisation, is essential in any interpretation of electron micrographs.

Results and Discussion

A comparison of the ancient charcoal fragments with reference material enabled secure identifications to be made in almost all cases. There seems little doubt, therefore, that opaque charcoal is a material particularly well suited to examination under SEM. A large amount of detail including such features as the structure of ray or vessel pitting or the nature and distribution of crystalline inclusions can be obtained from minute fragments, even those which fractured with an uneven surface (see Plate 2). Such detail is not always necessary, though in palaeoclimatic interpretations, identifications at a specific level may be essential and this is often impossible without the maximum anatomical information. Dark or light field incident light microscopy is frequently used as an alternative technique in charcoal analyses (Leney and Casteel, 1975). Identifications can often be made more quickly and cheaply using these techniques of optical microscopy, but since much of the fine detail cannot be seen, identification below generic level becomes extremely difficult.

Balekane and Mashila. Plate 1 demonstrates that the technique used for hardening the fragile Balekane charcoal did not cause any significant occlusion of structural details under SEM. Of the 29 charcoal fragments from Mashila, 14 originated from Terminalia sericea, 12 from Combretum imberbe (see Plate 1), 2 from legumes (Mimosoideae) and 1 from Olea europaea subsp. africana. The preponderance of the first two dense hardwoods at this smelting site is being investigated further in order to determine whether their anatomical structure or their chemical composition, in terms of alkaline and alkaline earth metal content, contributed more to their selection by Iron Age people (van der Merwe and Killick, 1979; Evers, 1982).

Siphiso. A summary of some of the charcoal analyses from Siphiso is shown in Table 1 and some of the details obtained using SEM are shown in Plate 2. Comparatively small assemblages of taxa persisted within each stratum. Careful selection on the part of successive inhabitants may be one interpretation of this situation but selective preservation of the ancient material is certainly a contributory factor. Preservation appears to be a function of anatomical structure of the original wood rather than the age of the charcoal. Androstachys johnsonii, the Lubombo ironwood for example, with small vessels, narrow rays and thick walled fibres is consistently well preserved, whereas some Acacia charcoal with large vessels, wide rays and extensive parenchyma is often fragmentary. In a recent paper, Zicherman and Williamson (1981), stress that both chemical and anatomical differences between species affect their response to heat. Further, they state that such interactions have so far been the subject of very little research.

Recent studies in Botswana by Jelenic and van Vegten (1981), indicate that dead branchwood still attached to the trees is the preferred fuelwood in areas where a choice is still available. A

Plate 1.

Fig. 1. Transverse fracture of hardened *Rapanea melanophloeos* charcoal from Balekane river terrace; x268.

Fig. 2. Transverse fracture of modern, charred *Rapanea melanophloeos*; x268.

Fig. 3. Transverse fracture of *Combretum imberbe* charcoal from smelting site at Mashila, showing few, solitary vessels and mass of fibres; x68.

Fig. 4. Transverse fracture of modern, charred *Combretum imberbe*; x68.

Fig. 5. Tangential longitudinal fracture of *Combretum imberbe* charcoal from smelting site at Mashila, showing numerous, uniseriate rays; x68.

Fig. 6. Tangential longitudinal fracture of modern, charred *Combretum imberbe*; x68.

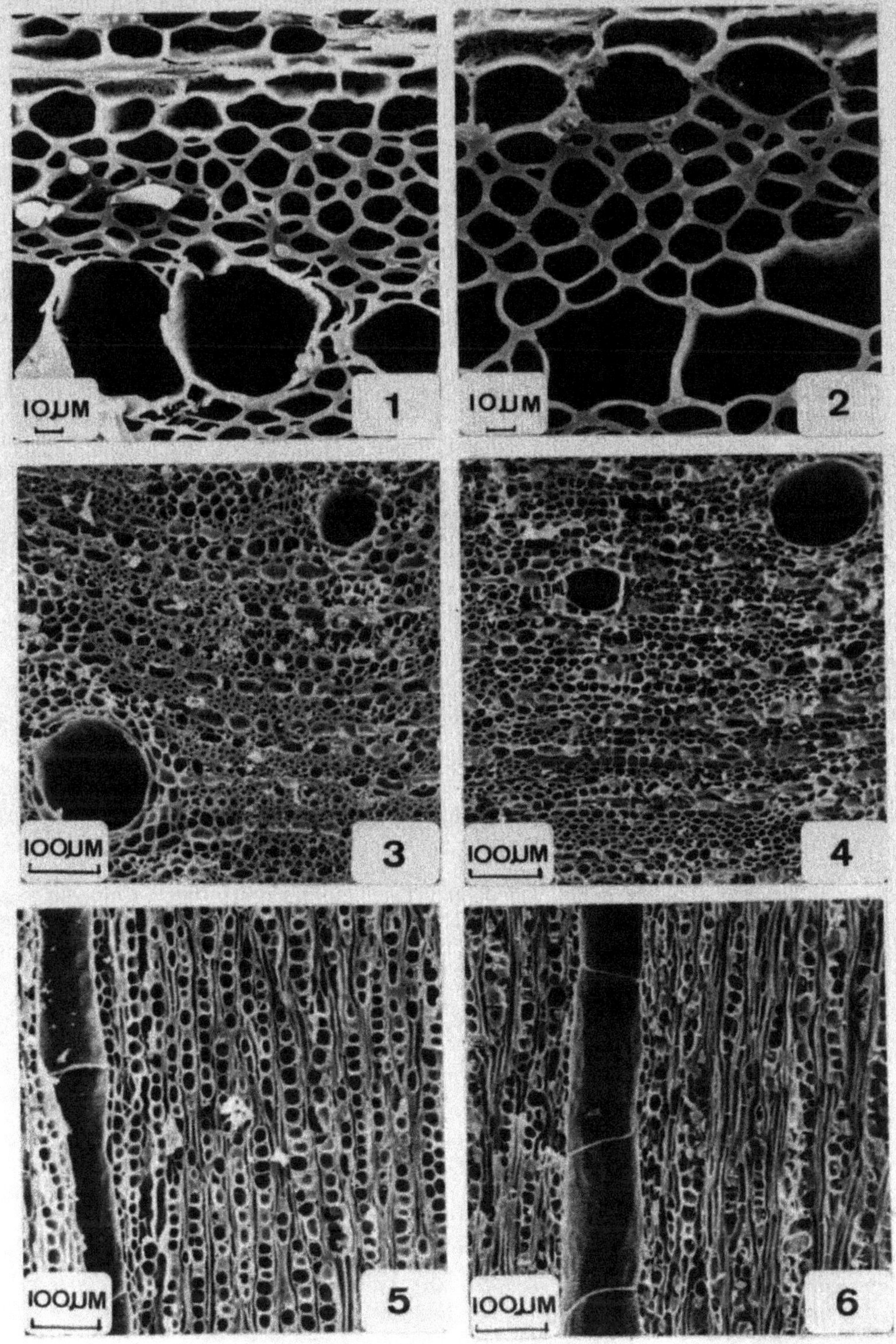
10µM
1
10µM
2
100µM
3
100µM
4
100µM
5
100µM
6

Plate 2.

Fig. 7. Transverse fracture of *Combretum apiculatum* charcoal from Stratum 3 at Siphiso, showing thick-walled, oval, solitary vessels and phloem island (top, left); x68.

Fig. 8. Transverse fracture of modern, charred *Combretum apiculatum*; x68.

Fig. 9. Tangential longitudinal fracture of modern, charred *Maytenus heterophylla*, showing crystals in ray parenchyma cells; x268.

Fig. 10. Transverse fracture of modern, charred *Dichrostachys cinerea*, showing gelatinous fibres. In these fibres, there is a large space between the thick, inner wall layers which have separated from the outer; x535.

Fig. 11 Inter-vessel pits in legume charcoal from Stratum 3 at Siphiso, showing remains of lignin-rich vesturing around the pit aperture; x2135.

Fig. 12. Transverse fracture of *Androstachys johnsonii* charcoal from Stratum 5 at Siphiso, showing homogenisation of inter-cellular material with resultant loss of structural detail; x268.

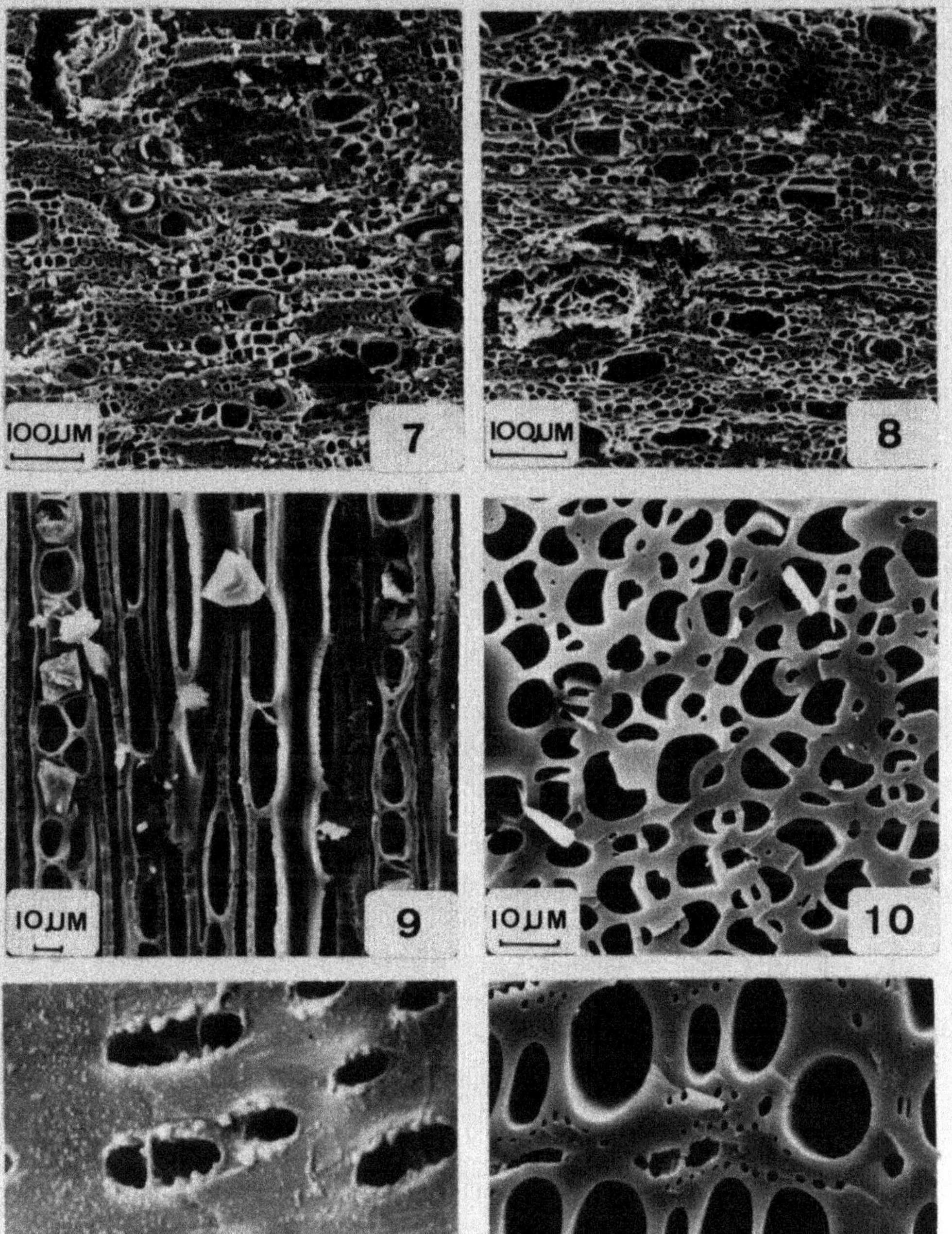
100µM
7
100µM
8
10µM
9
10µM
10
2µM
11
10µM
12

TABLE 1. Charcoals from Siphiso.

STRATA	APPROXIMATE DATE	TOTAL NO. FRAGMENTS	% LEGUMES (MIMOSOIDEAE)	% COMBRETUM APICULATUM
1+2	last 100 yrs.	261	42.9	27.6
3	Late Holocene	134	47.8	26.9
4	Mid Holocene	91	50.5	22.0
5	Early Holocene	119	63.0	10.1

TABLE 2. Fuelwoods Represented at Siphiso.

SPECIES RANKED IN ORDER OF DECREASING DENSITY	DENSITY VALUES $G.CM.^{-3}$	% FIBRE WALL *	CALORIFIC VALUES $KG.G.^{-1}$
DICHROSTACHYS CINEREA	0.88	52.9	20.7
ACACIA CAFFRA (D)	0.70	60.2	
ACACIA CAFFRA (M)	0.68	49.4	
COMBRETUM APICULATUM	0.68	40.5	18.6
ACACIA KARROO (D)	0.66	58.3	
ACACIA KARROO (M)	0.66	50.6	
ACACIA NIGRESCENS (D)	0.64	55.8	
ACACIA NIGRESCENS (M)	0.62	42.7	20.3
ACACIA TORTILIS	0.52	34.0	19.3

D=Drier Provenance
M=Moister Provenance
*=Average area of fibre wall x % total fibres in at least 50 high power microscope fields.

similar situation seems to have obtained at Siphiso in past times. Many charcoal fragments showed charred fungal hyphae within their cellular structure and since charred wood is totally inorganic, the fungi must be in primary deposition. Small twigs, presumably kindling, often form part of the charcoal spectra and though their anatomical structure may not be fully developed (Jane, 1970), this does not generally preclude secure identification.

Much information concerning the charring regime used on ancient sites can be gained from a comparative study of electron micrographs of ancient material and those from similar modern woods charred under known conditions. A comprehensive review of the thermal degradation of wood was published by Beall and Eickner (1970). Slocum et al. (1978), state that on charring, wood behaves as a mixture of its three polymers, which degrade at different temperatures and recombine into a carbonaceous structure. Many workers using SEM techniques have demonstrated cellular deformations which are distinct from those caused by simple shrinkage, and are the result of some degree of plastic flow. As Zicherman and Williamson (1981:247) state, "Such flow would be a response to the thermal stresses which are applied to the softened lignin in the degrading cell walls which no longer contain functional cellulose reinforcing networks. Thus, one can envision the observed deformations as a combined result of plastic flow during the combustion process and of shrinkage caused by volumetric loss of water and volatilization of cell wall constituents". In the Siphiso material, such deformations are particularly evident both in the lignin-rich vestured pits of legumes (Ohtani et al., 1984), and in the thick walled Androstachys fibres shown in Plate 2.

Application of Results to the Present Fuelwood Crisis

The type of detailed information obtained from ancient charcoals using SEM techniques has far wider potential (Prior, 1985). At Siphiso, the bulk of the fragments belong either to the Leguminosae (Mimosoideae) or to Combretum apiculatum (see Table 1 and Plate 2). Clearly therefore, it was these groups that were consistently selected as a fuel source by successive inhabitants. The Mimosoideae include Dichrostachys cinerea and a wide range of Acacia species. The acacias, a widespread and ecologically important group in arid and semi-arid areas of Africa, are difficult to distinguish solely on the basis of wood anatomy (Robbertse et al., 1980). In some instances, however, where large fragments were examined or where characteristic features had been particularly well preserved, specific identifications were possible. From these, it was evident that the range of Acacia species present during the Holocene included Acacia caffra, Acacia karroo, Acacia nigrescens and Acacia tortilis, all species occurring in the vicinity of the shelter today.

What then were the characteristics of these woods which resulted in their consistent selection and how did they withstand climatic change? All were dense woods with a high proportion of fibres and all had high calorific values, as shown in the preliminary results recorded in Table 2. They would, therefore, have burnt slowly with a considerable heat emission. One assumes too, that the smoke produced during their combustion was non-toxic and that they did not spark to excess. It is probable that the large number of calcium oxalate crystals present in axial and ray parenchyma cells, which were identified using X-ray diffraction techniques, affected the speed of combustion of the woods. As Shafizadeh (1984) and others have found, cations such as calcium reduce the ratio of carbon monoxide to carbon dioxide production, thereby altering the rate of heat release.

An unexpected and persistent anatomical feature present in the majority of the charcoal fragments from these groups was the presence of large amounts of gelatinous fibres (G-fibres), clearly visible under SEM (see Plate 2). These dense, modified fibres with an enhanced proportion of hydrophilic celluloses and hemicelluloses within their walls, are known to occur in tension wood as a response to stress (Hughes, 1965). In the present instance however, the large quantity present suggested an additional role, that of drought tolerance. Two lines of evidence support this hypothesis. Firstly, branchwood samples of _Acacia caffra_, _Acacia karroo_ and _Acacia nigrescens_ were collected from wet and dry provenances. In each case, preliminary studies show that the number of G-fibres and hence the percentage of fibre wall material, was enhanced by increased aridity (see Table 2). Secondly, the number of G and non-G fibres in samples of 30 ancient legume fragments from each of Strata 1 to 5 was analysed (see Fig. 13). The considerable increase in the ratio of the G to non-G variant in Stratum 4 suggests the possibility of drier conditions between about 4000 and 6000 bp.

Recent FAO and World Bank reports have frequently stressed the severity of the present fuelwood crisis, which is most acute in marginal, semi-arid lands where population pressures are greatest. Botswana, eastern Swaziland, Zambia and Zimbabwe are among those areas currently at risk. All the trees mentioned as fuelwoods at Siphiso are indigenous to these regions. As such, they are familiar to the Africans who, being greatly dependent upon oral traditions, select these same woods for fires today. Further, they also use the trees for fodder, shade, soil enrichment, magic and medicine. Of recent years, forestry projects in southern Africa have concentrated upon the growth of fast growing, non-indigenous species in areas of comparatively high rainfall. The Nairobi Plan of Action, adopted at the UN Conference of New and Renewable Energy Resources in August 1981, recommends a five-fold increase in the annual rate of tree planting in developing countries by the year 2000. During the next decade therefore, it would perhaps be appropriate if those indigenous species already shown by the archaeological record to be desirable fuels and to have drought tolerance, were used for wide-scale planting on a collective, as well as an individual basis.

Fig. 13. **Totals of Gelatinous (White) and Non-gelatinous (Shaded) Fibres In Samples of 30 Ancient Fragments**

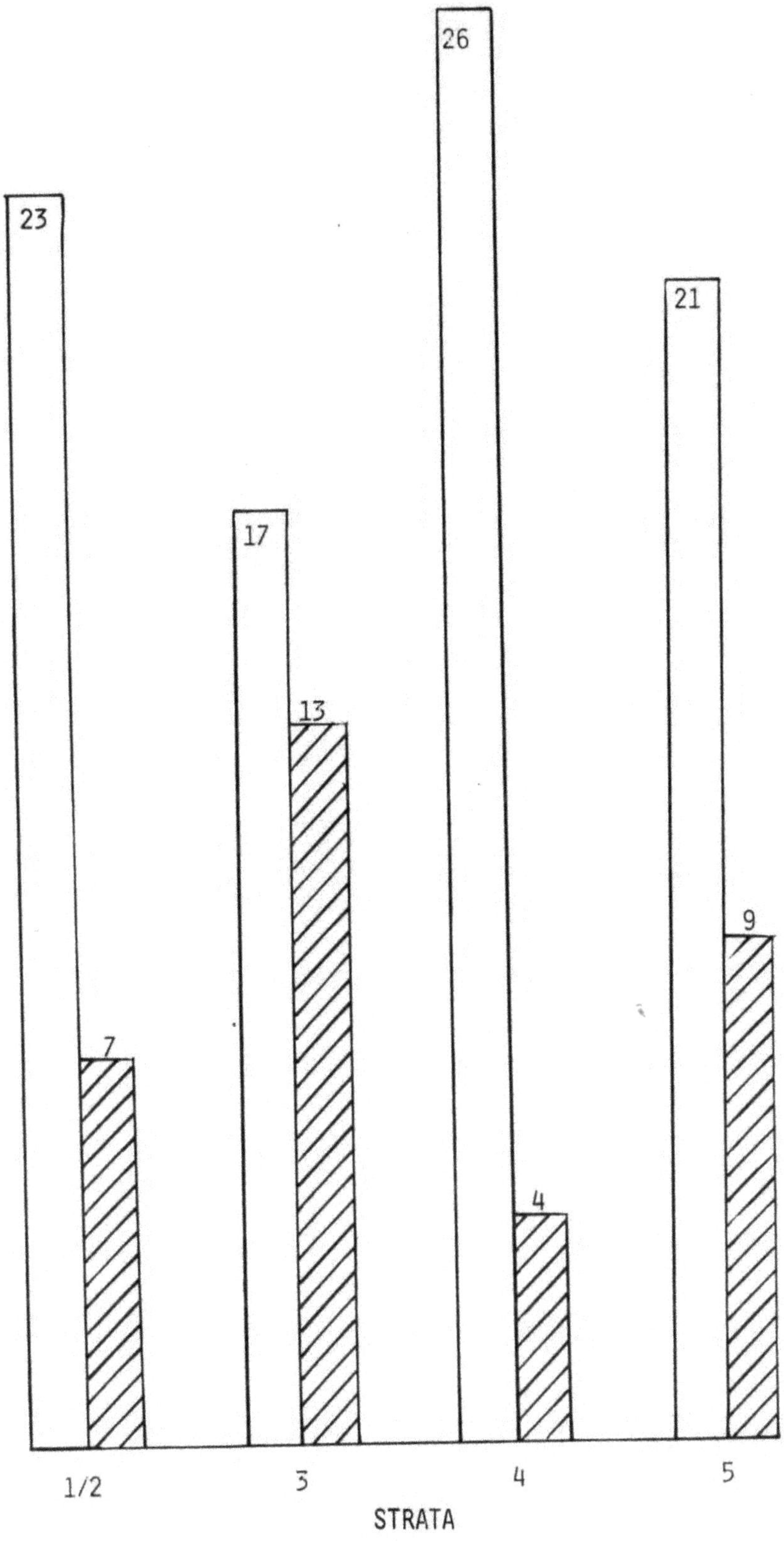

Acknowledgements

Funding for these studies is gratefully acknowledged from the Swaziland National Trust Commission, the Anglo American de Beers Chairman's Fund and the Central Research Fund of the University of London. Dr. D. Price Williams was responsible for excavation of the Siphiso site and Dr. K. Alvin collaborated in the work leading to the preliminary results recorded in Table 2. Many members of the Department of Pure and Applied Biology, Imperial College, particularly Professor J. Levy, Mr. N. Costa and Mrs. M. Docherty, provided help and advice during the course of the work.

References

Acocks, J.P.H. 1975. Veld types of South Africa. Memoirs of the Botanical Survey of South Africa 40: 1-128.

Beall, F.C. and Eickner, H.W. 1970. Thermal degradation of wood components: a review of the literature. U.S.D.A. Forest Service Research Paper FPL: 130.

Beall, F.C., Blankenhorn, P.R. and Moore, G.R. 1974. Carbonized wood-physical properties and use as an SEM preparation. Wood Science 6: 212-219.

Deacon, H.J. 1979. Excavations at Boomplaas Cave-a sequence through the Upper Pleistocene and Holocene in South Africa. World Archaeology 10: 241-257.

Deacon, H.J., Scholtz, A. and Daitz, L.D. 1983. Fossil charcoals as a source of palaeoecological information in the Fynbos region. In Deacon, H.J., Hendy, Q.B. and Lambrechts, J.J.N. (eds.), Fynbos Palaeoecology: a Preliminary Synthesis, pp. 174-182. South African National Scientific Programmes Report: 75.

Evers, T. 1982. Two later iron age sites on Mabete, Hans Merensky Nature Reserve, Letaba district, N.E. Transvaal. South African Archaeological Bulletin 37: 63-67.

Goldblatt, P. 1978. An analysis of the flora of southern Africa: its characteristics, relationships and origins. Annals of the Missouri Botanical Gardens 65: 369-436.

Goudie, A.S. and Price Williams, D. 1983. The Atlas of Swaziland. The Swaziland National Trust Commission. Occasional Paper no. 4.

Hughes, F.E. 1965. Tension wood: a review of literature. Forestry Abstracts 26: 2-9.

Jane, F.W. 1970. The Structure of Wood, second edition. London: A. and C. Black.

Jelenic, N.E. and van Vegten, J.A. 1981. A pain in the neck: the firewood situation in south-western Kgatleng, Botswana. NIR Research Note 5: Gaborone, Botswana.

Leney, L. and Casteel, R.W. 1975. Simplified procedure for examining charcoal specimens for identification. Journal of Archaeological Science 2: 153-159.

Murdoch, G. 1970. Soils and Land Capability in Swaziland. Mbabane: Swaziland Ministry of Agriculture.

Ohtani, J., Meylan, B.A. and Butterfield, B.G. 1984. Vestures or warts - proposed terminology. IAWA Bulletin n.s. 5: 3-8.

Price Williams, D. 1980. Archaeology in Swaziland. South African Archaeological Bulletin 35: 13-18.

Prior, J. 1983. Investigations on archaeological charcoals from Swaziland, using SEM techniques. In Vogel, J.C. (ed.), Late Cainozoic Palaeoclimates of the Southern Hemisphere, Proceedings of an International Symposium, Swaziland, 1983, pp. 353-360. Rotterdam: Balkema.

Prior, J. 1985. The identification and significance of woods represented in archaeological charcoals from Swaziland. In Forest Products Research International-Achievements and the Future, Proceedings volume 1, C.S.I.R., Pretoria, pp. 16-14, 1-16.

Prior, J. and Alvin, K.L. 1983. Structural changes on charring woods of Dichrostachys and Salix from southern Africa. IAWA Bulletin n.s. 4: 197-206.

Prior, J. and Alvin, K.L. 1986. Structural changes on charring woods of Dichrostachys and Salix from southern Africa: the effect of moisture content. IAWA Bulletin n.s. 7: 243-250.

Prior, J. and Price Williams, D. 1985. An investigation of climatic change in the Holocene epoch using archaeological charcoal from Swaziland, southern Africa. Journal of Archaeological Science 12: 457-475.

Robbertse, P.J., Venter, G. and van Rensburg, H.J. 1980. The wood anatomy of the South African acacias. IAWA Bulletin n.s. 1: 93-103.

Salisbury, E.J. and Jane, F.W. 1940. Charcoals from Maiden Castle and their significance in relation to the vegetation and climatic conditions in prehistoric times. Journal of Ecology 28: 310-325.

Scott, L. 1982. A late Quaternary pollen record from the Transvaal bushveld, South Africa. Quaternary Research 17: 339-370.

Shafizadeh, F. 1984. The chemistry of pyrolysis and combustion. In Rowell, R. (ed.), The Chemistry of Solid Wood. Advances in Chemistry Series 207: 490-529. American Chemical Society, Washington, D.C.

Slocum, D.H., McGinnes, E.A. Jr. and Beall, F.C. 1978. Charcoal yield, shrinkage, and density changes during carbonization of oak and hickory woods. Wood Science 11: 42-47.

Tusenius, M.L. 1986. The study of charcoal from some southern African archaeological contexts. M.A. thesis, University of Stellenbosch.

United Nations. 1981. Report of the UN Conference on New and Renewable Energy Sources, New York.

van der Merwe, N.J. and Killick, D.J. 1979. Square: an iron smelting site near Phalaborwa. South African Archaeological Society, Goodwin Series 3: 86-93.

van Zinderen Bakker, E.M. 1978. Quaternary vegetation changes in southern Africa. In Werger, M.J.A. (ed.), Biogeography and Ecology of Southern Africa. The Hague, Junk, pp. 133-143.

Zicherman, J.B. and Williamson, R.B. 1981. Microstructure of wood char. Part 1: Whole wood. Wood Science and Technology 15: 237-249.

SEM AND THE IDENTIFICATION OF MICRO-MORPHOLOGICAL INDICATORS OF DOMESTICATION IN SEED PLANTS

Bruce D. Smith

Introduction

Identifying Prehistoric Cultigens. The initial emergence and subsequent elaboration of prehistoric food production systems centering on crop plants were developments of profound importance in the cultural evolution of human populations in most areas of the world. Gaining an understanding of the incipient stages of prehistoric cultivation of crop plants by human populations has been hampered, however, by difficulties inherent in attempting to identify in the archaeological record the changing nature of prehistoric human-plant relationships. This difficulty in identifying incipient cultivation of plants is evident in the relative weakness of the indirect arguments offered in support of assigning "cultivated" status to plant species recovered in archaeobotanical assemblages. Plant species are usually designated prehistoric cultigens or "quasi-cultigens" on the basis of one or more of the following arguments (see Smith 1985a for references regarding the application of these criteria):

1. The existence of modern weed analogs for a plant species;
2. A prehistoric range extension for the plant species;
3. The archaeological abundance of a plant species relative to its modern occurrence;
4. The perceived economic importance of a plant species;
5. The cultivation of a similar species elsewhere;
6. A prehistoric association of the plant species with known cultigens;
7. Early historical descriptions of cultivation of a plant species;
8. A documented prehistoric increase in abundance of a plant species through time;
9. An increase in human population levels and sociopolitical complexity, implying an agricultural economy; and
10. A documented morphological change in reproductive propagules (e.g. seeds) associated with domestication.

Only the last of these arguments, however, allows the unequivocal identification of a plant species as a cultigen, as opposed to an intensively harvested wild plant or weedy camp follower. While these other arguments certainly serve to strengthen the case for cultivated status, the single unquestionable indicator of cultivation that can be observed directly in the archaeobotanical record remains morphological change in plant propagules (e.g. seeds) associated with domestication.

SEM and the Identification of Morphological Change Associated with the Adaptive Syndrome of Domestication. Resulting from strong and automatic selective pressures within an anthropogenic habitat, i.e. disturbed soil, such genetic-morphological changes reflect an adaptation to the harvesting and sowing of seeds by human cultivators.

The nature of these selective pressures and the morphological changes which they produce have been clearly established by Harlan and others (Harlan _et al._, 1973; deWet and Harlan, 1975). Seed bed competition, on the one hand, automatically selects both for increased seedling vigor (frequently resulting in increased seed size) and reduced germination dormancy (frequently resulting in a loss or reduction of seed dormancy mechanisms, such as the seed coat). In addition, harvesting, in association with successive deliberate sowing, selects for uniform ripening of the seeds, increased seed production, better "packaging" of seeds for the harvester (more terminalized and compact infructescences), and a loss of natural dispersal mechanisms.

Recognizing and documenting these domestication-related morphological changes in archaeobotanical assemblages, however, is often difficult. The plant parts that are necessary to study uniform maturation, an increased production of seeds, and increased compaction and terminalization of seedheads are rarely preserved. At the same time, even though seeds are frequently present in abundance in archaeobotanical assemblages, the changes in seed morphology associated with domestication, resulting from selective pressures for reduced seed dormancy and seed dispersal, and increased seedling vigor often involve such small scale modifications in seed structure that accurate description or quantifiable documentation using only a standard light microscope is extremely difficult.

It is only in the past decade, through the application of scanning electron microscopy, that the full research potential of micro-morphological analysis of archaeobotanical seed assemblages has been recognized. In Europe and Great Britain the application of SEM technology to the study of seed morphology began in the mid to late 1970's, as represented by the landmark work of Korber-Grohne (Korber-Grohne and Piening, 1980; Korber-Grohne, 1981) in distinguishing Old World cereal grains on the basis of surface microstructure. A similar interest in applying SEM technology to the study of the microstructure of archaeologically recovered seeds of New World domesticates developed in the United States in the late 1970's (Wilson, 1981).

To illustrate the value of scanning electron microscopy in the recognition and quantification of micro-morphological indicators of domestication, a North American example involving the prehistoric cultivation of _Chenopodium berlandieri_ will be briefly summarized (end note 1).

An Example from the Eastern Woodlands

Early Claims for Domesticated Chenopodium in the Prehistoric East. A debate regarding whether or not a domesticated form of chenopod was present in the prehistoric eastern woodlands of North America has extended over six decades.

As early as 1926 (Asch and Asch, 1977:12), Oliver Freeman identified chenopod seeds recovered from Bluff Shelters in the Ozarks of Arkansas as a probable cultivated form, due to their morphological similarities to _C. berlandieri_ ssp. _nuttalliae_, a modern Mexican cultivated chenopod. Ten years later, Volney Jones suggested that chenopod seeds recovered from poorly dated prehistoric deposits at Newt Kash Shelter, Menifee Co., Kentucky, represented a cultivated crop plant. His identification of these seeds as being from a domesticated crop was based on their large size in comparison to the seeds of wild species (Jones, 1936:149; Asch and Asch, 1977:14).

The Rejection of Early Claims for Domesticated Status. In the mid 1970's, David and Nancy Asch reanalyzed the Ozark and Newt Kash chenopod assemblages studied by Freeman and Jones and rejected both of the earlier cultigen status claims for prehistoric *Chenopodium* in the eastern woodlands of North America (Asch and Asch, 1977). The large Newt Kash "chenopod" seeds were found to actually be pokeweed (*Phytolacca americana*). Restudy of the Ozark Bluff Shelter chenopod seeds showed them apparently differing morphologically from wild seeds in two respects (Asch and Asch, 1977:19):

"cross-sectional shape - a tendency toward truncate margins in the archaeological specimens as opposed to published descriptions of *C. bushianum* as biconvex in cross section.

prominence of markings on the seed coat - less prominent on most archaeological specimens than indicated by published descriptions."

The Asches, however, also observed these morphological traits (truncate margins and variable seed coat markings) in a significant percentage of the fruits produced by a modern, late season stand of wild *C. bushianum* which they studied in the mid-1970's. Because of the relative abundance of such "red morph" fruits in the modern wild stand studied, Asch and Asch concluded that rather than indicating domestication, these morphological characteristics of the prehistoric Ozark specimens could be explained as reflecting phenotypic polymorphism in wild populations of *C. bushianum* (1977:20-22). No specific mention was made of observed variation in seed coat (testa) thickness, and this characteristic was not discussed as a possible indicator of domesticated status. The Asches concluded that they had "... found no convincing evidence for genetic changes [domestication] in Eastern North American *Chenopodium*" (1977:24). This position was reiterated in a 1980 article (Asch and Asch, 1980:158).

The First Application of SEM to the Problem. In the late 1970's, however, Hugh Wilson looked again at the Ozark chenopod materials and convincingly demonstrated the prehistoric presence of a domesticated form of *Chenopodium* in Ozark Bluff Shelter archaeobotanical assemblages. Employing scanning electron microscopy, Wilson compared cross-sectioned seeds of: a) wild *Chenopodium berlandieri*; b) the modern Mexican cultivar *C. berlandieri* ssp. *nuttalliae* cv. 'huauzontle', and c) archaeological specimens from Holman Bluff Shelter, Madison County, Arkansas. Wilson's comparative SEM analysis showed that while the wild fruit had a thick, black outer epiderm seed coat layer (testa), this tough lignified structure, essential for seed dormancy in the wild, appeared to be entirely absent in both the modern Mexican cultivar and the Holman Bluff Shelter specimens, due to strong selective pressure for reduced germination dormancy. This apparent elimination of the outer epiderm, in addition to providing unequivocal evidence for domesticated status and giving the archaeological and modern Mexican "naked" seeds a light coloration as opposed to the characteristic black coloration of wild fruits, also appeared to result in the truncate margin morphology previously recognized by the Asches (Wilson, 1981:237).

In comparing the micromorphological structure of prehistoric and modern seeds with the aid of scanning electron microscopy, Wilson demonstrated, for the first time, the critical role that SEM could play in elucidating long standing questions regarding prehistoric plant husbandry in the Eastern Woodlands.

Although it was generally thought at the time that the domesticated pale-colored chenopod recognized by Wilson was a late prehistoric (post A.D. 1200) introduction from Mexico, direct accelerator dating of Ozark specimens has recently shown that the pale-seeded variety of domesticated Chenopodium was present in Eastern North America by ca. A.D. 300 (Fritz, 1986a, 1986b; Fritz and Smith, 1988).

Initial Recognition of a "Thin-Testa" Chenopod. In his 1981 article Wilson also provided a brief, but important, reference to another Mexican cultivar variety of chenopod (Chenopodium berlandieri ssp. nuttalliae cv. 'chia') which retained an outer epiderm (testa) structure that was substantially reduced in thickness in comparison to that of wild species. Within a few years, several researchers would suggest the prehistoric presence of a second domesticated form of chenopod in the Eastern Woodlands of North America, a thin-testa, 'chia'-like taxon. The supporting arguments that would eventually be developed in support of the presence of a prehistoric thin-testa chenopod in the East would, once again, rest on comparative SEM micromorphological analysis of archaeological and modern seeds.

In a paper presented in 1982 at the annual meeting of the Society for American Archaeology, Asch and Asch briefly reported on chenopod seeds from the Newbridge and John Roy sites in west-central Illinois. The seeds were described as being: "...morphologically dissimilar to wild C. bushianum. Specifically, they have a truncate seed margin, they have almost entirely smooth seed coats, and they appear to have a thinner seed coat than modern C. bushianum or the other prehistoric chenopods."

The Asches here identify additional morphological characteristics that appear to distinguish the prehistoric specimens from modern wild seeds of Chenopodium berlandieri, namely an apparently thinner and smoother seed coat.

In addition, the Asches drop their earlier position that the "red morph" characteristic observed in prehistoric specimens of chenopod reflect variation within wild populations and agree with Wilson that: "...change of morphology through domestication is the most probable explanation for prehistoric samples diverging from typical C. bushianum seed characteristics (Asch and Asch, 1982:11)."

This change of opinion, and the recognition of "apparently domesticated forms from Woodland sites" (Asch and Asch, 1982:11) was based on the fact that production of thin-testa truncate margin "red morph" fruits in wild stands of C. bushianum was extremely rare: "...in a previous paper, we suggested (Asch and Asch, 1977) that phenotypic variability in C. bushianum could possible account for similar variation in cave and rockshelter collections. Since then, we have found no natural C. bushianum populations with more than a small percentage of seeds approximating the prehistoric material." (Asch and Asch, 1982:11).

The following year, in a paper presented at the annual meeting of the Society for American Archaeology, Richard Yarnell (1983) noted the presence of chenopod seeds in human fecal samples from Salts Cave, Kentucky, which were: "lacking the hard, black outer covering." Yarnell went on to say: "These finds and others provide us with evidence that Chenopodium bushianum was in some stage of the domestication process beginning during the terminal Archaic in Illinois, Kentucky, and Tennessee..."

Later the same year, the Asches (1983:692-695) recount the presence, in roughly equal frequency, of two types of Chenopodium seeds at the Middle Woodland (ca. A.D. 100-250) Smiling Dan site, in Illinois. In addition to the deeply alveolate, thick seed coat fruits produced by wild stands of Chenopodium berlandieri, the Asches recognize a second chenopod type having: "a thin seed coat with a truncate cross-section, as well as a much smoother seed coat surface." The Asches again consider the possibility that these thin-testa seeds represented the thin-testa "red morph" type produced in low frequency by wild plants, but reject this explanation because of the abundance of the thin-testa form in the Smiling Dan assemblage: "The conditions under which the thin seed-coat morph predominated in the wild are unusual ones. If archaeological specimens were collected from wild plants, one would not expect the predominance of thin seed coats which characterize Woodland period Chenopodium..." (Asch and Asch, 1983:694).

While in their 1982 paper the Asches refer to "apparently domesticated" Woodland period Chenopodium, by 1983 they take a stronger position - that the thin-coated type at Smiling Dan was "either the product of indigenous domestication of C. bushianum or an introduction from Mexico of a closely related domesticate." (Asch and Asch, 1983:695).

Building a Case for Domestication. By 1983 chenopod fruits appearing to have a thin testa had been recognized in prehistoric contexts at four eastern sites (Salt's Cave, ca. 500 B.C.; Smiling Dan, ca. A.D. 100-250; Newbridge, ca. A.D. 480-660; and John Roy, ca. A.D. 550). It had been proposed that these fruits reflected the prehistoric presence of a domesticated form of chenopod in the Eastern Woodlands during the Early, Middle, and Late Woodland periods. It was still necessary, however, to control for potential sources of bias, rule out alternative explanations, and establish a convincing case for the presence of a domesticated thin-testa form of chenopod in prehistoric garden systems of the Eastern Woodlands. A detailed comparative micro-morphological analysis of modern wild, modern domesticated, and prehistoric chenopod fruits following Wilson's general research design (1981) was clearly required.

Such a program of analysis was carried out over a four year period (1982-1985). Critically dependent upon scanning electron microscopy, it focused on describing and comparing chenopod fruits of four general kinds: 1) modern thick testa fruits from eastern North American wild stands of *Chenopodium berlandieri*; 2) modern thin-testa "red-morph" fruits from wild stands of *Chenopodium berlandieri*; 3) modern thin-testa fruits of the Mexican cultivar *C. berlandieri* ssp. *nuttalliae* cv. 'chia'; and 4) archaeobotanical assemblages of the thin-testa chenopod type (Smith, 1984, 1985a, 1985b; Smith and Funk, 1985; Smith and Cowan, 1987; Fritz and Smith, 1988).

Methods

This comparative analysis was conducted at the Scanning Electron Microscope Laboratory of the National Museum of Natural History, Smithsonian Institution, on Cambridge Stereoscan 100 instruments. Both modern and prehistoric seed specimens were first attached to glass SEM stub slides with DuPont's Elvace 1874. This liquid adhesive was found to provide a better and longer lasting attachment to SEM stubs than double sided tape, resulting in less charging of specimens and less specimen loss. Once mounted, specimens were first given a base coat of carbon, which insured a more uniform application of a second, overlying coating of gold-palladium. An accelerator voltage of 10 Kv produced the best results within the range of magnification employed (rarely greater than 2000-2500X). Built-in measurement capabilities of the Cambridge Stereoscan 100 allowed rapid and accurate determination of testa thickness values for chenopod specimens.

One obvious alternative explanation for prehistoric thin-testa chenopod fruits, suggested in the early 1980's, was that testa thickness varied in wild populations of *C. berlandieri*, either along geographical clines, with southern populations producing fruits having thinner seed coats, or in response to habitat variability, with stands growing in shady or wet settings producing fruits having thinner seed coats. In order to evaluate this possibility, *Chenopodium berlandieri* stands were harvested from a wide range of different habitat settings across the eastern United States (Smith, 1987).

Testa thickness values of samples of the collected modern wild fruits, determined through SEM analysis, ranged from 40-80 microns (Smith, 1985a, 1985b). Thickness values of the measured wild stand fruits were determined to be uninfluenced by either latitude or habitat setting.

In dramatic contrast, when cross-sectioned and measured, seed coats of fruits of the Mexican cultivar *Chenopodium berlandieri* ssp. *nuttalliae* cv. 'chia' (provided by Hugh Wilson), yielded testa thickness values in the 10-20 micron range.

Similar SEM analysis of samples of "red morph" fruits from Illinois, Michigan, and South Carolina stands of *C. berlandieri* showed testa thickness values ranging from 7-30 microns, overlapping considerably with the 10-20 micron range documented for the Mexican chenopod cultivar "chia".

SEM analysis also suggested that "red morph" fruits could be distinguished from the thin-testa domesticate fruits on the basis of greater pitting of the testa surface of the former. In addition, "red morph" fruits were shown to be produced only in very low frequencies, comprising 1-3% of the total fruit production of wild C. berlandieri stands (Smith, 1985b:122). Nevertheless, the overlap in the ranges of "chia" and "red morph" testa thickness values underscored the necessity of locating and analyzing large, unmixed, thin-testa archaeobotanical chenopod assemblages in order to establish a convincing case for the prehistoric presence of a "chia"-like domesticated chenopod in the Eastern Woodlands.

Two previously unanalyzed museum collections of Chenopodium fruits provided an excellent opportunity to document, through SEM analysis, the micro-morphology of the prehistoric thin-testa chenopod of the Eastern Woodlands.

The Russell Cave and Ash Cave Assemblages. A grass-lined storage pit excavated in Russell Cave, Alabama, in 1956, contained a large volume of carbonized Chenopodium seeds. A sample of approximately 50,000 uncatalogued and unanalyzed Russell Cave chenopod seeds was located in NMNH collections in 1982. Subsequent detailed SEM analysis of the Russell Cave chenopod assemblage (Smith, 1984, 1985a) demonstrated three important facts. First, there was a complete absence of any thick-testa chenopod seeds. Second, testa thickness values of 7-16 microns were comparable to "chia" and considerably lower than the 40-80 micron range of the fruits of modern wild stands of C. berlandieri. Third, there was a consistent occurrence of a distinctive set of micro-morphological features, notably a rectanguloid fruit cross-section, a truncate margin, and a smooth testa. When combined with a reduction in testa thickness, these features represent an adaptive response by a chenopod cultigen to strong seed bed selective pressure for reduced germination dormancy (Figures 1, 2). A sample of the Russell Cave seeds provided a particle accelerator AMS) direct date of 2,340 +/- 120 B.P.

Excavated in 1876 and curated for over a century by the Peabody Museum, Harvard University, the Ash Cave, Ohio, chenopod assemblage yielded an AMS date of 1,720 +/- 100 B.P (Smith, 1985b). Detailed SEM analysis of the well-preserved, uncarbonized assemblage of approximately 25,000 seeds documented the same set of micro-morphological features associated with an adaptive response to selection for reduced germination dormancy seen in the Russell Cave assemblage. Testa thickness values for Ash Cave truncate-margin seeds ranged from 11-21 microns.

Conclusions

In addition to underscoring the research potential of museum collections, SEM analysis of the large unmixed Russell Cave and Ash Cave chenopod collections allowed, for the first time, a detailed description of the micro-morphological structure of the fruits of the

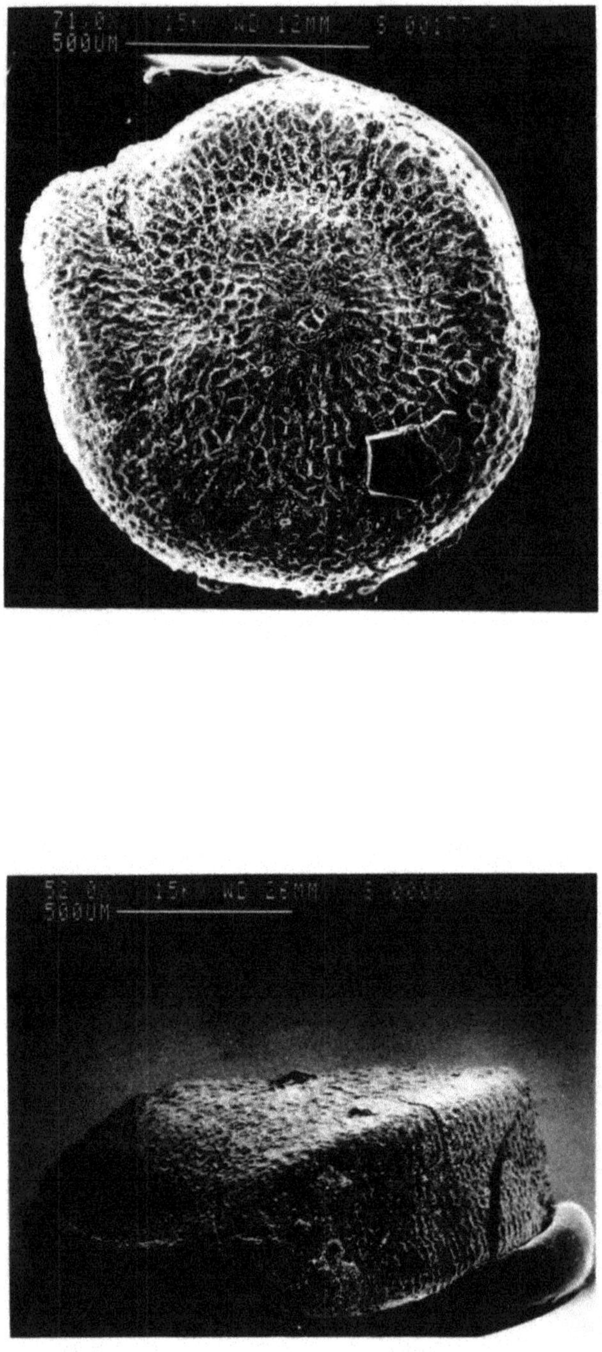

Fig. 1. Scanning electron micrograph of Russell Cave _Chenopodium_ seeds, showing the distinctive reticulative-alveolate dorsal pericarp pattern and the truncate margin seed configuration. From Smith, 1985a.

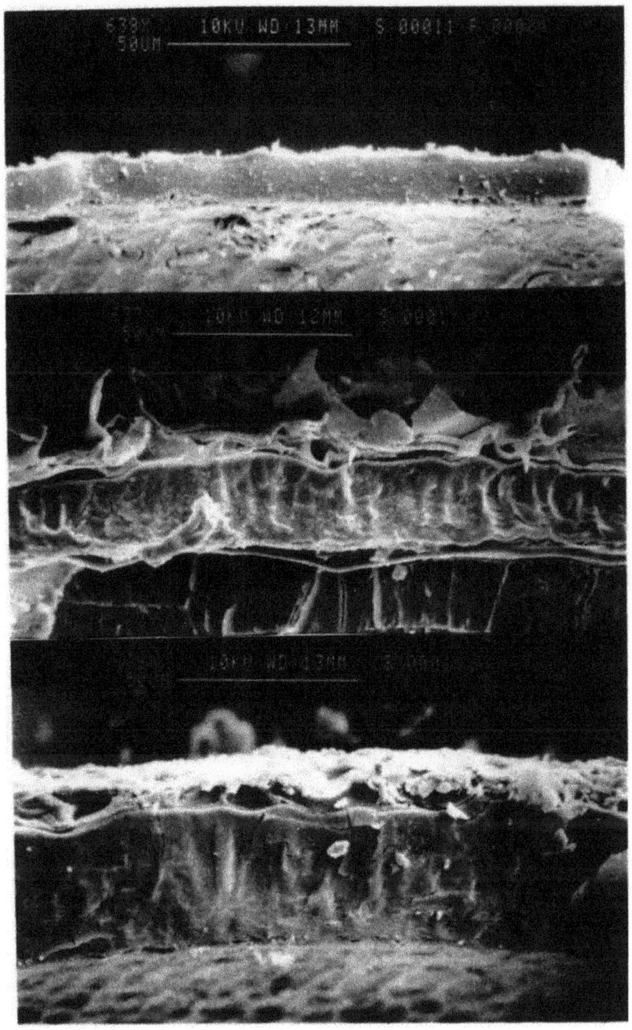

Fig. 2. Scanning electron micrographs showing outer seed coats in cross section: (top) Russell Cave seed (14 microns thick); (middle) *C. berlandieri* ssp. *nuttalliae* cv 'chia' seed (20 microns thick); (bottom) *C. berlandieri* seed (41 microns thick). From Smith, 1985a.

thin-testa chenopod cultivar of prehistoric garden systems of the Eastern Woodlands. This accurate description and measurement of extremely small fruit structures (e.g. pericarp and epiderm patterning and thickness) allowed rather vague descriptive labels such as "thin-testa" and "thick-testa" to be replaced by specific numerical values. In addition, the detailed comparison between the Russell Cave and Ash Cave seed assemblages and modern wild and domesticated forms, provided, for the first time, a basis for the confident assignment of domesticated status and a formal taxonomic subspecies designation (C. berlandieri ssp. jonesianum (Smith and Funk, 1985) to the prehistoric thin-testa chenopod of the eastern United States.

This specific example serves to demonstrate the significant potential that exists for expanding and enriching archaeobotanical research through the application of SEM technology. SEM micro-morphological description and comparative analysis of archaeobotanical seed assemblages should become a routine procedure, and be acknowledged as a necessary prerequisite to any proposed assignment of domesticated status to archaeobotanical plant materials.

Scanning electron microscopy clearly holds a number of distinctive advantages for the study of the micromorphology of prehistoric seeds. Often quite delicate, particularly if carbonized, archaeobotanical seed specimens selected for study can easily be mounted to SEM cover slides using "Elmers" glue. Once mounted and coated, such delicate seed specimens can be repeatedly studied intact through a wide range of magnification levels and viewing angles, and with good depth of field. Photographic capabilities are potentially excellent, with on-line Polaroid and 4x5 negative formats allowing extensive and easy documentation of previously unrecognized and potentially important micromorphological characters. Ironically, however, it is often still difficult, even in large research university settings, for archaeologists and archaeobotanists to take advantage of SEM technology. Many times, scanning electron microscopes have either restricted use within a particular academic unit, or are operated on a paying customer basis, which may seriously limit the application of this valuable user-friendly technology.

End Notes

1. A second chenopod taxon (Chenopodium bushianum) is frequently mentioned in quoted passages. Although still technically recognized as distinct species, C. berlandieri and C. bushianum are very closely related. Once detailed genetic research on present day stands of these two eastern North American taxons are carried out, Chenopodium bushianum is likely to be relegated to subspecific status within C. berlandieri.

References

Asch, D. and Asch. N. 1977. Chenopod as cultigen: a re-evaluation of some prehistoric collections from Eastern North America. Midcontinental Journal of Archaeology 2: 3-45.

Asch, D. and Asch, N. 1982. A chronology for the development of prehistoric horticulture in West-Central Illinois. Paper presented at the 47th Annual Meeting, Society for American Archaeology, Minneapolis, Minnesota.

Asch, D. and Asch, N. 1983. Archaeobotany. In Stafford, B. and Sant, M.B. (eds.), Excavations at the Smiling Dan Site, Center for American Archaeology, Contract Archaeology Program, Report of Investigations No. 137. Submitted to the Illinois Department of Transportation.

Asch, D. and Asch, N. 1985. Archaeobotany. In Conner, M.D. (ed.), The Hill Creek Homestead, Kampsville Archaeological Center Research Series 1: 115-170.

Asch, N. and Asch, D. 1980. The Dickson Camp and Pond Sites: Middle Woodland Archaeobotany in Illinois. In Cantwell, A.M., Dickson Camp and Pond, Illinois State Museum, Reports of Investigations 36: 152-160.

deWet, J.M.J. and Harlan, J.R. 1975. Weeds and domesticates: evolution in the man-made habitat. Economic Botany 29: 99-107.

Fritz, G. 1986a. Prehistoric Ozark Agriculture, the University of Arkansas Rockshelter Collections. Unpublished Ph.D. dissertation, University of North Carolina at Chapel Hill.

Fritz, G. 1986b. Starchy grain crops in the Eastern United States: Evidence from the desiccated Ozark plant remains. Paper presented at the 51st Annual Meeting of the Society for American Archaeology, New Orleans, Louisiana.

Fritz, G. and Smith, B. 1988. Old collections and new technology: documenting the domestication of *Chenopodium* in Eastern North America. Midcontinental Journal of Archaeology 13(1).

Harlan, J.R., deWet, J.M.J. and Price, E.G. 1973. Comparative evolution of cereals. Evolution 27: 311-325.

Jones, V. 1936. The vegetal remains of Newt Kash Hollow Shelter. In Webb, W.S. and Funkhouser, W.D. (eds.), Rock Shelters in Menifee County, Kentucky. University of Kentucky, Reports in Archaeology and Anthropology 3(4): 147-167.

Korber-Grohne, U. 1981. Distinguishing prehistoric cereal grains of *Triticum* and *Secale* on the basis of their surface patterns using the scanning electron microscope. Journal of Archaeological Science 8: 197-204.

Korber-Grohne, U. and Piening, U. 1980. Microstructure of the surfaces of carbonized and non-carbonized grains of cereals as observed in scanning electron and light microscopes as an additional aid in determining prehistoric findings. _Flora_ 170: 189-228.

Smith, B.D. 1984. Chenopodium as a prehistoric domesticate in Eastern North America: evidence from Russell Cave, Alabama. _Science_ 226: 165-167.

Smith, B.D. 1985a. The role of _Chenopodium_ as a domesticate in pre-maize garden systems of the Eastern United States. _Southeastern Archaeology_ 4: 51-72.

Smith, B.D. 1985b. _Chenopodium berlandieri_ ssp. _jonesianum_: evidence for a Hopewellian domesticate from Ash Cave, Ohio. _Southeastern Archaeology_ 4: 107-133.

Smith, B.D. and Funk, V.A. 1985. A newly described subfossil cultivar of Chenopodium _(Chenopodiaceae)_. _Phytologia_ 57: 445-448.

Wilson, H.D. 1981. Domesticated _Chenopodium_ of the Ozark Bluff dwellers. _Economic Botany_ 35: 233-239.

Yarnell, R.A. 1983. Prehistoric plant foods and husbandry in Eastern North America. Paper presented at the 48th Annual Meeting of the Society for American Archaeology, Pittsburgh, Pennsylvania.

THE SEM AND SEED IDENTIFICATION, WITH PARTICULAR REFERENCE TO THE VICIEAE

E. Ann Butler

Introduction to the Vicieae

Legumes have been important economic resources since earliest times and are thought to have played a role similar to that of the cereals in the beginnings of agriculture (Renfrew, 1973). However, differential preservation acts against the seeds of legumes and charred material from these plants in early archaeological contexts is usually sparse in relation to the remains of cereals.

Amongst the Leguminosae the Vicieae, a tribe represented by some 300 species, has contributed from its four main genera a number of well-known cultivars common throughout the Old World, including the pea, chickling pea, lentil and broad bean amongst many lesser known species.

Identification by Optical Microscopy

The identification of the remains of charred pulses has always been difficult. Commonly the material consists of fragments of the cotyledons which yield only estimations of the dimensions of the original seed as criteria by which to make the identification. Seed shape and size are known to be highly variable within species in the Vicieae, and such characters may even be unreliable indicators by which to separate genera (Gunn, 1970a). However, frequently some of the seedcoat remains, and even the hilum may be retained on the seed. The degree of rough- or smoothness of the testa, and the dimensions and shape of the hilum recorded under low magnifications have added to the criteria that have been employed for identification. The very limited depth of focus and poor resolution obtained by optical microscopy, even using epi-illumination, have severely restricted the observation of morphological detail, thereby making identifications tenuous.

SEM Analysis

The application of SEM to the study of seed surfaces has already shown rewarding results in archaeobotany, charred remains often showing very good preservation of morphology (Conolly, 1976). The recent work by Kislev and Hopf (1985) on the identification of Lathyrus sativus and L. cicera from the 11th century BC site of Tell Qasile in Israel illustrates particularly the observations achievable by SEM on seeds of the Vicieae from antiquity.

It is now ten years since Brisson and Peterson (1976) published a critical review of the use of SEM in the study of the seedcoat, and some of the points raised in that work are of particular relevance to archaeobotany. Of the several problems highlighted, a major cause for concern was the lack of consideration of the variation within species of the morphology and anatomy of the seedcoat; most frequently single specimens of a species had been examined and the records thus displayed little evidence for the stability of the characters describing the different species. The need was expressed for more illustrations of seeds under SEM that might be used as comparatives and provide a knowledge of the breadth of the range of variation.

A decade later, the situation remains largely unresolved. Many publications still contain data recorded by SEM from very few specimens per species and few studies have examined the potential range of variation in features under investigation. Part of the problem lies in the different objectives for the use of SEM. Where the purpose is taxonomic, broad trends in morphology may be sought rather than the range of morphological expression in individual characters. Yet for those engaged in the identification and interpretation of what may be very small amounts of ancient plant material, some appreciation of the potential range of variation in individual characters is important.

This variation in seed anatomy and morphology may be due to a number of factors, including differences in the genotype, phenotypic plasticity, the state of maturity, pathologies and other causes. That the results of these factors have not been a major focus of study by SEM may be explained by the fact that this technique does not readily lend itself to large scale surveys of replicated material. The operation of the instrument and necessarily associated photography may be time-consuming and labour intensive; the equipment is expensive to buy, maintain and run; and access for the borrowing or hiring of SEM facilities is thus often very restricted. Furthermore, reference material may be in short supply so that the preparation of a large number of SEM samples may involve the permanent removal of too much of the reserve of a seed collection. In the case of charred archaeological material, the irreversible removal of part of a sample has to be very carefully justified.

Given the constraints described above, a partial solution to the problem may lie in the maximisation of information that may be obtained from publications of other authors in order to extend the database. This requires the inclusion of full details of method, accurate records of the numbers of observations, and descriptions of the populations from which specimens have been drawn.

SEM is a technique that has some great advantages in the examination of seeds. In general, relatively little preparation of the specimen is necessary prior to scanning. Where the seeds are dry, as is the case in mature seeds of the Vicieae, fixing and critical point drying are not required. Charred seed material is particularly suitable for scanning and may even be scanned without coating (Feindt and Mesterharm, 1980). The depth of focus allows for good photographic records of such subjects as spherical seeds at magnifications as low as x12 on certain instruments, and the high resolution provides for detailed observation of anatomy at magnifications up to x10,000 of opaque charred seeds that could not be studied by other means.

The Present Studies

In order to evaluate characters that may be applied to the identification and interpretation of charred material of archaeological derivation, a study has been undertaken of the morphology and anatomy of the seedcoat in members of the Vicieae. Fresh seeds have been examined from 94 taxa, including 15 species of *Lathyrus*, 5 species of *Lens*, as classified by Ladizinsky (1986), 2 subspecies of *Pisum* and 14 species of *Vicia*, a number of different populations of the same species being studied where possible (Table 1). Charred material from three archaeological sites has also been examined.

Table 1. Species of the tribe Vicieae examined in the study of seedcoat anatomy.

Lathyrus aphaca
Lathyrus annuus
Lathyrus cicera
Lathyrus clymenum
Lathyrus heterophyllus
Lathyrus hirsutus
Lathyrus inconspicuus
Lathyrus laxiflorus
Lathyrus niger
Lathyrus nissolia
Lathyrus ochrus
Lathyrus pratensis
Lathyrus sativus
Lathyrus sylvestris
Lathyrus tuberosus

Lens culinaris
Lens orientalis
Lens odemensis
Lens nigricans
Lens ervoides

Pisum sativum elatius
Pisum sativum sativum arvense

Vicia bithynica
Vicia cracca
Vicia ervilia
Vicia faba var minor
Vicia faba var paucijuga
Vicia lutea
Vicia melanops
Vicia narbonensis
Vicia noeana
Vicia orobus
Vicia peregrina
Vicia sativa
Vicia sativa nigra
Vicia sepium
Vicia sylvatica
Vicia tetrasperma

Methods

The primary technique employed has been SEM, using a Philips PSEM 500 at 12 and 25 kv, and a JEOL JSM 35 at 15 kv, under magnifications of between x12 and x10,000. Specimens have been coated with gold by a Polaron E5000 sputter coater to a depth of 400-500A.

Stubs have been prepared with whole seeds mounted to show the hilum and testa surface both adjacent to the hilum and at midseed. Seeds have been halved by pressure fracture at right angles across the hilum and mounted to show the hilum in transverse section, and the testa in longitudinal section around the seed circumference, after the manner of Stirton (pers. comm.). Since the seeds are hygroscopic and unfixed, they can take up atmospheric water or dry out while attached to the stub. Consequently, the use of hard-setting glues such as Araldite, or conducting paints such as silver dag for the adhesive, has been avoided, due to the risk of losing the specimen under the differential stresses imposed by the vacuum of the sputter coater or the SEM itself. Generally glues that retain a slight plasticity when dry, such as Durafix or UHU, have been found to be the most satisfactory, though double-sided sticky tape has been effective for small, light seed fragments or charred material.

Towards a greater understanding of the anatomy of certain features revealed under SEM, thin sections of the seedcoat cut transversely across the hilum have been observed under light microscopy after staining, as well as by SEM in three species of *Lathyrus*.

Characters traditionally used to identify seeds of the Vicieae under low magnifications have been recorded under light microscopy at magnifications of less than x50.

Observations have been qualitatively assessed and compared with those of other authors. Where appropriate, data have been numerically expressed and analysed by cluster analyses and principal coordinate analysis.

Results

External Testa Surface. Previous publications have described the testa surface in the Vicieae as consistently papillose (Lersten, 1979). While usually this was found to be the case in this study, 2 species of *Vicia* and 6 of *Lathyrus* were seen with other pattern types, as exemplified by *Lathyrus aphaca* (Plate 1). The general pattern type is consistent within each species but variations in detail have been observed within different populations of the same species that are sometimes greater than those of different species. Similarly, seeds from different genera may exhibit strikingly similar testa morphology, exemplified by *Vicia ervilia* and *Lathyrus sativus* (Figs. 2, 3).

It has been concluded that testa morphology alone cannot be used to identify species in the Vicieae in most cases (Butler, in press).

Testa pits described by a number of authors in several genera of the Leguminosae and reported in 3 species of *Lathyrus* (Lersten and Gunn, 1982) have been recorded here in 4 species of *Lathyrus* (Butler, 1986) and one of *Vicia*. The pits are of occasional occurrence and are not a constant feature within species or even within populations; their presence or absence is therefore considered to be unstable and unsuitable for use in the separation of species.

The Hilum. The external morphology of the hilum has yielded a number of characters that have been used traditionally as a basis by which to identify species of the Vicieae, particularly when applied to the separation of weedy species from cultivars (Gunn, 1970b). It was anticipated that a study of the internal anatomy of the hilar region in transverse section might be of value. It is considered by some taxonomists that while seed morphology may be subject to the influence of environmental factors, plant cells and tissues have more conservative characteristics. Vascular tissue has been a particularly useful focus of interest in this respect (Bell, no date). Consequently, the tracheid bar, a structure derived from the ovule vascular bundle (Lersten, 1982) lying beneath the hilum, has been

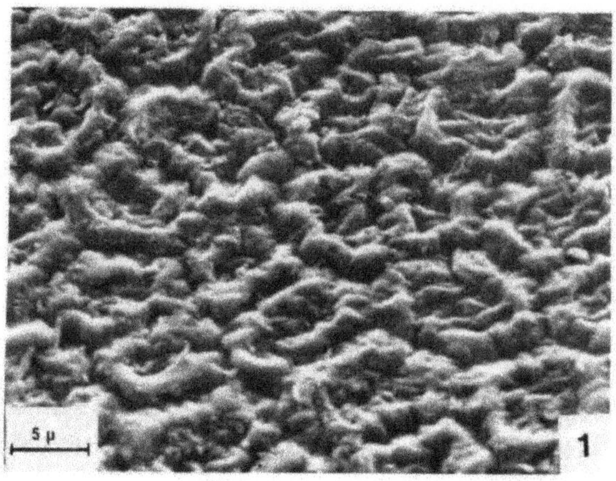

Fig. 1. *Lathyrus aphaca*. Testa surface, non-papillose type.

Fig. 2. *Vicia ervilia*. Testa surface, papillose type.

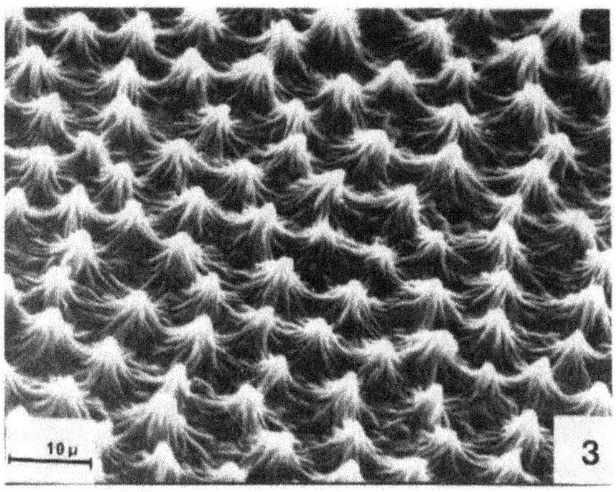

Fig. 3. *Lathyrus sativus*. Testa surface, papillose type.

carefully examined in this study. Figure 4 shows the hilum in transverse plane in Lathyrus cicera.

The tracheid bar is composed of tracheoids, intercommunicating by means of perforations similar to the pits found in wood elements. The size and shape in section of this tracheid bar appear to remain relatively constant within species, and to be criteria that might prove to be useful for identification. However, the tracheoid pits have not conformed to those described by Lersten and Gunn (1982) as typical in the Vicieae, and have not been uniformly regular, and non-vestured. Only one species of Lathyrus, no Vicia and not all species of Lens had smooth, plain-bordered pits. All other taxa showed at least some tracheoids with warty or even elaborately vestured pits (Fig. 5). Most commonly within a species a wide range of pit ornamentation has been seen, and variations in pit distribution and shape and aperture shape have been the rule. Clearly a greater understanding of the implications behind these variable features is necessary before they can be incorporated in data used for the separation of species. Serial sections along the hilum in 3 species of Lathyrus have been studied, and no correlation has been found between types of pitting and the position of pits relative to the situation of the tracheoid elements in the tracheid bar.

Observations in the seedcoat of members of the Vicieae have shown even at very high magnifications the breadth of variation encountered during studies by light microscopy under low magnifications. Individual characters of morphology and anatomy usually cannot be reliably employed for the identification of legume seeds. Thus cluster analysis and principal coordinates analysis have been applied to an array of criteria to seek separation of species.

Multivariate Analysis. The results of a pilot study of 13 species of Lathyrus using 11 characters of the external testa morphology and hilum in transverse section, have shown that some taxonomically significant separations can be achieved, and that useful characters can be found in the width and depth of the hilum and associated tracheid bar (Butler, 1986).

Currently, analyses are in progress of 60 populations of Lathyrus and Vicia using 47 characters. Preliminary results indicate that, while intrageneric separations can be achieved, those between genera are more difficult. This is in agreement with the work of Lersten and Gunn (1982), using traditional criteria visible under low power microscopy. Scattergrams of the eigen values of axes I and II show overlap of groupings of species from both genera. Again, however, the parameters relating to the depth and width of the hilum and tracheid bar are responsible for most of the separations (Butler, in press). Further analyses will be necessary before definite conclusions can be drawn, and these results together with full data will be the subject of a future publication.

Applications to Archaeobotany

Material from three archaeological sites has been examined by SEM, in order to establish the relevance of the study of fresh material to

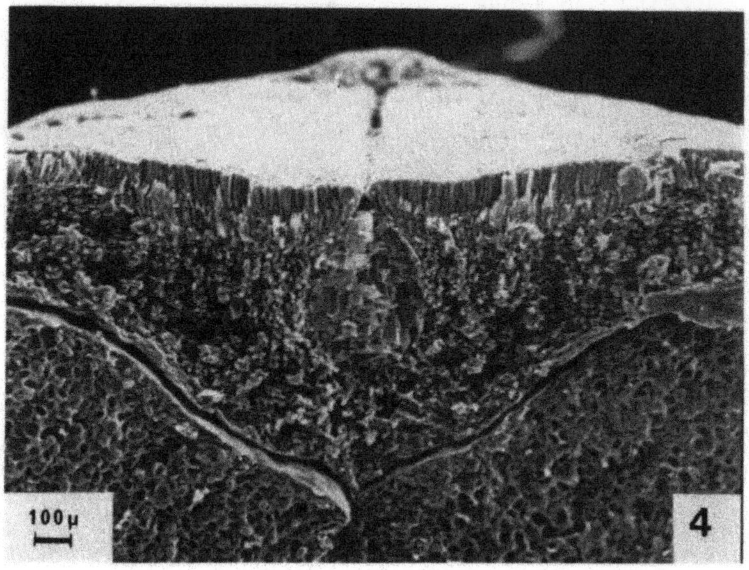

Fig. 4. Lathyrus cicera. TS hilum.

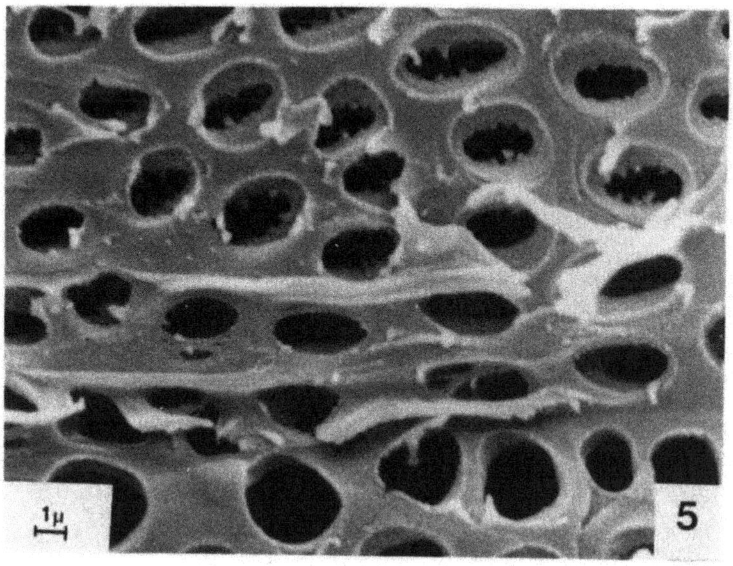

Fig. 5. Vicia faba. Tracheoid pits of mixed type.

archaeobotany. Charred legume seeds from sites at Tell Brak, Iraq and Manerba, Italy, from the Bronze Age and Chalcolithic respectively and from Gordion, Turkey, of 7th century BC date have retained little of the testa. In addition, very few have had the hila preserved _in situ_. However, careful sorting of fractions retained in small mesh sieves has shown that detached fragments of testa bearing the hilum can be separated. These very small remains often show under SEM clear patterning of the topography of the exterior surface of the testa, and when pressure-fractured across the hilum reveal well-preserved anatomical characters that would permit records of the same criteria that have been collected in fresh material (Figs. 6, 7). The use of these detached hila for SEM study, rather than the whole seeds, would allow for the conservation of the major specimens in the archaeobotanical sample, and thus remove an important obstacle to the application of the SEM in archaeobotany.

Conclusions

The need for a greater understanding of the anatomical features revealed by SEM of plant surfaces has been expressed by Heywood (1985). During this project it has been found that while SEM together with light microscopy can contribute towards the collection of an array of characters of anatomy and morphology, the variation in and significance of a number of features that have been observed and recorded remain insufficiently understood to allow for their inclusion within the database that may be used for the identification of seeds of the Vicieae. Establishment of the stability of characters employed in seed identification is important.

The application of the SEM to the collection of an array of criteria from features of the seedcoat of fresh material may form the basis for more certain interpretations of archaeobotanical samples of seeds of the Vicieae.

Acknowledgements

Seed samples have been kindly supplied by Mr. G.C. Hillman, Institute of Archaeology, London; by Dr. F.A. Bisby, Taxonomic Unit, Biology Department, University of Southampton; Professor G. Ladizinsky, the Hebrew University of Jerusalem; and the N.I. Vavilov All-Union Institute of Plant Industry, Leningrad. SEM facilities have been made available at the Biology Department, Imperial College and the Crystallography Department, Birkbeck College London. Using an unpublished program for cluster analysis written by Professor R.G. Davis and his own program for principal coordinate analysis, the multivariate analyses were carried out by Dr. D.H. Dalby of Imperial College London, whose help is gratefully acknowledged. Archaeological charred seed material was supplied by Mr. G.C. Hillman (Gordion), Mr. M. Charles (Tell Brak) and Miss S. Colledge (Manerba) of the Institute of Archaeology London. The advice of Dr. C. Stirton, Royal Botanic Gardens, Kew, has been much appreciated. The work was funded by a grant from the Science and Engineering Research Council.

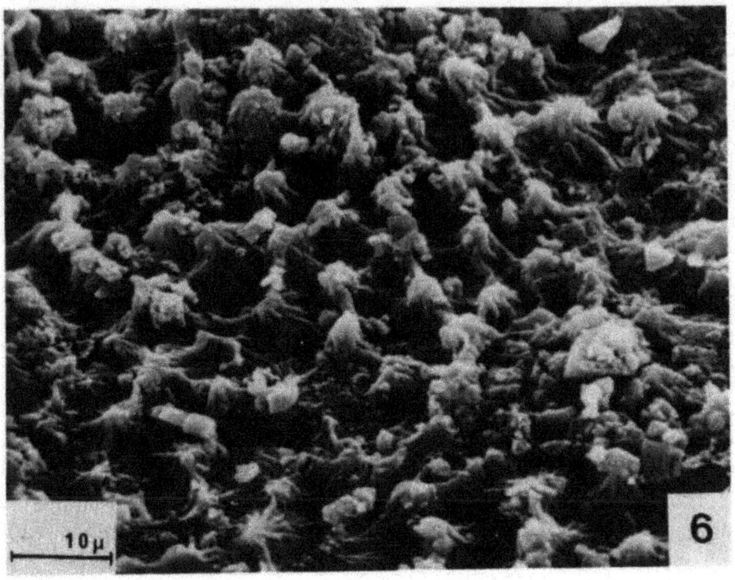

Fig. 6. Pisum cf. sativum arvense. Charred testa surface, from Manerba, Italy. Chalcolithic site.

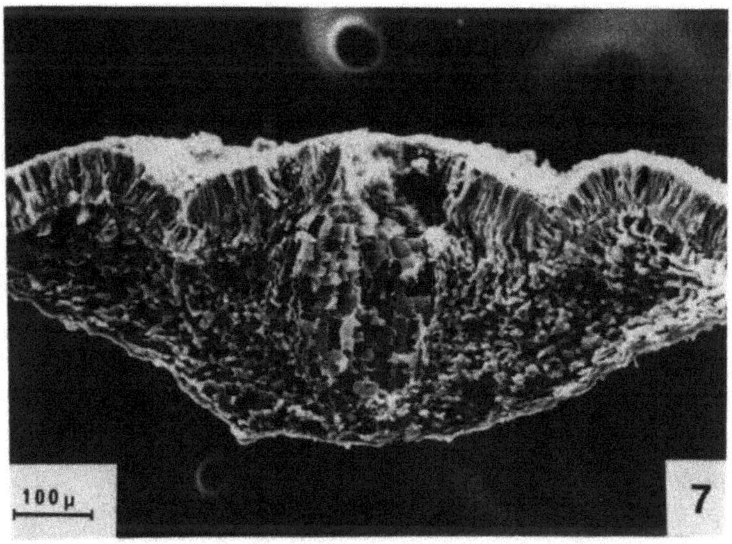

Fig. 7. Pisum sp. Charred TS hilum, from Tell Brak, Iraq. Bronze age site.

References

Brisson, J.D. and Peterson, R.L. 1976. A critical review of the use of the scanning electron microscope in the study of the seedcoat. Scanning Electron Microscopy Part VII/1976.

Butler, E.A. 1986. Studies in the seedcoat of Lathyrus. In Kaul, A.K. and Combes, D. (eds.) Proceedings of the Colloque Lathyrus, Pau 1985, pp. 25-38. New York: Third World Medical Research Foundation.

Butler, E.A. In press. The Vicieae: problems in identification. Proceedings of the 7th Symposium of the International Work Group for Palaeoethnobotany. Cambridge 1986.

Bell, C.R. No date. Plant variation and classification. London: Macmillan and Co.

Conolly, A. 1976. The use of the scanning electron microscope for the identification of seeds, with special reference to Saxifraga and Papaver. Folia Quaternaria 47: 29-32.

Feindt, F. and Mesterharm, G. 1980. Rasterelekronenmikroskopie in der Palaoethnobotanik. Mitteilung der Institut für Allgemeine Botanik, Hamburg, 17: 135-143.

Gunn, C.R. 1970a. A key and diagram for the seeds of 100 species of Vicia (Leguminosae). Proceedings of the International Seed Testing Association 35: 773-790.

Gunn, C.R. 1970b. Seeds of the tribe Vicieae (Leguminosae) in North American Agriculture. Proceedings of the Association of Official Seed Analysts 60: 48-70.

Heywood, V.H. 1985. Scanning electron microscopy: perception, character recognition and systematics. Proceedings of the Royal Microscopical Society 20: 22-23.

Kislev, M.E. and Hopf, M. 1985. Food remains from Tell Qasile with special reference to Lathyrus sativus/cicera. Qedem 20: 140-147.

Ladizinsky, G. 1986. A new Lens from the Middle East. Notes from the Royal Botanic Garden Edinburgh 43: 489-492.

Lersten, N.R. 1979. A distinctive seed coat pattern in the Vicieae (Papilionoidae: Leguminosae). Proceedings of the Iowa Academy of Science 86: 102-104.

Lersten, N.R. 1982. Tracheid bar and vestured pits in legume seeds (Leguminosae: Papilionoidae). American Journal of Botany 69: 98-107.

Lersten, N.R. and Gunn, C.R. 1982. Testa characters in the tribe Vicieae, with notes about tribes Abreae, Cicereae and Trifolieae (Fabaceae). USDA Technical Bulletin No. 1667.

Renfrew, J.M. 1973. Palaeoethnobotany. New York: Columbia University Press, p. 104.

SCANNING ELECTRON STUDIES OF THE CELL PATTERNS OF THE PERICARP LAYERS OF SOME WILD WHEATS AND RYES. METHODS AND PROBLEMS

Susan Colledge

Introduction

The work outlined in this paper was carried out during the period of funding of the Science and Engineering Research Council sponsored project number GR/B/80796, "Early Manipulations of Plant Resources in Near Eastern Steppe and Riverine Forest" (hereafter referred to as "The Abu Hureyra Project"). The project was initiated and directed by Mr. Gordon Hillman and Professor David Harris at the Institute of Archaeology, London.

The high powered magnification which can be achieved with scanning electron microscopy was required for this study to facilitate the investigation of the microstructure of the outer cell layers of cereal and grass seeds. Figure 1 shows the composition of these layers. In the family Gramineae (which includes grasses and cereals) the pericarp (fruit wall) and testa (seed coat) are fused together to form the covering layers of the seeds. Of particular interest in this investigation are the longitudinal, transverse and tube cell layers. Figure 2 gives the configuration of these layers on a grain of wheat.

Important pioneer work carried out by Professor Korber-Grohne (Korber-Grohne and Piening 1980; Korber-Grohne 1981) provided the basis for this study. She realised the potential of using morphological details of the microstructure of the seed pericarp layers in identification of cereal species. Her work was a direct response to one of the greatest problems in archaeobotany: the identification of ancient cereal grains where the gross morphological characteristics of the seeds are not necessarily diagnostic. The verification of species of wild and domestic cereals found within archaeological deposits, from all periods, is vital in order to understand the development, emergence and spread of the various races of crop plants.

As part of her investigation Korber-Grohne prepared modern, identified grains of wheat, barley and rye species and used a scanning electron microscope, at magnifications of between 200 and 2,000 times, to survey the layers of the pericarp. Much of her work was concentrated on investigation of the transverse cell layers. It was necessary to quantify certain morphological features to allow for an objective assessment of the cell patterns. Korber-Grohne included the length and breadth measurements of the cells (both absolute and relative), the thickness of the cell walls, the amount of pitting in the cell walls and the overall shape and arrangement of the cells in the transverse layer. To complete the characterisation of the patterns she also noted the number of cell layers within the longitudinal layer and the presence of the tube cells. Korber-Grohne found a constancy of these features within species, such that, on the basis of comparison of these quantitative parameters for modern specimens, she was able to suggest the possibility of differentiation between wild ancestral forms and their equivalent domesticated species. She was, however, careful to stress the inadequacy of scientific testing in her work. The diagnosis of morphological features in the pericarp layers was applied

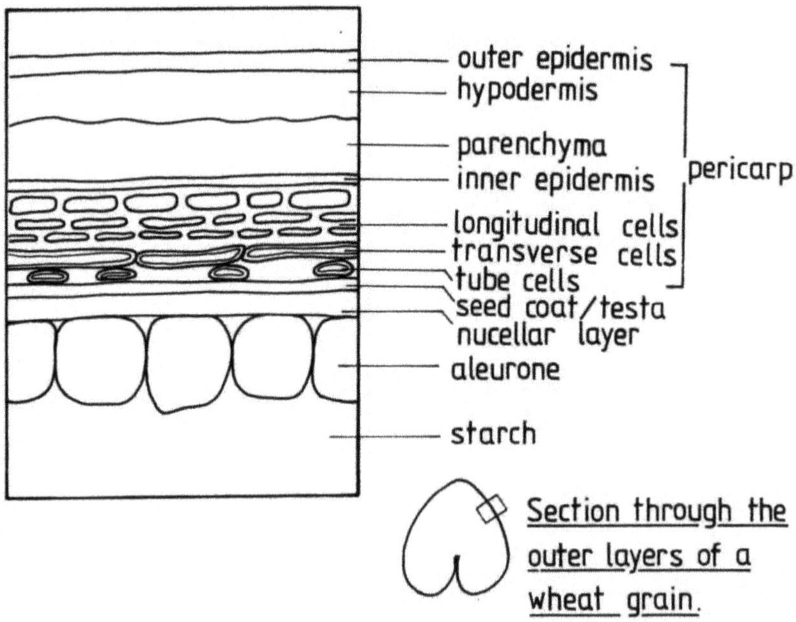

Fig. 1. Diagrammatic representation of the outer cell layers of a wheat grain.

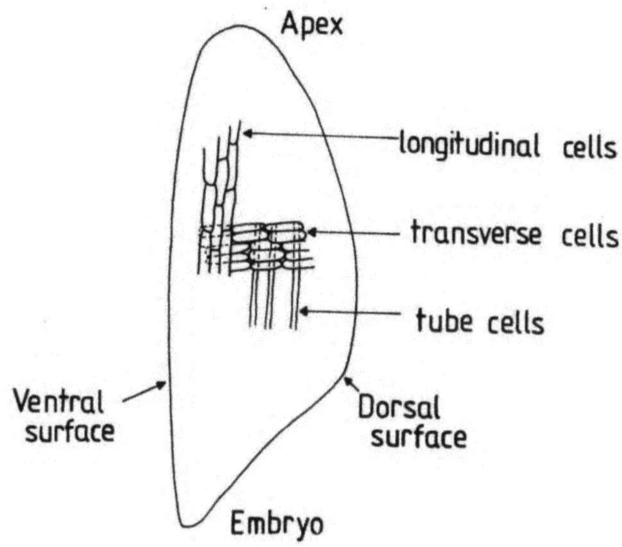

Fig. 2. Diagrammatic representation of the lateral surface of a wheat grain showing the configuration of longitudinal, transverse and tube cell layers.

to SEM photomicrographs of ancient charred grains and, by comparison, identifications were made.

This was the state of knowledge at the time of the Abu Hureyra Project. Archaeobotanists were beginning to use the diagnostic criteria in the identification of ancient cereal grains from archaeological sites of varied distribution, both geographically and chronologically, and their interpretations were made accordingly.

The Abu Hureyra Project

Tell Abu Hureyra is a site in northern Syria on the southern bank of the River Euphrates. Occupation at the site extended from Epipalaeolithic through to Aceramic and Ceramic Neolithic periods (c. 9000 bc to 6000 bc). The site was excavated during two seasons in 1972 and 1973 (Moore, et al. 1975). The initial recovery of the ancient, charred plant remains from within the occupation deposits and the subsequent archaeobotanical research, carried out during the three year funding of the project, were undertaken with the aim of answering very specific questions relating to the beginnings of cultivation and the emergence of fully domestic crops in that area of the Near East. The preservation of the charred plant material from Epipalaeolithic and Neolithic levels within the tell was excellent. Hillman recognised the potential for investigation of the transition from gathering wild plant foods to the first stages of agriculture with managed crops. He wanted, in particular, to trace any signs of early manipulation of wild plant foods in the Epipalaeolithic period and, thus, to attempt to define the circumstances and processes which gave rise to fully domesticated cereal species in the Neolithic. Hillman (pers. comm.) considered there was a possibility that histological changes in the seed (i.e. changes in the microstructure of the cells of the pericarp layers) might have occurred prior to any gross morphological changes in rachis structure and grain shape once the wild plants were under man's influence. Domestic cereals have a tough rachis which does not shatter when the ear is mature, in wild species the rachis is 'brittle' and seed dispersal occurs automatically at maturity. Alteration of grain shape and size accompanies domestication of the species.

The research was designed to look for these histological changes using the scanning electron microscope on grains of wild einkorn, Triticum boeoticum, and wild ryes, Secale spp., from Epipalaeolithic occupation deposits at Abu Hureyra (Figs. 3c, d). If these changes could be isolated it was obvious that the implications would be far reaching.

SEM Studies of Modern Grains

Prior to the examination of the ancient grains it was necessary to undertake a thorough survey of modern specimens of the equivalent species. Table 1 lists the species which were used and the habitats where they were collected. For the wild cereal species the primary habitats refer to the natural wild stands and the segetal habitats refer to when the plants were found growing in a field of crops. The latter forms might well mimic the situation of the Epipalaeolithic

TABLE 1: Modern Populations of Wheats and Ryes used in the SEM Reference Study*

1. Triticum boeoticum: wild-type einkorn
   - A. Derived from primary (wild) habitats
     - 1. Turkey, GCH 3091 - 1^g and 2^g spikelets
     - 2. Turkey, GCH 3771 - " " " "
     - 3. Turkey, GCH 3219 - 1^g spikelets only
   - B. Derived from segetal habitats (i.e. growing as weeds of crops)
     - 4. Turkey, GCH 18176 - 1^g and 2^g spikelets
     - 5. Turkey, DHF 420 - " " " "
     - 6. Turkey, GCH 3070 - " " " "
     - 7. Turkey, GCH 2264 - " " " "
     - 8. Turkey, GCH 3705 - " " " "
   - C. Derived from intermediate habitats (disturbed, but uncultivated)
     - 9. Turkey, GCH 3341 - 1^g and 2^g spikelets
     - 10. Turkey, PBI - " " " "
2. Triticum boeoticum var. urartu: single-toothed wild einkorn
   - A. Derived from primary habitats
     - 11. Turkey, ICARDA 55130 - 1^g and 2^g spikelets
     - 12. Syria, ICARDA 20049 - " " " "
     - 13. Syria, ICARDA 20033 - " " " "
     - 14. Syria, ICARDA 20125 - " " " "
   - B. Derived from segetal habitats
     - 15. Syria, ICARDA 20186 - 2^g spikelets only
3. Triticum monococcum: fully domesticated einkorn
   - A. Grown as crops
     - 16. Turkey, PBI - 1^g spikelets only
     - 17. Turkey, Zohary - " " "
     - 18. Turkey, JBB - " " "
4. Secale montanum: wild mountain rye (all rye spikelets are 2 grained)
   - A. Derived from primary habitats
     - 19. Turkey, GCH 3094 - 2^g spikelets
     - 20. Turkey, GCH 3215 - " "
     - 21. Turkey, GCH 2618 - " "
5. Secale sylvestre: wild sand rye
   - A. Derived from primary habitats
     - 22. USSR, via URBG - 2^g spikelets
   - B. Derived from segetal habitats
     - 23. USSR, via UMBG - 2^g spikelets
6. Haynaldia villosa: hairy wild rye
   - A. Derived from primary habitats
     - 24. Crete, van Zeist
7. Secale cereale: cultivated rye
   - A. Derived from segetal habitats (subsp. segetale)
     - 25. Turkey, GCH 18186 - 2^g spikelets
   - B. A cultivated crop in its own right (subsp. cereale)
     - 26. Turkey, GCH 2921a - 2^g spikelets
     - 27. Turkey, GCH 3544a - " "
     - 28. United Kingdom, PBI - " "
     - 29. Turkey, GCH 2872b - " "

* 1^g refers to the grain from a one-grained spikelet, 2^gS refers to the smaller of the two grains from a two-grained spikelet, and 2^gL refers to the larger of the two grains from a two-grained spikelet.

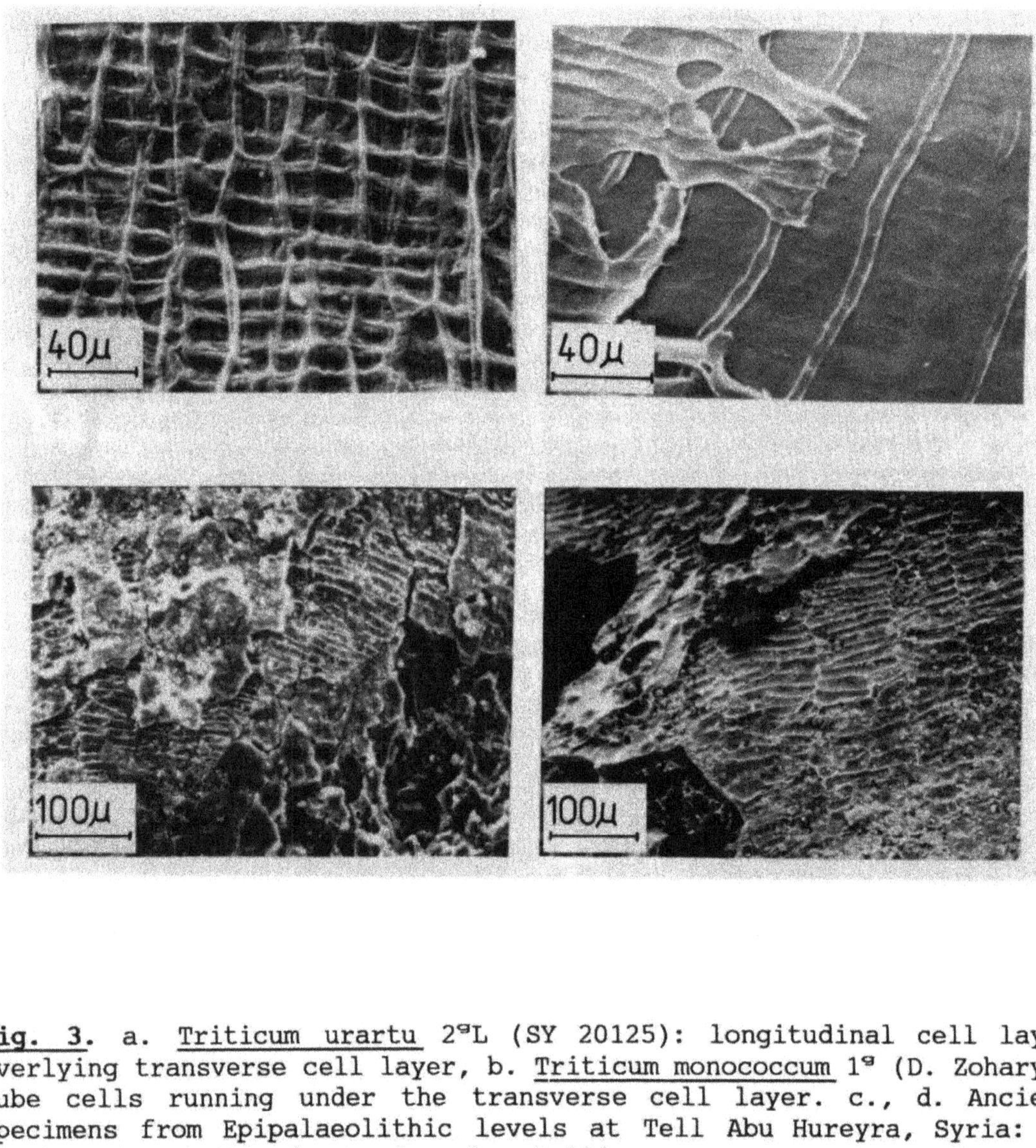

Fig. 3. a. Triticum urartu $2^{g}L$ (SY 20125): longitudinal cell layer overlying transverse cell layer, b. Triticum monococcum 1^{g} (D. Zohary): tube cells running under the transverse cell layer. c., d. Ancient specimens from Epipalaeolithic levels at Tell Abu Hureyra, Syria: c. Level number E 281, d. Level number E 326.

manipulation of wild grasses and cereals. The first stage of the research was, therefore, to determine whether there were differences in the transverse cell patterns between grains of the same species from primary and segetal populations (Fig. 4). Table 2 lists the features and measurements of the cell patterns which were considered for comparison.

Methods

The modern grains were soaked in water for 1-2 hours to facilitate the removal of cell layers above the transverse cells. A probe was used to secure the grain on its side and with fine forceps it was possible to peel off the layers by gently pulling at the hairs at the apical end. Usually the layers on the two lateral surfaces came away without tearing. The grains were then put in an oven for about 16 hours at a temperature of 230°C to replicate the charring of the ancient remains. All specimens were drawn before and after charring, to take account of any size changes which might affect the cell patterns. No other preparation was required and the grains were mounted on the SEM stubs with double sided sticky tape on their lateral surfaces and were sputter coated with gold to a thickness of c. 40 nm.

In her investigations Korber-Grohne described an area on the cereal grains which showed least distortion of transverse cells, and she based all her quantitative analysis on the patterns in this area. For this study the area of the grain was divided into three sections (Figure 5) and each part was scanned separately.

Results

It must be stressed that the investigation is far from complete. As yet, no detailed quantitative analyses have been applied to the SEM photomicrographs. The results given here are, therefore, preliminary.

The study highlighted many more problems than were initially expected. The research indicated that:

i) there was no constancy of cell patterns of the transverse layer on one grain (Figs. 6a-c);
ii) there was no constancy of cell patterns of the transverse layer on grains from a one-grained spikelet and a two-grained spikelet of the same ear (Figs. 6d-f);
iii) there was no constancy of cell patterns of the transverse layer on grains of the same species from different populations (Figs. 7a, b);
iv) there were, however, striking similarities of cell patterns of the transverse layer on grains of wild ancestral forms of wheat and rye and on their equivalent domesticated species (Figs. 7c, d);
v) and there were similarities of cell patterns of the transverse layer on grains of different genera (Figs 7e, f).

From this study it is obvious that the use of micromorphological features of the pericarp layers of cereals cannot be applied in such

TABLE 2: Criteria Examined in the Cell Patterns of Layers of the Pericarp on Modern Cereals

Longitudinal Cells *:	number of layers cell shape cell size wall thickness visibility of pitting
Transverse Cells:	number of layers cell length cell width length/width ratio presence of intercalated cells shape of intercalated cells end wall curvature end wall thickening end wall overlap with adjacent rows of cells side wall thickening side wall pitting: prominence : distribution, regularity occurrence of angular lacunae at cell ends
Tube Cells:	presence distribution over the whole grain regularity, density width of tubes thickness of tube walls

* examined under optical light microscope only

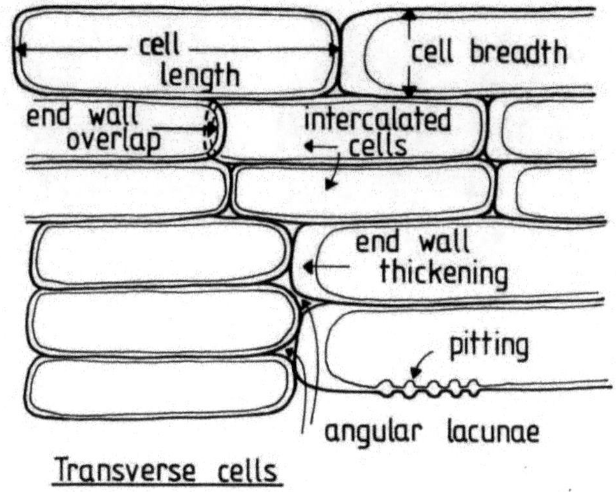

Fig. 5. Diagrammatic representation of the lateral view of a wheat grain showing the division into areas for scanning.

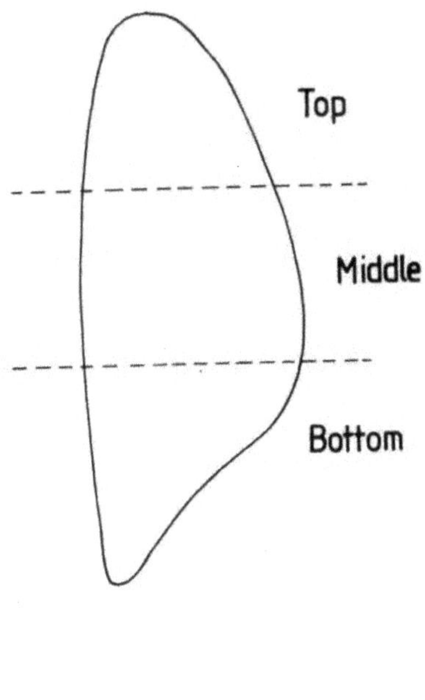

Fig. 4. Diagrammatic representation of the transverse cell layer, showing the features used for the comparison of the cell patterns.

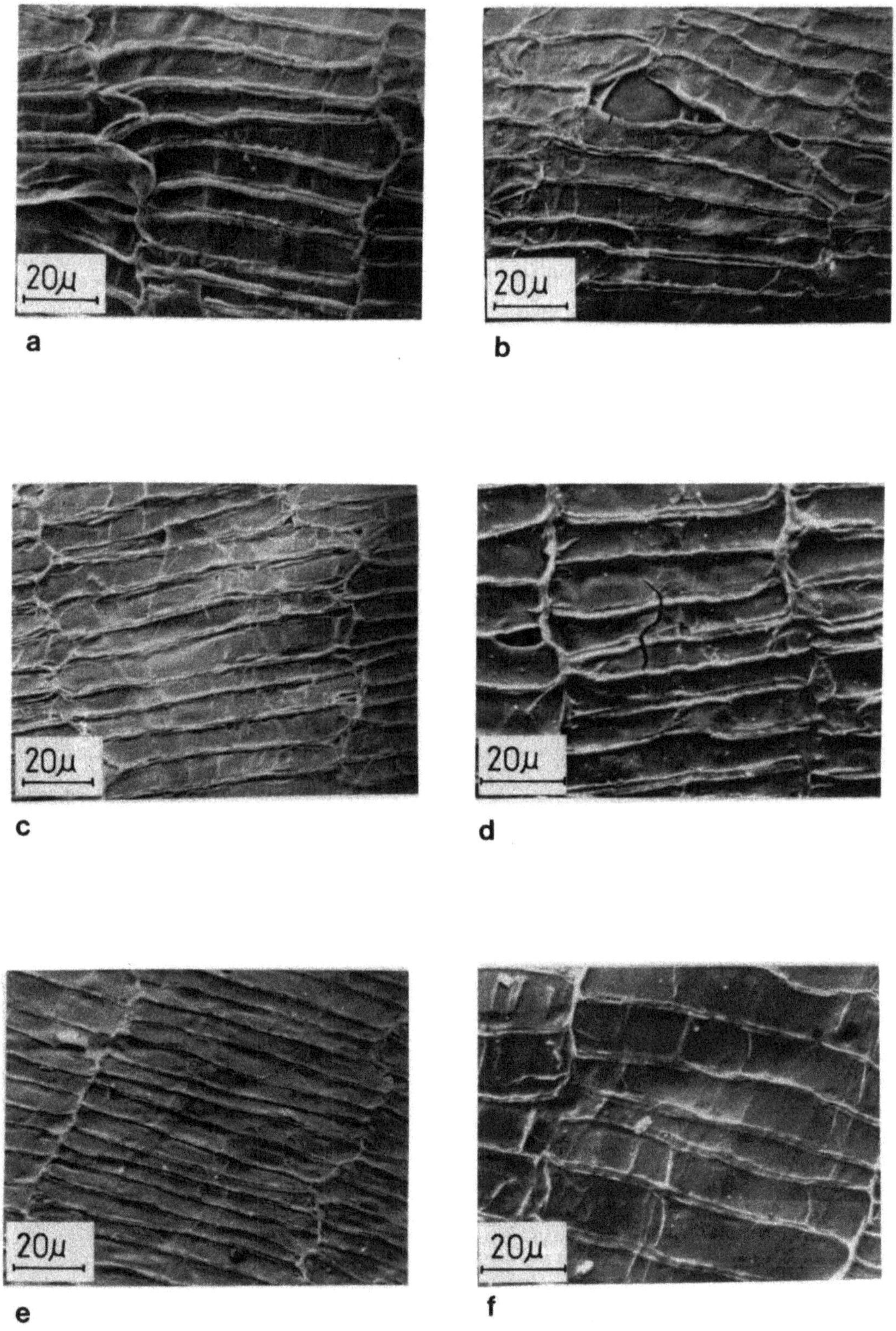

Fig. 6. a-c. Triticum monococcum 1^{g} (PBI, c): different cell patterns in different areas of the same grain. a. apex, b. middle, c. bottom. d-f. Triticum monococcum (D. Zohary): differences in cell patterns on grains from one-grained and two-grained spikelets of the same ear: d. 2^{g}S, e. 2^{g}L, f. 1^{g}.

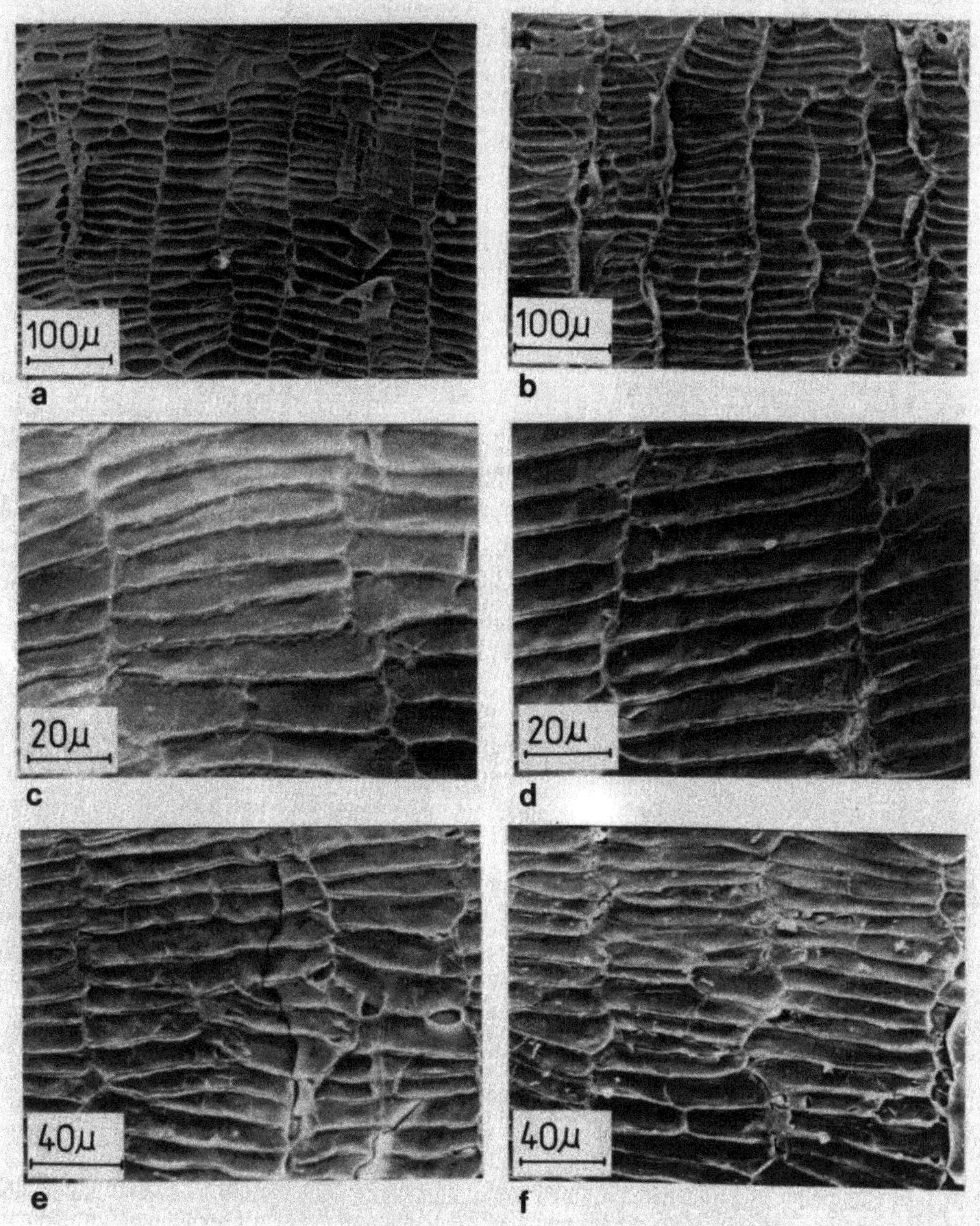

Fig. 7. a, b. Secale cereale: differences in the cell patterns on the grains of the same species from different populations: a. (GCH 2921a), b. (PBI). c, d. Similarities of the cell patterns on grains of different species: c. Triticum boeoticum, 2ᵍL (PBI), d. T. monococcum, 2ᵍL (D. Zohary). e, f. Similarities of the cell patterns on grains of different genera: e. Triticum monococcum' 2ᵍS (D. Zohary), f. Secale cereale, (GCH 3094).

critical analyses until, as Korber-Grohne herself admits, there has been a much more thorough investigation of a range of modern species. The temptation, in the past, has been to use scanning electron microscopy in the final confirmation of the identification of the crop species represented in the ancient remains. This investigation has highlighted the limitations imposed both by the preservation of the archaeological specimens and also by the lack of characterisation amongst the modern species.

References

Korber-Grohne, U. and Piening, U. 1980. Microstructure of the surfaces of carbonised and non-carbonised grains of cereals as observed in scanning electron and light microscopes as an additional aid in determining prehistoric findings. Flora 170: 189-228.

Korber-Grohne, U. 1981. Distinguishing prehistoric cereal grains of Triticum and Secale on the basis of their surface patterns using the scanning electron microscope. Journal of Archaeological Science 8: 197-204.

Moore, A.M.T. et al. 1975. The excavation of Tell Abu Hureyra in Syria: a preliminary report. Proceedings of the Prehistoric Society 41: 50-77.

PART VI:

ANIMAL MATERIALS

THE DETECTION OF DIAGENETIC CHANGES IN NON-MARINE SHELLS PRIOR TO THEIR SUBMISSION FOR ^{14}C DATING

Tim Yates

Abstract

The use of shells from terrestrial and freshwater molluscs for ^{14}C dating has in the past been viewed with skepticism because the results are prone to distortion from post-depositional diagenesis, or the incorporation of material with a low $^{14}C/^{12}C$ ratio whilst the mollusc was alive, or both. This paper deals with three aspects of a larger project: diagenetic changes, their detection and their effect on ^{14}C dating.

Diagenesis was studied by comparing the structure and chemical composition of modern specimens with those of fossil samples. Scanning electron microscopy showed that five structures were present in freshwater shells and one in terrestrial shells; three forms of post-depositional structural change were observed. The effect of diagenetic alteration on ^{14}C dates is considered using dates obtained on fossil samples using Accelerator Mass Spectrometry (AMS) techniques.

Introduction

The chronology of developmental processes in environment, landscape or society often calls for ^{14}C dating of some form of calcium carbonate, such as shells, carbonate nodules, calcite or tufa. Where preservation conditions are suitable these materials may be available in quantities that allow conventional ^{14}C dating (Evin et al., 1980; Thorpe, Otlet and Sweeting, 1980; Callen, Wasson and Gillespie, 1983), and will certainly be available in the quantities required for AMS techniques; they are, however, often used only as a last resort because the ages they yield are considered unreliable for accurate dating.

Earlier studies on marine molluscs (Gillespie and Polach, 1979; Mangerud, 1972) have suggested that the associated errors may be quantifiable within statistical limits, and the work of Vita-Finzi (1980; Vita-Finzi and Roberts, 1984) has emphasized the importance of sample preparation in obtaining reliable results. Olsson, Göksu and Stenberg (1968) have also emphasized the risk for contamination by adsorption of CO_2 by storing in the atmosphere. Land and freshwater molluscan shells are even more suspect than marine shells since they contain material which may have been severely depleted in ^{14}C when taken from the environment, besides being subject to recrystallisation and contamination since death. This paper builds on the original work of Chappell and Polach (1972) and shows that, once the original shell structure has been recognised, diagenetic changes can readily be identified using the SEM and often the light microscope. It goes on to show the effect that undetected diagenetic material can have on ^{14}C dates measured by AMS.

Methods and Techniques

Procedures based on stable isotope analysis ($^{13}C/^{12}C$, $^{18}O^{16}O$) and chemistry of shells and their environment of deposition allowed selection of samples in which apparent age effects were at a minimum (Yates 1986). These samples were then examined for diagenetic material and contamination. In order to make a reliable assessment of the extent and position of diagenetic changes and areas of contamination it was necessary to establish the form of unaltered structures, the levels at which diagenetic material and contamination could be detected, and their potential for removal. The techniques necessary to achieve these aims were integrated to give a consistent and reliable method and are shown in Figure 1.

After cleaning, the mineralogy of individual shells was established. Analysis of modern samples confirmed that all the species studied were pure aragonite unless they contained regenerated shell material (Saleuddin, 1971). Each shell was sampled at five to ten points to provide a minimum sample of 50 mg for x-ray diffraction (XRD) analysis. The analyses were undertaken on a Philips PW 1010 diffractometer with a CuK α source fitted with a Ni filter and calibration showed the minimum detectable limit to be 0.6% calcite.

All samples containing calcite and a number of those that were pure aragonite were examined under a scanning electron microscope. Samples were coated with gold/palladium in an argon atmosphere and examined using a Jeol 35CF and an Hitachi S-530, both operating at 15-25KV; uncoated samples were studied at 2-4KV. The working range was 30-80,000 times, so that single aragonite crystal units could be resolved.

If observation showed the contamination or recrystallisation to be a surface deposit, further cleaning was undertaken followed by further XRD analysis. When the shell was judged to be clear of calcite the whole shell was etched in 5% HCl, rinsed in distilled water, dried and ground. The sample was then subsampled and again analysed for calcite. Samples in which none was detected were considered suitable for ^{14}C analysis.

Structures Present in Freshwater and Terrestrial Shells

Details of the structures of 18 freshwater bivalves were already available (Taylor, Kennedy and Hall, 1969; Mackie, 1978), and the observations on Unio species were confirmed by further study. Eight species of freshwater and 15 species of terrestrial gastropods were examined, the structure of only one species being known from earlier research. The nomenclature used is based on that of Taylor, Kennedy and Hall (1969).

Five structures were found to be present: crossed-lamellar, complex crossed-lamellar, sheet nacreous, lenticular nacreous and prismatic, the last four being present only in the freshwater bivalves.

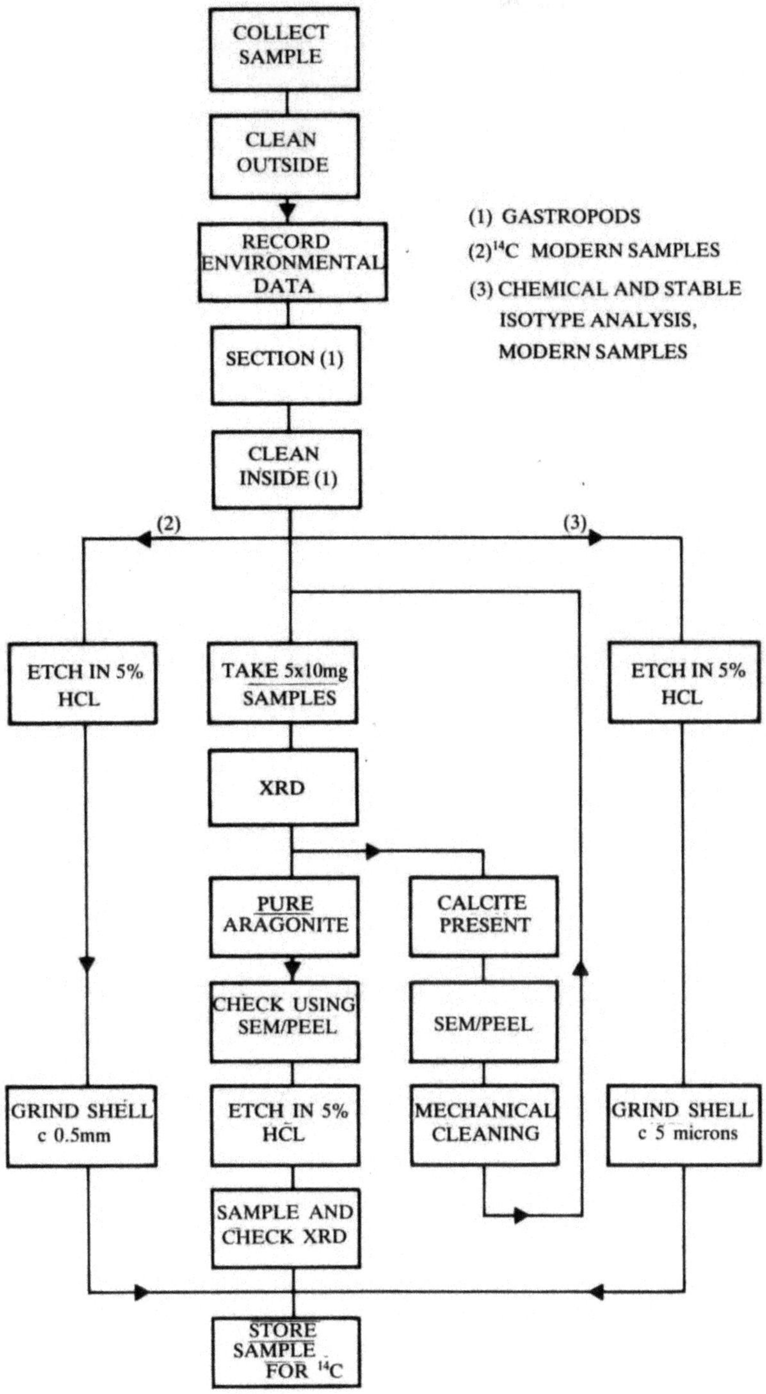

Fig. 1. Pretreatment procedure for mollusc shells.

Sheet Nacreous Structure. This is distinguished by the regular arrangement of layers parallel to the shell interior. The overall effect is of a brick wall, with the interlamellar and intercrystalline organic matrices representing the mortar (Fig. 2).

Lenticular Nacreous Structure. In this form the euhedral or rounded tablets, formed of piles of thin layers, are stacked in uneven groups. In general, the surfaces are distorted, leading to a more irregular form than that of sheet nacre.

Prismatic Structure. The prisms that form the basic structural unit are generally pentagonal cells divided from one another by thick conchiolin walls. The cells are narrow near the periostracum, giving rise to a large number of small prisms. Geometric selection eliminates many of these in growth and the characteristic prismatic form is produced.

Crossed-Lamellar Structure. This is the only structure found in non-marine gastropods, though its exact form shows wide variation. The structure is formed from a three-fold hierarchy consisting of first-, second-, and third-order lamellae (Fig. 3). MacClintock (1967) and Philippon (1974) describe two types of crossed-lamellar structure, concentric and radial, that differ in their orientation to the shell margin. Both types occur in most gastropod shells.

Complex Crossed-Lamellar Structure. This form is built of the same structural units as the crossed-lamellar structure but with four orientations of crystallites instead of two.

Diagenetic Structures

The present study of fossil shells (3000 BP - 9000 BP) from the British Isles revealed three forms of post-depositional changes: surface etching, formation of surface crusts, and recrystallization. Associated with these is the exfoliation of shell layers which accounts for the "flaky" texture frequently found on non-marine gastropod shells. The gaps caused by exfoliation are susceptible to both contamination and the growth of secondary calcite.

Etching. This appears to represent the earliest stages of dissolution; in the terrestrial gastropods examined the attack was generally confined to the surface and had penetrated only a few microns. In some cases, however, etching of individual third-order lamellae was occurring in association with secondary deposition; this is discussed below. The surfaces available for dissolution are increased by the exfoliation processes because it gives access to the first-order lamellae interfaces which are not protected by organic matrices.

Surface Crusts. This form of diagenetic change was observed on many shells. A similar structure was recorded by Evans (1972), who suggested it was a secondary deposit on the original shell surface. The present study suggests that it can also be formed by recrystallisation of original shell material (Fig. 4). It was observed on both inner and outer surfaces, but only occasionally exceeded 10μ in thickness. The thickness is limited by the crust breaking away from

Fig. 2. Detail of a sheet nacreous structure in a freshwater bivalve (*Unio* sp.).

Fig. 3. Crossed-lamellar structure in *Helix aspersa*.

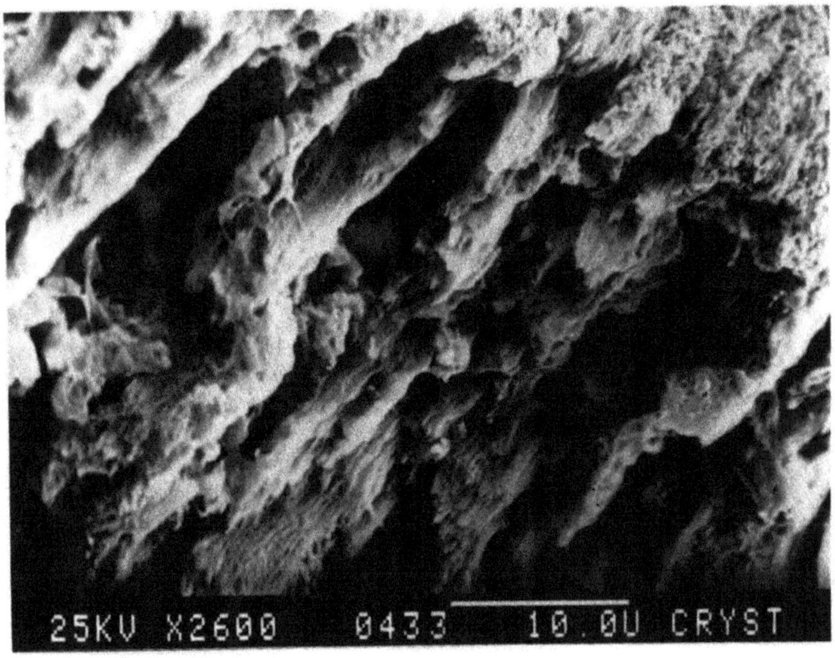

Fig. 4. Weathered surface of a fossil shell of <u>Cepaea hortensis</u> from Caerwys, Clwyd (7880+150BP). Note the surface crust (top right).

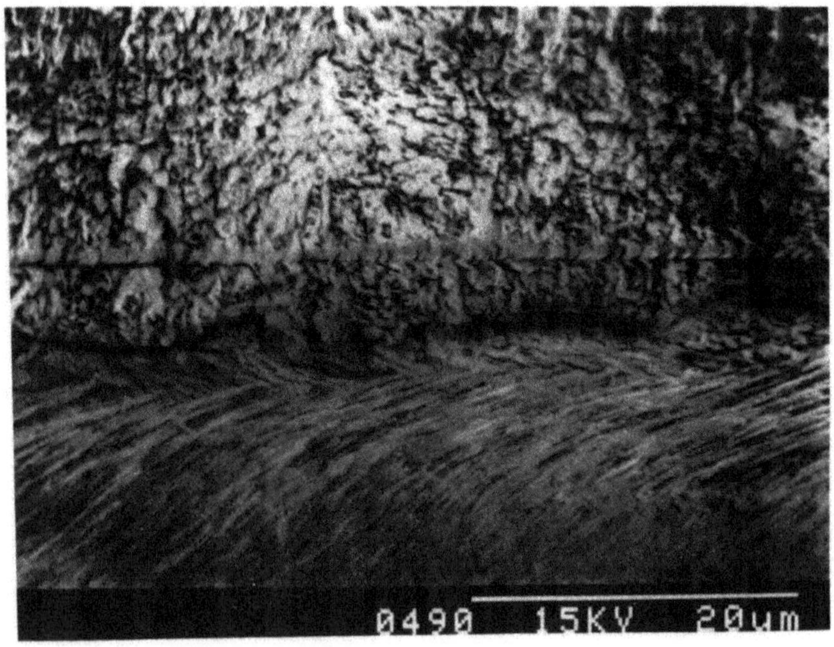

Fig. 5. Boundary between crossed-lamellar structure and recrystallised calcite (top). Fossil shell of <u>Cepaea nemoralis</u> from Caerwys, Clwyd (7880+150BP0.

the shell surface, as was clearly seen in a number of specimens. This is also seen as exfoliation and is comparable with the loss of the hydration layers observed on obsidian. The exfoliation of crusts may owe something to differential expansion of the shell aragonite relative to the calcite crust. Such crusts are believed to form from the deposition of secondary calcite from groundwaters and, therefore, to contain carbonate that is older than the context.

Replacement of Aragonite by Calcite. Where the surface crusts had not peeled off, or where the recrystallisation had taken place from an inner layer surface, larger areas of replacement calcite were found (Fig. 5). In the shells studied the areas affected did not exceed 5% of the shell's volume. This still represents, however, an unacceptably high level in older samples, as it usually incorporates post-depositional CO_2.

The recrystallization process appears to conform to the model proposed by Pingitore (1976; 1982), with a narrow diagenetic front separating the calcite from the unaltered aragonite. Although the front is narrow (1-2 microns), no apparent retention of the original structure was observed.

The Effect of Diagenetic Changes on ^{14}C Dates

Validation of the pretreatment and selection procedures required the dating of samples from a context for which other dating was available. Samples were collected at an extensive tufa deposit at Caerwys in Clwyd, Wales for which ^{14}C dates were already available. The shells were subjected to the selection methods outlined above and five samples submitted for ^{14}C dating by the AMS technique. Access to this technique allowed small sample sizes to be prepared and therefore be subjected to a thorough selection procedure. Samples were generally 100 mg in weight but two were only 10 mg. Details of the samples are given in Table 1.

Samples OX4-6 were unaltered aragonite shell. The dates for OX4 and OX6 were not significantly different from the date obtained on charcoal. Sample OX5 was younger than the accepted date by 350 years, but further SEM analysis showed that it contained small areas of recrystallised material, particularly between individual lamellae. The presence of secondary carbonate could explain the age error and its direction. Samples OX13 and OX14 were diagenetic deposits; OX13 was recrystallised material (calcite) and OX14 a surface crust. As predicted they were in error and in the expected direction.

Conclusion

The work dealt with in this part of the project has shown that one of the two major problems associated with the use of non-marine shells for ^{14}C dating can be reduced to low levels and in the case of selected samples confined to within the statistical errors associated with counting. To achieve this requires a rigorous approach to the cleaning and preparation of material for dating and in this the SEM plays an important role.

Table 1. Results of AMS Analysis of Shells from Caerwys, Clwyd

No.	Code	Species of Shell	Shell date* (years b.p.)	^{13}C	^{18}O
OX4	CW35	*Cepaea nemoralis*	7950 +/- 95 (OxA 704)	-8.94	-2.53
OX5	CW30	*C. hortensis*	7530 +/- 110 (OxA 705)	-8.06	-1.58
OX6	CW8	*Arianta arbustorum*	7860 +/- 100 (OxA 706)	-11.39	-0.74
OX13	CW31	*C. nemoralis*	6360 +/- 600 (OxA 713)	-8.00**	-
OX14	CW9	*C. hortensis*	8990 +/- 650 (OxA 714)	-8.00**	-

* compared to a charcoal date of 7880 +/- 150 (BM 1736).

** assumed value

Acknowledgements

I should like to thank C. Vita-Finzi, the Department of Crystallography in Birkbeck College, London, the Department of Geology in University College London, and the Accelerator Unit, Oxford for help and advice. The work was funded by the NERC.

References

Callen, R.A., Wasson, R.J. and Gillespie, R. 1983. Reliability of radiocarbon dating of pedogenic carbonate in the Australian arid zone. Sedimentary Geology 35: 1-14.

Chappell, J. and Polach, H.A. 1972. Some effects of partial recrystallisation on ^{14}C dating of late Pleistocene corals and molluscs. Quaternary Research 2: 244-252.

Evans, J.G. 1972. Land Snails in Archaeology. London and New York: Seminar Press.

Evin, J., Marechal, J., Pachiaudi, C. and Puissegue, J.J. 1980. Conditions involved in dating terrestrial shells. In Stuiver, M. and Kra, R.S. (eds.), International ^{14}C Conference Tenth Proceedings: Radiocarbon 22: 2: 545-555.

Gillespie, R. and Polach, H.A. 1979. The suitability of marine shells for radiocarbon dating of Australian prehistory. In Berger, R. and Suess, H.E. (eds.), Radiocarbon Dating International ^{14}C Conference, Ninth Proceedings, pp. 404-421. Berkeley: University of California Press.

MacClintock, C. 1967. Shell structures of patelloid and bellerophontoid gastropods (Mollusca). Peabody Museum of Natural History Bulletin 22: 1-140.

Mackie, G.L. 1978. Electron microscopy of Pisidiiae. Canadian Journal of Zoology 56: 1: 1-6.

Mangerud, J. 1972. Radiocarbon dating of marine shells, including a discussion of apparent age of recent shells from Norway. Boreas 1: 2: 143-172.

Olsson, I.U., Göksu, Y. and Stenberg, A. 1968. Further investigations of storing and treatment of Foraminifera and molluscs for ^{14}C dating. Geologiska Foreningens i Stockholm Forhandlinger 90: 417-426.

Philippon, J. 1974. Structure et composition mineralogique de la coquille de Gastropodes actuels et fossiles. Notes et Contributions, Centre pour Etudies et Recherche Paleontologie et Biostratigraphie 7: 1-84.

Pingitore, N.E. 1976. Vadose and Phreatic diagenesis: processes, products and their recognition in corals. Journal of Sedimentary Petrology 46: 985-1006.

Pingitore, N.E. 1982. The role of diffusion during carbonate diagenesis. Journal of Sedimentary Petrology 52: 27-39.

Saleuddin, A.S.M. 1971. Fine structures of normal and regenerated shell of Helix. Canadian Journal of Zoology 49: 37-41.

Taylor, J.D., Kennedy, W.J. and Hall, A. 1969. The shell structure and mineralogy of the Bivalvia. I. Introduction. Nuculacea-Trigonacea. Bulletin of the British Museum (Natural History) Zoology Supplement 3: 1-125.

Thorpe, P.M., Otlet, R.L. and Sweeting, M.M. 1980. Hydrological implications from ^{14}C profiling of UK tufa. In Stuiver, M. and Kra, R.S. (eds.), International ^{14}C Conference, Tenth Proceedings, Radiocarbon 22: 3: 897-908.

Vita-Finzi, C. 1980. ^{14}C dating of recent crustal movements in the Persian Gulf and Iranian Makran. In Stuiver, M. and Kra, R.S. (eds.), International ^{14}C Conference, Tenth Proceedings, Radiocarbon 22: 3: 763-773.

Vita-Finzi, C. and Roberts, N. 1984. Selective leaching of shells for ^{14}C dating. Radiocarbon 26: 1: 54-58.

Yates, T.J.S. 1986. The selection of non-marine Molluscan shells for radiocarbon dating. Unpublished Ph.D. thesis, University of London.

THE SCANNING ELECTRON MICROSCOPE AND THE STUDY OF ANCIENT TEETH

Simon Hillson

Ancient teeth form a prominent part of the mammal remains from archaeological sites and dental tissues are often preserved relatively intact. Usually, the enamel, which coats the crown of the tooth is the best-preserved. Dentine, which forms the main internal structure of the tooth and cement, which coats the root surfaces (and crown in some mammals), may be less well-preserved, but it is often possible to examine the detailed microscopic structure of all three tissues.

The polarising microscope has been the standard piece of equipment for studying the histology of dental tissues. A section of the tooth, polished to a thickness of less than 100 microns, is examined in transmitted light. Polarising filters are arranged above and below the microscope stage, so that the brightness and colour of the image reflects the mineralogy and orientation of crystals, collagen fibres and pore spaces in the tissue. As dental tissues are largely composites of mineral crystals (the apatite family of calcium phosphate minerals) and collagen fibres, this makes it possible to see many details of structure which cannot be discerned by other means. For this reason, much work in dental histology and its archaeological applications has been carried out using the polarising microscope (Schmidt and Keil, 1971; Rose et al., 1981; Rudney, 1983; Hillson, 1986). Transmitted light microscopy of this kind does, however, have a number of disadvantages. First of all, it cannot resolve details less than 1 micron or so across. In addition, it is necessary to prepare polished thin sections of teeth, which is time consuming, destroys valuable specimens and can be difficult with fragile archaeological material. Finally, the material being examined must be translucent and, whilst this is no problem with most fresh dental material, archaeological specimens are frequently rendered opaque by the processes of preservation. Such difficulties vary from site to site and, although it is often possible to produce usable sections of dental enamel, dentine and cement are often too fragile, too altered, or too opaque.

Use of the scanning electron microscope can avoid many of these problems. It has a high resolution and large depth of focus, and structures much less than 1 micron across can be examined routinely. It looks at surfaces, not sections so that the specimen does not need to be translucent. Preparation techniques can be very simple. Small fragments of tissue, already broken and lying in the bottom of a box, can be utilised. Replicas can be made using silicone rubber or other methods (see Claugher, this volume) without damaging the specimen at all, whilst still showing sufficient structural detail.

Scanning electron microscopy does not demonstrate well all the features that can be seen in polarising microscopy, but it is able to show some features that polarising microscopy cannot; either by virtue of its greater magnification, or because it is possible to look at surfaces which have been fractured along the "grain" of the tissue. This paper presents three examples of ways in which the scanning electron microscope can yield useful information for archaeologists; one from each of the dental tissues.

Enamel and Identification

The tiny apatite crystals of dental enamel are arranged into small bundles, about 5 microns across, called prisms (Fig. 1). Three main patterns of enamel structure can be distinguished from the way in which the prisms are arranged (Boyde, 1965):

Pattern 3; which is the most common form in man and the apes, carnivores and elephants, produces a characteristic "keyhole" tessellation when the prisms are seen end-on (Fig. 2).

Pattern 2; the main pattern of hoofed mammals, monkeys and marsupials, and which is the basis of enamel structure in rodents (below), has "inter-row sheets" of enamel arranged between rows of prisms (Fig. 3).

Pattern 1; found in the toothed whales, bats, insectivores and some of the lemurs, is the most simple form, with cylindrical prisms running through tunnels in the "interprismatic" enamel.

These patterns vary greatly and most teeth have at least some of all three distributed through their enamel, but there are some general differences between groups of mammals. Boyde and Martin (1982) were able to distinguish between man and groups of great apes in the relative proportions of Patterns 3 to Patterns 1 and 2. This has important implications for the study of hominid fossils.

There are, however, further possibilities for distinguishing between different groups of mammals on the structure of their enamel. The prisms only rarely pass straight through the enamel. In most mammals, they bend from side to side and are woven into one another (Fig. 5). This is called prism decussation. The weave is complex and variable, but it seems possible to make some general distinctions between the following mammal groups (Kawai, 1955):

1. Toothed whales, insectivores and bats.
2. Carnivores. Pinnipeds (seals and sea lions) fit into the carnivore pattern, but diverge toward the primate form.
3. Primates.
4. Ungulates (hoofed mammals). There seem to be some grounds for distinguishing between even-toed (cattle, deer, etc.) and odd toed (horses and rhinoceros) ungulates on the basis of their decussation pattern.
5. Rodents and lagomorphs (rabbits and hares) have the most complex patterns of all, so complex that it may be possible to distinguish between rodent families and even genera on the

Fig. 1. Prisms in medieval human enamel from York, England. Scanning electron micrograph of fractured surface. Scale bar represents 10 microns.

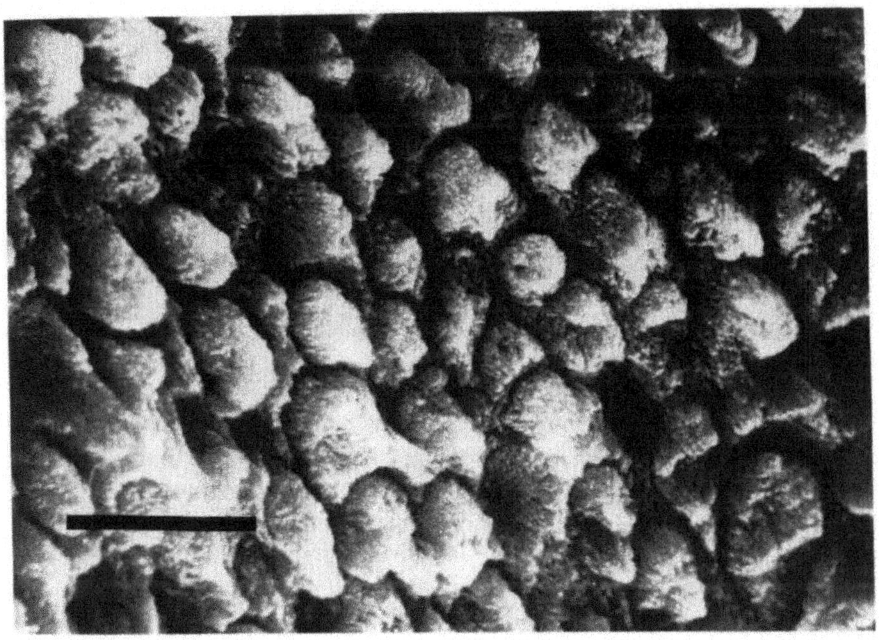

Fig. 2. Pattern 3 enamel in the same specimen as Figure 1. The prisms are seen end on, and show a roughly "keyhole" outline. Scanning electron micrograph of fractured surface. Scale bar represents 10 microns.

basis of their enamel (Korvenkontio, 1934; Yamakawa, 1959; Boyde, 1978).

The large rodent incisor shown in Figure 4 is an example of the use of enamel structure as an aid to identification. From its morphology, it is an upper left incisor, but because most rodent incisors are built along similar lines there are few definite clues as to its species. It is very large - the worn surface is about 5 mm across - and this narrows down the choice. It comes from a Neolithic site in Tunisia and the only Old World Holocene rodents which would be large enough are beaver Castor or porcupine Hystrix (Hillson, 1986). Because it comes from the North African coast, it is almost certain to be porcupine, but it is still desirable to confirm the identification. Fortunately, the weave of prisms in beaver enamel is very different from that of the porcupine. It conforms to a pattern known as sciuromorph, whereas the porcupine fits into an hystricomorph pattern. Scanning electron microscope examination of a small fragment of the Neolithic specimen (Fig. 5) confirms quickly that it is hystricomorph in form, arranged into alternate bands of different prism orientation, each 3 to 5 prisms across.

Similar studies may be carried out in other rodent teeth. For example, the scanning electron microscope has been used in the study of evolution in fossil vole cheek teeth (von Koenigswald, 1980). In general, however, the technique has been little applied in archaeology.

Age Determination and Root Dentine Translucency

There are several forensic methods for determining age at death from human teeth. Some of them could potentially be applied to archaeological material, but care is needed. Conditions of preservation may cause changes in the dental tissues which mask the effects used in age determination, or confuse the issue. This is where the scanning electron microscope may help.

One potentially useful method involves examining a section of the tooth root and looking for an area of translucency in the dentine. No magnification is required - the effect can be clearly seen with the naked eye on a lightly polished slice through the tooth (Fig. 6). In most individuals the size of the translucent zone is correlated with increasing age and, if it is measured, then standard formulae can be used to calculate an estimate of the age at death (Bang and Ramm, 1970). The method is particularly useful in determining the age of older adult people and this is its main attraction in archaeology, where individuals over 25 or so years are difficult to age reliably by other means.

Under the scanning electron microscope, dentine (Fig. 7) is characterised by closely spaced tubules which run through the tissue. These are 4 microns or less in diameter and their structure is difficult to see under the polarising microscope. The dentine becomes translucent as the tubules are progressively infilled over time. In archaeological specimens, poor preservation may make it difficult to see the limits of the area of translucency with the naked eye. The dentine may be opaque because the collagen fibres and apatite crystals

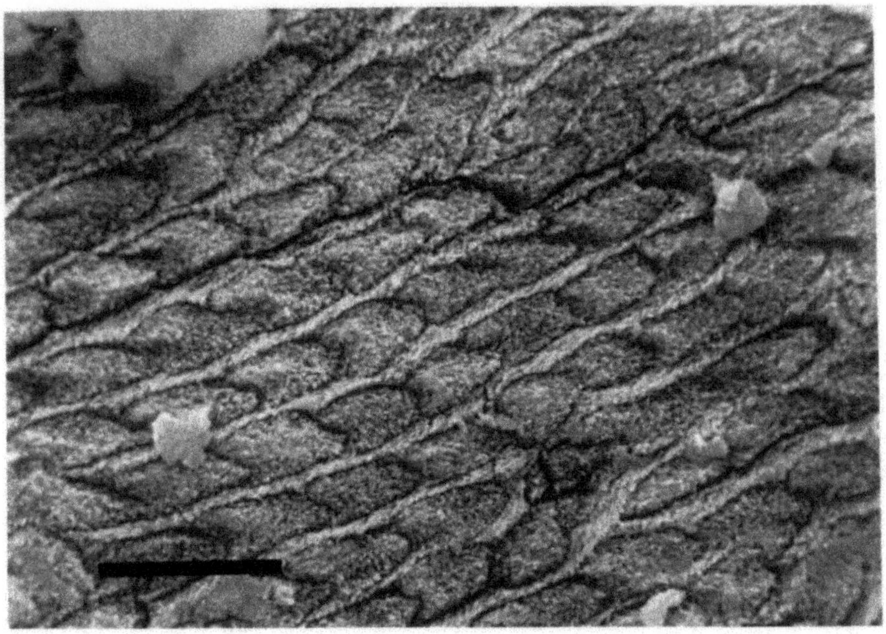

Fig. 3. Pattern 2 enamel in medieval pig enamel from York, England. The prisms are seen end on, arranged in rows with sheets of interprismatic enamel between them. Scanning electron micrograph of fractured, etched surface. Scale bar represents 10 microns.

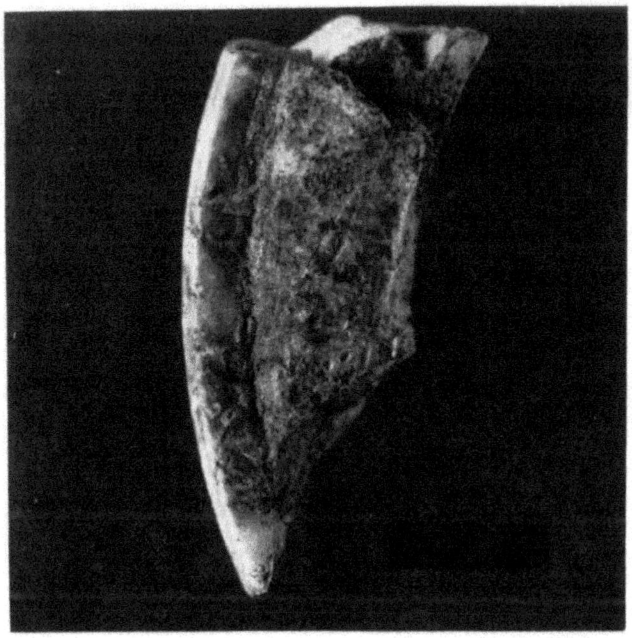

Fig. 4. Large rodent incisor (upper right) from a Neolithic site in Tunisia. Scale bar represents 0.5 cm.

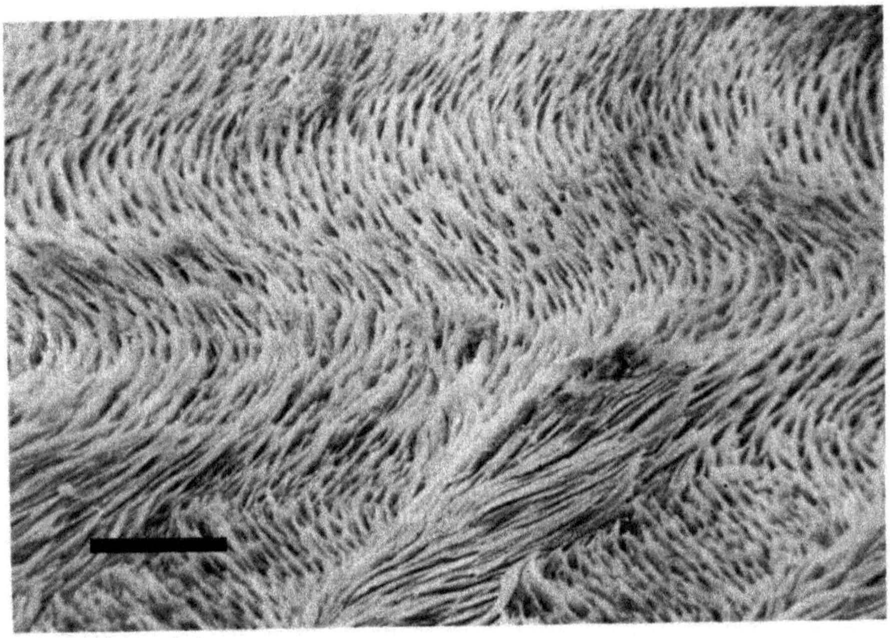

Fig. 5. Enamel from the specimen in Figure 4, showing hystricomorph structure. In the lower half of the photograph, the prisms are arranged into band, 5 or so prisms wide, which present themselves to the plane of section at different angles. Scanning electron micrograph of polished, etched surface. Scale bar represents 25 microns.

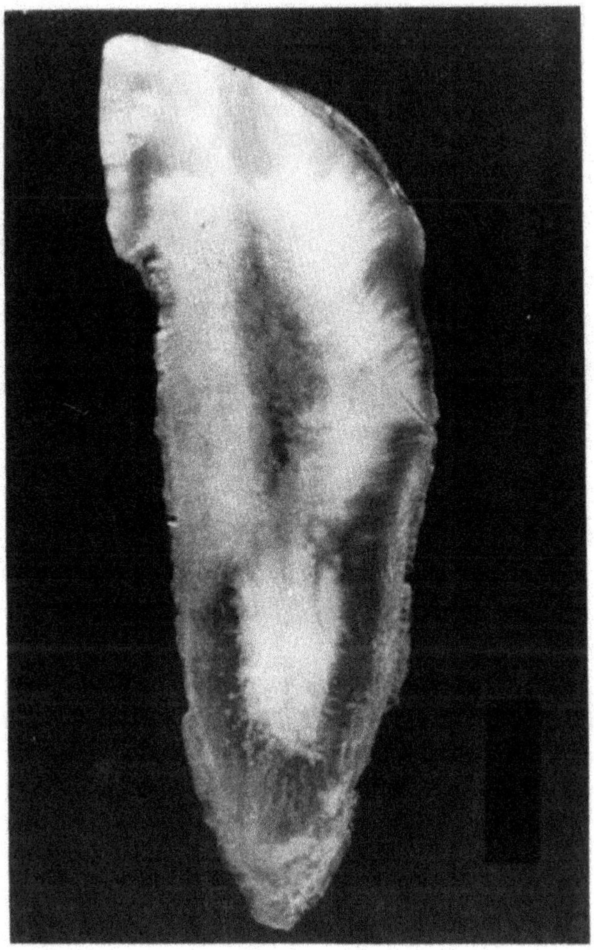

Fig. 6. Polished section through Neolithic human incisor (upper central) from Tunisia. The specimen is arranged so that the root of the tooth extends downwards, and the zone of translucent dentine can be seen as a darker area extending back from the root tip. Scale bar represents 0.5 cm.

have been replaced by other minerals which crystallise out from the ground water, or the tooth roots may have been invaded by mycelium-producing microorganisms. These invasions carve out irregular canals (Fig. 8), up to 20 microns or so in diameter, which greatly disrupt the dentine structure and can obscure the edge of the translucent zone. The scanning electron microscope can be used to examine the dentine which remains in between the canals of the invading mycelium, and check the infilling of tubules. For example, parts of the translucent zone shown in the tooth sectioned in Figure 6 show clearly, but towards the edge of the zone, there is a large pale patch of dentine which has been invaded and where the detail is obscured. Under the scanning electron microscope (Fig. 8) it was possible to examine the tubules in between the damaged areas and to delimit the area with infilled tubules. Specimen preparation was simple; the half of the tooth which remained after the slice had been cut was polished, etched and then coated with gold. Another method would be to produce a replica after etching. The average length of the translucent zone in this specimen yields an age at death estimate of 68 +/- 11 years if the standard formulae for modern man are applied (Bang and Ramm, 1970).

Cement Layers and Age

The cement coating of the roots is frequently layered. In many mammals, these layers appear to be laid down seasonally or annually. This is particularly so in the marine mammals such as toothed whales or seals and sea lions (Perrin and Myrick, 1980), but also seems to be the case in terrestrial hoofed mammals, hibernating mammals (Klevezal and Kleinenberg, 1967; Morris, 1972; Casteel, 1976) and even man himself (Stott et al., 1982). In many mammals, these layers are broad enough to be seen in simple slices through the tooth root, examined under modest magnification. Another method is to decalcify the tooth (usually with acids), leaving the organic component (mostly collagen fibres) which can then be sectioned with a microtome and stained before being examined in a conventional optical microscope. Counts of the patterns of layers are thought to yield estimates of the age at death, although there are difficulties with the procedure (Hillson, 1986). The layers are not always clearly defined and different observers often produce different counts. In addition, the seasonal basis of the layering is not clearly understood, and the relationship between layer pattern counts and age is not proven in all mammals. Cautiously applied, however, the method does seem to have promise and it would be very useful to apply it in archaeology.

As in the case of root dentine transparency (above) a simple polished slice through the tooth will often suffice to see the coarser layering such as is found in the hoofed mammals. It is more difficult to see the finer layering shown in other mammals, and the details may often be obscured by poor preservation. The scanning electron microscope can be used to examine cement layers in root fragments which have been etched to bring out their structure in relief. Figure 9, for example, is a specimen of cement from a cave bear (<u>Ursus arctos spelaeus</u>) canine, found at a Mousterian cave site in Yugoslavia. The layers are not sharply defined, but it is possible to count some 20 major layer groups through the thickness of the cement. Experimental

Fig. 7. Dentinal tubules in medieval human dentine (the same specimen as Figure 1). Scanning electron micrograph of fractured surface. Scale bar represents 10 microns.

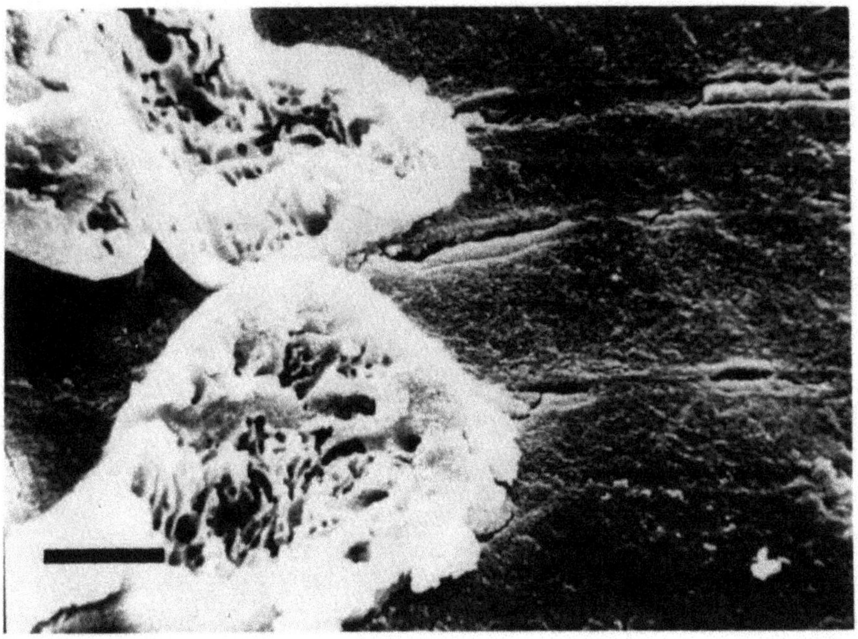

Fig. 8. "Canals", probably due to invasion by a mycelium-producing microorganism, in the same specimen as Figure 6. The canals (on the left of the photograph) are filled with a roughly textured deposit of secondary minerals, deposited from ground waters percolating through the tooth during burial. The dentine is from the transparent zone and the tubules, which can be seen in between the canals, are filled with later dentine deposits which have etched out to show as cores. Scanning electron micrograph of polished, etched surface. Scale bar represents 10 microns.

work needs to be done on modern U. arctos teeth from individuals of known age, but there does seem to be some potential for the method.

It may also be possible to apply similar techniques to human remains. Human cement frequently has finely spaced layers, and the layer count seems to be related to age. Stott et al. (1982) demonstrated this using polarising microscopy of polished thin sections. This technique rarely yields much detail of cement layering in archaeological specimens of human teeth, but etched cement surfaces under the scanning electron microscope frequently do show layers. Not all teeth are suitable; for example, the specimen in Figure 6 did not have clear layering, so that counts could not be compared with the root dentine transparency method above. In another specimen, however, the second molar of a young man from Nubia (c. 1700 BC), it is possible to make a count (Fig. 10). Thirteen to 15 layers are visible in many places on the root and, since the second molar roots are completed around 14 years of age, this would seem to suggest an age estimate of 27 to 29 years. This compares favourably with independent evidence from the skeleton of this individual.

Much more work needs to be done on these techniques, but there do seem to be real possibilities for using the scanning electron microscope to expand the range of information which can be extracted from ancient dental remains.

References

Bang, G. and Ramm, E. 1970. Determination of age in humans from root dentine transparency. Acta Odontologica Scandinavica 28: 3-35.

Boyde, A. 1965. The structure of developing mammalian dental enamel. In Stack, M.V. and Fearnhead, R.W. (eds.), Tooth Enamel, International Symposium on the Composition, Properties and Fundamental Structure of Tooth Enamel, pp. 163-167. Bristol: Wright.

Boyde, A. 1978. Development of the structure of the enamel of the incisor teeth in the three classical subordinal groups of the Rodentia. In Butler, P.M. and Joysey, K.A. (eds.), Development, Function and Evolution of Teeth, pp. 43-58. London: Academic Press.

Boyde, A. and Martin, L. 1982. Enamel microstructure determination in hominoid and cercopithecoid primates. Anatomy and Embryology 165: 193-212.

Casteel, R.W. 1976. Incremental growth zones in mammals and their archaeological value. Papers of the Kroeber Anthropological Society (Berkeley) 47: 1-27.

Hillson, S.W. 1986. Teeth. Cambridge Manuals in Archaeology. Cambridge University Press.

Kawai, N. 1955. Comparative anatomy of the bands of Schreger. Okajimas Folia Anatomica Japonica 27: 115-131.

Klevezal, G.A. and Kleinenberg, S.E. 1967. Age determination of mammals from annual layers in teeth and bones. Academy of Sciences USSR, translated 1969 Department of the Interior and National Science Foundation, US Department of Commerce, Clearinghouse for Federal Scientific and Technical Information, Springfield.

Korvenkontio, V.A. 1934. Mikroskopische untersuchungen an nagerincisiven unter hinweis auf die schmelzstruktur der backenzahne. Annal. Zool. Soc. Zool. -Bot. Fenn. Van. 2: 1-274.

Morris, P. 1972. A review of mammalian age determination methods. Mammal Review 2: 69-104.

Perrin, W.F. and Myrick A.C. Growth of odontocetes and sirenians: problems in age determination. Proceedings of the International Conference on Determining Age of Odontocete Ceteans (and Sirenians), La Jolla, California, September 5-19, 1978. Cambridge: International Whaling Commission.

Rose, J.C., Boyd, L.F. and Condon, K.W. 1981. Histological enamel indicator of childhood stress in prehistoric skeletal samples. American Journal of Physical Anthropology 49: 511-516.

Rudney, J.D. 1983. The age-related distribution of dental indicators of growth disturbance in ancient Lower Nubia: implications for hypotheses of X-group origins. Journal of Human Evolution 12: 535-543.

Schmidt, W.J. and Keil, A. 1971. Polarization Microscopy of Dental Tissues. Oxford: Pergamon Press.

Stott, G.G., Sis, R.F. and Levy, B.M. 1982. Cemental annulation as an age criterion in forensic dentistry. Journal of Dental Research 61: 814-817.

von Koenigswald, W. 1980. Schmeltzstruktur und morphologie in den molaren der Arvicolidae (Rodentia). Abhandlungen der Senckenbergische Naturforschende Gesellschaft 239: 1-29.

Yamakawa, K. 1959. Comparative anatomical studies on the enamel structure of the rodents (in Japanese). Acta Anatomic Nipponica 34: 852-866.

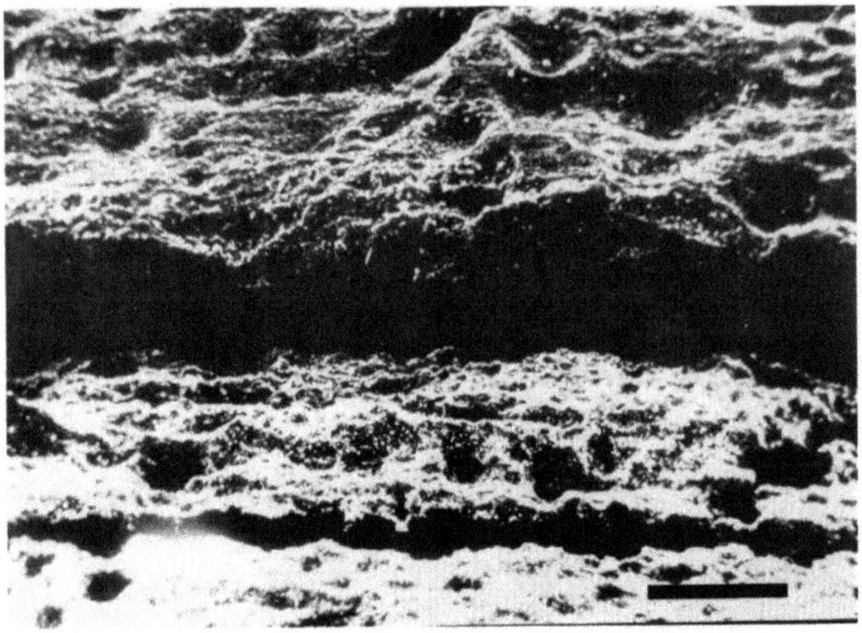

Fig. 9. Layers in cement from a cave bear canine, found at a Mousterian cave site in Yugoslavia. Scanning electron micrograph of fractured, etched surface. Scale bar represents approximately 500 microns.

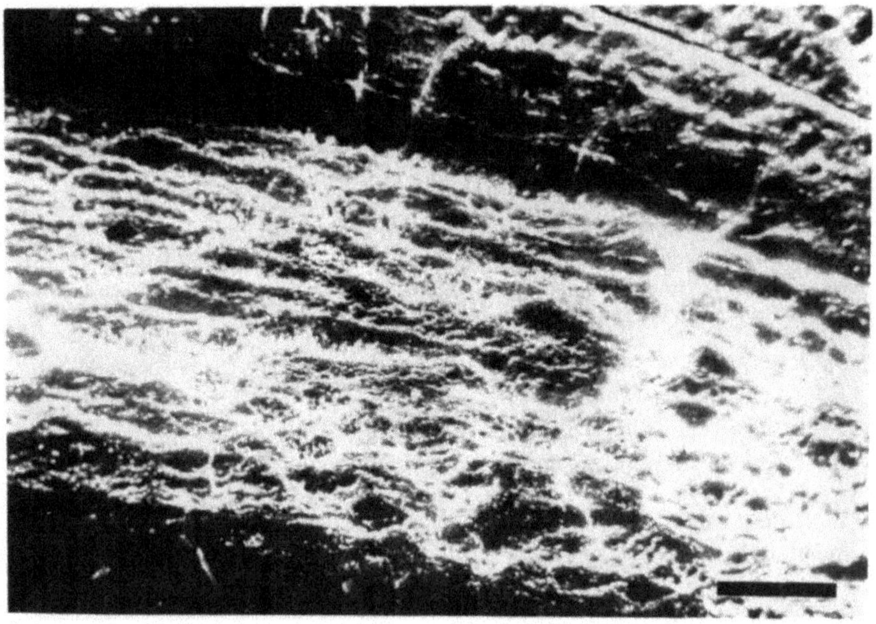

Fig. 10. Layers in human cement from Kerma, Nubia (c. 1700 BC). Scanning electron micrograph of polished, etched surface. Scale bar represents 50 microns.

ACTUALISTIC STUDIES OF ANIMAL RESOURCES AND HOMINID ACTIVITIES

Pat Shipman

Theoretical Framework

Although the ostensible purpose of this paper is to review some of my work in using the SEM to resolve archaeological problems I shall begin by addressing a more crucial issue: what is the meaning of the answers? Until this issue is clearly understood, and explicitly stated, there is little point in discussing the types of questions that can be answered. It is all too common for researchers to lose sight of the significance and limits of resolution of the answers in the struggle to obtain any answers at all.

In 1978, I first began exploring the usefulness of the SEM as a tool that would provide new types of data about fossil bones. My interests were primarily in finding new and more powerful means of deducing the taphonomic history of specific sites and to this end I began collecting bones of known taphonomic history, either naturally or experimentally produced. Structurally, the problems I faced at the time were ones that are undoubtedly familiar to most or all of the participants in the conference on the SEM in Archaeology that produced this book. That is, I was faced with a vast virgin territory. To my knowledge, no one prior to that time had considered that bones might encode in the microscopic morphology of their surfaces the set of taphonomic events which they had undergone. There was no data base from which to operate, no comparative sample to which unknowns (fossils) could be compared and then diagnosed (their history deduced). I drew up a list of major taphonomic events and began conducting experiments and collecting "wild bones", that is, those that had been exposed under natural conditions to different taphonomic events, and documenting the surface changes that had occurred. As this work proceeded, the list of events that I felt were important grew geometrically, as did the collection of replicas of controls. The first synthesis of this exploratory work can be found in Shipman (1981a); additional studies relevant to archaeological problems are also in print and some are summarized below (Shipman, 1981b, 1986a, b, and in press; Shipman and Rose, 1983, and Chapter 21, this volume; Shipman et al., 1984a, b).

The theoretical underpinnings of this work, as in all actualistic studies, is fairly straightforward. Although I have discussed it in detail elsewhere (e.g., Shipman, 1981a: 359-360; Shipman, 1981b: 10-12), it is worth repeating here because it is of such fundamental importance. Archaeology, taphonomy, and palaeontology are among the historical sciences in which the task is not to predict the future outcome or effects of a given set of events but rather to retrodict the set of events that gave rise to a particular effect. Historical scientists are forever "discovering" the results of experiments and then trying to deduce what the experiment itself was. The consequence of this is that, structurally, historical sciences must operate by a unique set of rules. Simpson (1970) proposed that the investigation of questions in the historical sciences must rely upon three stages. First, an understanding of the causal links between modern processes and their effects must be achieved; then the historical data (effects of ancient processes) must be gathered; finally, the historical data can be confronted and interpreted using the understanding of the process-effect linkage gained from the study of modern data. This approach is fundamental to all actualistic studies, including those in which the SEM is used to resolve archaeological problems.

Theoretically sound though this approach may be, in practice it is difficult. One cannot, reasonably, study and comprehend all modern processes that might by any stretch of the imagination have affected a fossil or archaeological assemblage. Even worse, if one could, the study would still be limited by what one imagined was relevant or likely to have happened, and one would still face tremendous difficulties in dealing with those aspects of the past that are not replicable today. A simple example will demonstrate what I mean. Suppose the question concerned the effects that animals' teeth have on bones. The experiments would seem to be simple: go to a zoo, feed animals bones, collect and study the bones. Then, for good measure, try to find "wild bones" of animals killed or processed by carnivores, just in case the zoo animals were behaving atypically. But what about extinct carnivores? What marks did a saber-toothed cat leave on bones, if any? And what about ungulates, which have been persistently reported to chew on bones to obtain calcium, phosphates, or trace elements? Would one imagine that giraffes' tooth marks ought to be studied?

In short, the results of actualistic studies are of limited, but not trivial, value. As I have observed elsewhere (Shipman, 1981a:360):

> ...care must be taken in interpreting the fossil record so that the appropriate level of resolution of the results is not exceeded. Conclusions must take the form, for example, that 'this fossil shows marks that best match those produced by the gastric juices of owls' rather than 'this fossil was swallowed by an owl.' Given the constraints of retrodiction inherent in working in an historical science, this is an acceptable and, in fact, optimal level of resolution.

This point has been reiterated by myself and others who have been careful to define what they mean when they conclude that their actualistic studies replicate what they observe in the archaeological materials. Why, then, are the results of these studies received either with complete acceptance or with deep skepticism? At the risk of oversimplification, I will attempt to characterize the acceptors and skeptics in terms of the issues that underlie their positions.

The acceptors, I believe, fall into two groups. There are those familiar with the SEM and with the microstructure of the type of object being studied who can inspect and thoughtfully evaluate the images published or shown. If their judgment about the resemblances and differences between the controls and the unknowns agrees with that of the experimenter, then strong acceptance results. The other group of acceptors are those unfamiliar with the SEM who are so impressed by the extraordinary detail it can reveal that they uncritically regard the SEM as a direct route to the truth. This group underestimates the variability within and among samples and the subjectivity of the judgment that two images are, or are not, fundamentally the same. Their acceptance relies on a combination of faith and their feeling of being unable to judge the study critically.

The skeptics fall into three groups. The first of these are those who are familiar with the SEM and the microstructure of the object studied and whose evaluation of the resemblances between the controls and the unknowns differs from that of the experimenter. I suspect this group is relatively small in numbers. The second group is unfamiliar with the SEM and, at an intuitive level, dislikes and mistrusts evidence arising from a technology that they do not understand or possess. This group, in sum, is making judgments on an emotional, nonrational basis that is highly similar to the second group of acceptors described above. The third group is skeptical because of the major, inherent problem with actualistic studies: the universe of possible causes can never be fully known. They are unquestionably correct in this assertion. All actualistic studies can be seen as arguments by elimination (e.g., Binford, 1981:40, 83), which are structurally weak for this reason. As a result, some skeptics would discount all actualistic studies as inadequate or insufficient.

What I think has been missed by this group is that actualistic studies must be understood in the broader context of Popperian science. I do not believe that actualistic studies are <u>only</u> arguments by elimination. One common source of skepticism is a misunderstanding of the intent of the researchers (or what I hope is their intent) in concluding that ancient specimen X shows features that match those of experimental specimen Y, which was modified, produced, or affected by event Z. Such conclusions are not meant to be definitive, unalterable truths to be written in stone and never re-examined thereafter. Such conclusions are, at best, working hypotheses, carefully formulated statements of "greatest similarity" framed within the context of the experimental sample available to that researcher.

Another point is relevant to the issue of argument by elimination; it is a more common and widely accepted procedure in science than is sometimes realized. Because my training is in physical anthropology, not archaeology, the example that I think best demonstrates my point has to do with the identification of skeletal remains. Taxonomy is classically no more and no less than argument by elimination. An unknown is systematically compared with the available and relevant knowns in search of a point by point matching of features; when this is found, given the limitations of the material and its preservation, the fossil is identified as belonging to species X. In reality, such identifications directly parallel those in actualistic studies. The researcher is, in fact, stating that he or she finds the features of the unknown fossil more similar to those of species X than to those of any other species considered and sufficiently similar to be accepted as a "match". If the unknown fossil is not judged sufficiently similar to any other known species, then the researcher creates a new taxon and must state what the diagnostic and distinctive features of that taxon are (i.e., how it can be recognized from other similar taxa). Frequently, as a group of species becomes better known, the initial diagnosis must be revised; species have been known to emerge, submerge, sink, rise again, split and join as the initial understanding is refined. Why should we expect a different pattern of progress with actualistic studies? Further, since taxonomy and actualistic studies share the same basic methodology of defining and comparing relevant features between knowns and unknowns, and since in taxonomy this approach is so universally accepted, can actualistic studies be criticized or discounted for using argument by elimination? It seems to me that there is a good reason for healthy skepticism and controversy about the conclusions of actualistic studies. Have the criteria or diagnostic features been clearly and carefully defined? Are they the correct features? Have some been ignored? There is little reason, however, for condemning such studies _a priori_, simply because of their methodology.

A second common misunderstanding concerns what has been accomplished by reaching such an outcome or conclusion. If a similarity has been established between the effects of a given event and those observed on an unknown, the researcher has failed to refute the null hypothesis, which can be generally stated as: "The effects on this unknown specimen are dissimilar in nature and cause from those produced on this experimental specimen." Thus, the known event must still be included in the list of potential explanations for the observed effects on the unknown. If, in addition, a _dissimilarity_ has been established between the effects of a known event and those observed on an unknown, then at least one known event can be eliminated from further consideration. This is the fundamental geometry of Popperian science: two or more alternative explanations (hypotheses) are posed, implications or predictions are derived from those, and then new data are gathered to test those predictions. As Popper points out, using this logic it is impossible ever to prove that an hypothesis is true; it can only be proved that an hypothesis is either false or not-yet-falsified. Nonetheless, over the course of time and through clever and rigorous testing and re-testing of hypotheses, it begins to be convincing that an hypothesis repeatedly unable to be falsified is likely to be true. In its essence, of course, the Popperian method _is_ an argument by elimination. It is also apparent that an hypothesis can be unfalsifiable because it is either so vague or so trivial that no serious testing can be conducted, but I trust that most scientific researchers are capable of recognizing and discounting inappropriate or insignificant hypotheses.

As both argument by elimination and Popperian hypothesis-testing, actualistic studies have a great potential to resolve or at least whittle down archaeological problems to manageable size. If the most that can be said of a set of studies is that some of the blind alleys and false hypotheses have been identified and the universe of potential explanations diminished, then those studies have made a significant contribution.

Finally, as my work clearly shows, I believe it is both legitimate and important to explore the implications of an hypothesis that has not yet been falsified. In other words, we must work on working hypotheses. Of course, the testing of alternative hypotheses should not be stopped or stalled once an initial round of testing has been carried out. One of the best ways to generate further hypotheses to be tested is to take an explanation that remains potentially true and think, talk and write seriously about what it will mean to our broader understanding of the past if that hypothesis is true (remains unfalsified). Actualistic studies will be doomed to a sterile role as either "spoilers" (Yellen, 1977) or "stamp collecting" - the unproductive gathering of ever-growing piles of data that never flower into applications or conclusions - unless we strive to see the broader significance of our testable hypotheses.

Actualistic Studies

I have investigated in depth four main areas of actualistic study using the SEM. These are: 1) identifying linear grooves, including cutmarks, carnivore toothmarks, rodent gnawing, root etching, and vascular grooves, which can be used to identify the agent responsible for altering a particular bone; 2) distinguishing between immediate and delayed processing of bones, which may be useful in deducing the objective of that processing; 3) recognizing burnt bones or teeth and determining the temperature range to which they have been heated, which may help show whether the heating occurred in natural or manmade fire; 4) recognizing utilization wear on bone tools, which can be useful if the bones were little modified prior to use and which may indicate, in general terms, the use to which those bones were put.

Cutmarks. The identification of linear grooves on bones has been much discussed in the literature, primarily in the context of identifying cutmarks as evidence of hominid activities. An initial set of diagnostic criteria for distinguishing between cutmarks and carnivore tooth scratches was offered by Potts and myself (Potts and Shipman, 1981; Shipman, 1981a, b), with subsequent refining of those criteria to eliminate features that were not truly diagnostic (Shipman, 1983), to distinguish more clearly between cutmarks and vascular grooves (Shipman and Rose, 1984), and to document more fully the variability in cutmarks and to include a better description of rodent gnawing marks (Shipman and Rose, 1983).

Work by others has shown that the issue of recognizing cutmarks is more complex than I initially thought. Andrews and Cook (1985), Behrensmeyer et al. (1986), Oliver (1984), and Fiorillo (1984) have all shown that trampling of bones on sandy or, especially, rocky substrates may produce marks that to some extent mimic cutmarks. The most thorough treatment of trampling marks is found in Andrews and Cook (1985). Several aspects of the trampled assemblages reported in these studies and in an extensive set of experiments carried out recently (Olsen and Shipman, in press) are important for distinguishing pseudo-cutmarks (trampling marks) from genuine ones.

First, the reported incidence of marked bones in trampled assemblages is very high. For example, Andrews and Cook (1985:682) found 70% of the bone bore marks and Oliver (1984:27) observed that trampling marks "...are so numerous that no attempt was made to quantify their occurrence." Olsen and I have made similar findings in an independent study of various bone assemblages trampled under a wide range of circumstances (Olsen and Shipman, in press). Since no assemblage of bones known to have been butchered has more than 20-25% of cutmarked bones (Bunn, 1982), no analyst should feel comfortable attempting to identify cutmarks on a heavily marked assemblage. Trampled bone is commonly heavily marked, bearing multiple linear grooves (see Fig. 3a, b in Andrews and Cook, 1985:681); Andrews and Cook report 510 marks on 11 bones, or an average of 46 marks per bone. In contrast, bones that have been genuinely processed by hominids rarely show such extensive marking; in my experience, I would estimate that most bones show 1-6 individual cutmarks. The only exception which I have seen occurs when an attempt has been made to strip meat from areas of extensive muscle attachment, or remove tendon or periosteum, activities that leave abundant cutmarks, especially scraping marks. Certainly a heavily-marked assemblage bearing many heavily-marked individual bones is more likely to have been trampled or exposed to other forms of sedimentary abrasion than butchered or processed by hominids, all other things being equal.

Second, the most commonly mimicked type of cutmark is a scraping mark (Potts and Shipman, 1981:577),

> ...formed by drawing an edge across a bone surface in a direction roughly perpendicular to the long axis of the edge. This action results in multiple, fine, parallel striations across a broad area of bone rather than confined to a single, elongated main groove....

I would now add to this description that the broad area bearing striations is depressed below the unaltered surface of the bone. In addition, some scraping marks show perpendicular chattermarks that have not been observed in trampled bones (Olsen and Shipman, in press). This new criterion should help distinguish scraping marks from those produced by many small, individual particles of sand or rock which may leave parallel striations but which generally will not remove surface bone over a broad area. Caution must be observed, however, and assemblages with many scraping marks must be closely inspected to determine whether those marks are cutmarks (produced by tools) or effects of various types of sedimentary abrasion.

Third, sedimentary context is an important factor and more caution must be exercised in interpreting assemblages found in rocky or sandy substrates than those found in fine sediments. My own experimental data suggest that fine, silty substrates themselves do not produce cutmark mimics. This conclusion is supported by the famous <u>Homo erectus</u> boy's skeleton from Nariokotome, west of Lake Turkana, in Kenya. That skeleton was recovered from fine, silty, lacustrine sediments and was almost certainly trampled:

> The top of the fossiliferous horizon shows many signs of bioturbation [including hippo footprints] and several of the bones were found broken or lying in positions suggesting they had been trampled by large mammals. (Brown et al., 1985:788).

Nonetheless, none of the bones show any signs of sedimentary abrasion from trampling (Walker, pers. comm.), suggesting that the fine sediments perhaps could not produce obvious linear grooves.

It is difficult to know the appropriate treatment of bones found in fine substrates accompanied by stone tools. The tools themselves may or may not have been a part of the substrate at the time of any hypothesized trampling. Perhaps the best approach is to consider as much evidence as is available, including information about the weathering stage and presence or absence of rodent gnawing marks that might indicate how rapidly the bones became buried and were thus removed from the dangers of trampling.

Fourth, the orientation and location of marks may reveal whether they are due to trampling or hominid activities. It has been observed by several researchers that trampling marks occur more frequently on shafts than near joints. Unfortunately, my suggestion that a predominance of near-joint cutmarks indicates hominid butchery was based on analysis of an assemblage, Prolonged Drift, with a highly biased representation of skeletal elements. Most of the cutmarks on the Prolonged Drift assemblage are near-joint, but so are most of the bones preserved (Gifford et al., 1981). In fact, the identifiable elements from this site are remarkable in having very few and very short shaft fragments attached to the articular ends. A less biased assemblage, Ngamuriak, that was also certainly butchered, does not show a strong predominance of near-joint cutmarks (Marshall, 1986a, b; Shipman, 1986b). Thus it seems likely that hominids, carnivores, and trampling all produce many midshaft marks, although the frequencies may still be statistically different. In terms of orientation, Andrews and Cook (1985:682-684) have shown that trampling marks are often transverse to the long axis of long bones and oblique to the scapular margins. Of course, hominids might produce similarly placed and oriented marks, depending on their intent and skill. But, in circumstances like those in Bed I, Olduvai, where I have documented that more than 70% of the cutmarks on long bones are oriented oblique to the bone's long axis, it can be concluded that these are relatively unlikely to have been caused by trampling. Further, the frequency of cutmarks at these Bed I sites is significantly correlated with the number of flakes, the tool type most likely to have left cutmarks (Shipman, 1986b). This finding strengthens the interpretation of these marks as cutmarks, since there is no readily discernable reason why the frequency of trampling marks should be correlated with the frequency of any particular tool type.

Another possible source of cutmark mimics was presented at the SEM Conference by Friederman Schrenk, who showed hyena tooth scratches that included fine linear striations within the main groove. Since he and I have each looked at bones chewed by spotted and striped hyenas, and since he has seen striations frequently and I have never observed them, this problem obviously calls for a collaborative effort. We hope to discover distinctive features of his striated toothmarks that can be used to identify them reliably. It is my subjective impression, having seen only a few micrographs briefly, that they are not close cutmark mimics. I can offer at present two hypotheses about how such striated toothmarks are produced. First, perhaps these marks were caused by repeated gnawing in the same spot, similar to rodent gnawing which frequently produces marks that show fine striations (see Shipman and Rose, 1983). If this hypothesis is correct, then (like rodent gnawing marks), the striations ought to be raised, not incised. Second, perhaps the hyena producing these marks had broken off a tooth, leaving a rough edge that would produce such striations. Otherwise, I cannot see how the normal tooth morphology of hyenas could produce such an effect without repeated gnawing. Tooth breakage must be a relatively infrequent occurrence, considering the number of times a wild carnivore chews and does not break its teeth.

Although the foregoing paragraphs have focused on the growing complexities and cautions of identifying cutmarks, the potential usefulness of data about which bones were actually processed by hominids is so great that I believe it is worth persisting in trying to refine the methodology to enhance its certainty. There are several significant types of analyses that can be performed, once a sample of cutmarked bones has been isolated.

First, and most obviously, the presence of cutmarked bones provides a causal and temporal link between bones, stone tools, and hominids. Many sites have complex taphonomic histories, so it is necessary to find solid evidence for the interpretation that spatially associated bones and stone tools were deposited at the same time by hominids.

Second, the patterns of skeletal representation (which bones are cutmarked?) and species representation (which species are cutmarked?) may reveal a great deal. I would hesitate to jump from the presence of cutmarks on some bones in an assemblage to the conclusion that all of the bones in the assemblage were collected or processed by hominids. Nonetheless, analyzing only the cutmarked subset of an assemblage will enable the analyst to answer questions like: did systematic butchery, i.e., a highly patterned, repetitive processing of carcasses, occur? are cutmarks concentrated on bones that bear large muscle masses, suggesting meat was the focus of the processing? or do cutmarks occur in locations suggestive of skin, tendon or fat harvesting? are butchered species restricted to some size range or some habitat type? does the mortality profile of the butchered species look like a catastrophic or attritional assemblage?

Third, additional information may be derived from careful inspection of the cutmarks themselves. Bromage and Boyde (1984) have shown that directionality can sometimes be inferred from cutmark morphology. This information could potentially be used to determine when handedness and the accompanying lateralization of the brain occurred (Shipman and Rose, 1983).

Time of processing. Another aspect of studying cutmarks themselves may prove highly useful; cutmark morphology can reveal the timing of the processing of a carcass relative to the death of the animal involved. In a collaborative effort (Russell et al., 1985), Russell, Villa and I have shown that we can distinguish, on the basis of microscopic morphology of the marks, between immediate and delayed processing of carcasses. By immediate processing we mean processing of a carcass which occurs at or shortly after death, when the flesh is still a viable source of food. In contrast, delayed processing occurs later in time, after the flesh is dried or decayed. It seems likely that the time period intervening between immediate and delayed processing will be more difficult to recognize. Table 1 shows the types of activities that might be expected to occur in each type of processing; it is important that the objectives are different in immediate and delayed processing because different resources have different temporal lives. It is also significant whether the subject of the processing is human or nonhuman, since the former raises the prospect of either cannibalism (immediate processing) or secondary burial (delayed processing), both of which have interesting implications for understanding the system of beliefs of a group.

This approach has been applied to the Neolithic assemblage from Fontbregoua, France, in which both human and nonhuman remains show cutmarks in similar places (see Villa et al., 1986, for a more extensive discussion). This work constituted a test of the hypothesis that dietary cannibalism (i.e., eating human remains) was involved; by establishing that both the human and faunal materials were subjected to immediate processing, the hypothesis was supported.

Methodologically, the problem of distinguishing between immediate and delayed processing depends upon two factors: the age (state of weathering) of the bone surface and the age of the cutmark. Figure 1 shows four theoretical combinations of bone age and mark age, although one combination (fresh bone, old mark) cannot actually exist. In our experiments, Russell independently prepared a series of replicas of slicing marks on bones (N=8 specimens) that represented three states: fresh marks on fresh bone; old marks on old bone; and fresh marks on old bone. One specimen was an aged bone bearing old marks over which Russell created a new generation of marks. These replicas were assigned letter codes and sent to me without further data. My diagnoses of these specimens was 100% accurate. Although this is a small sample size, the results are modestly impressive. Since each replica could have borne any of three states, the probability of identifying all of the specimens accurately by chance were 6561:1. Some additional studies have been carried out since then, but more are still needed to explore the range of variability in mark morphology and to confirm these initial results.

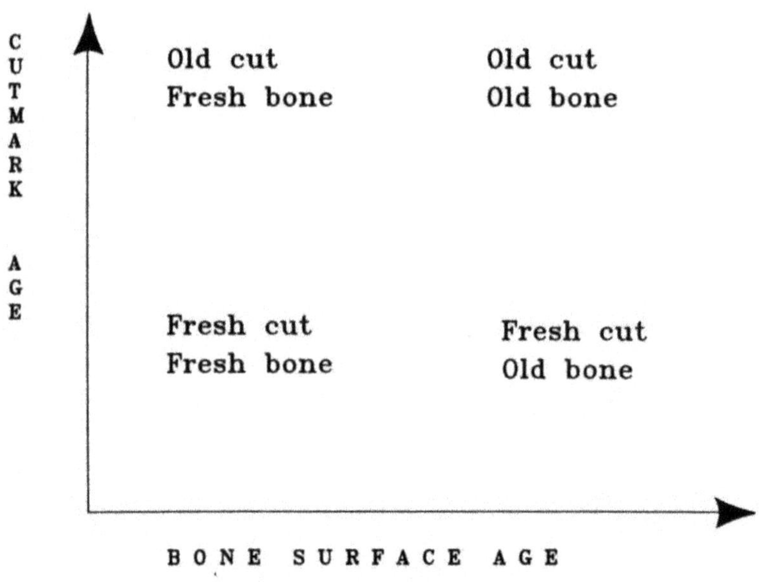

Fig. 1. Theoretical model of the different combinations of cutmark and bone age resulting from carcass-processing. Note that the alternative in the upper left (old cut, fresh bone) cannot actually occur.

TABLE 1. Some of the behaviors and objectives associated with immediate and delayed carcass-processing.

Subject	Timing	Behavior	Objective
Animal	Immediate	Hunting	Obtain meat, marrow, fat
		Early scavenging	Obtain meat, marrow, fat
		Manufacture	Obtain skin, tendon, bone
		Consumption	Obtain meat, marrow, fat
	Delayed	Late scavenging	Obtain marrow
		Marrowing	Obtain marrow
		?Manufacture	Obtain skin, tendon, bone
Human	Immediate	Ritual cannibalism	Appease gods
		Dietary cannibalism	Obtain meat, marrow, fat
		Scalping	Obtain trophy
		Mutilation	Obtain trophy, vengeance
		Sacrifice	Appease gods
	Delayed	Secondary burial	Finish burial ritual
		Cemetery relocation	Facilitate transport

At present, the following descriptions serve to characterize the two states. Immediate processing is characterized by fresh, intact bone surfaces with normal ultrastructure bearing slicing marks. These marks are elongate, linear grooves with multiple fine, parallel striations cut into their walls and floor. Often, a single nadir is distinguishable in such slicing marks. Bone smears, oblique faulting and oblique chipping (Bromage and Boyde, 1984) are often visible. Fresh cutmarks show much fine detail that is distinctly different from normal bone texture (Fig. 2a, b).

When a bone subjected to immediate processing becomes weathered and aged, these features alter. The bone surface develops cracks along lines of structural weakness. Since the cutmark itself has disrupted the normal structural organization of the bone, it serves as a focus for crack development. It is possible that the immediate area of the

<u>Fig. 2.</u> Experimental studies of delayed and immediate processing. a) Fresh cutmarks on a fresh bone with an intact surface. b) A close-up of the same slicing marks, showing the fine striations, bone smears and oblique chipping characteristic of fresh slicing marks. c) A series of intersecting, old slicing marks, originally made on fresh bone that is now weathered and old. Note the network of weathering cracks both within and without the cutmarks. d) A close-up of the same set of marks. Note the loss of detail within the cutmarks and the development of fine longitudinal and perpendicular cracks within each mark.

cutmarks develops cracks earlier than adjacent areas, though this is not yet certain. Cracks frequently develop along the nadir of the slicing mark; in addition, a series of short cracks may develop along the walls of the mark. These cracks often have two intersecting arms, one parallel to and one perpendicular to the mark's long axis. In our sample, such cracks do not extend outside the mark itself, although larger scale cracks parallel to the predominant fiber orientation of the bone surface may intersect and cross the slicing mark. Although the fine striations remain visible inside the main groove, they become much more difficult to identify as they have lost both detail and depth. As a result, they are often visible only when the specimen is highly tilted (angles of close to 45° to the surface) and with a high contrast image (Fig. 2c, d).

Delayed processing presents a different set of features. Typically, in delayed processing the bone surface is weathered and cracked. Work by Behrensmeyer (1978) and Gifford (1977) has shown that weathering begins at about 6 months to one year after death, depending upon microenvironment and climate. However, because the slicing mark itself is made after weathering and cracking has occurred, the mark appears fresh. It possesses all the detail, depth, and highly visible striations found in fresh marks on fresh bone. Thus, it is the difference in weathering state between the outermost bone surface and the cutmark that identifies delayed processing. Cutmarks resulting from delayed processing have no regular relationship to crack orientation; apparently cracks have not yet had time to develop within fresh marks. The distinction between an old mark and a new mark on the same weathered bone (Fig. 3a, b) is striking.

Burnt bones and teeth. The summary of this work presented here is drawn from Shipman et al. (1984b), which should be consulted for more extensive information about methodology and results. Based on controlled heating in a muffle furnace, manually cleaned sheep and goat mandibles (bearing teeth) and astragali show a series of changes in color, microscopic morphology, and crystal structure. Documenting changes permits the analyst to deduce the range of temperatures to which the skeletal element itself has been heated. It is important to realize that the maximum temperature reached by the bone or tooth represents the minimum temperature of the heating device (fire, pyre, heating oven, etc.). In other words, the heating device was at least as hot as the bones or tooth heated in that device. However, various circumstances, such as the extent of fleshy covering of the bones, may protect them from reaching as high a temperature as the heating device itself reaches.

The changes in microscopic morphology that accompany heating to different temperatures can be divided into five stages, whether the tissue monitored is bone, dentin or enamel. All three tissues show analogous changes in morphology that we have attributed to, first, the breakdown of the nonmineral component of bone or tooth and, second, to changes in the hydroxyapatite component of those tissues. Figures 4-6 (reproduced from Shipman et al., 1984b:315-318) show these stages, which can be described in general terms as follows:

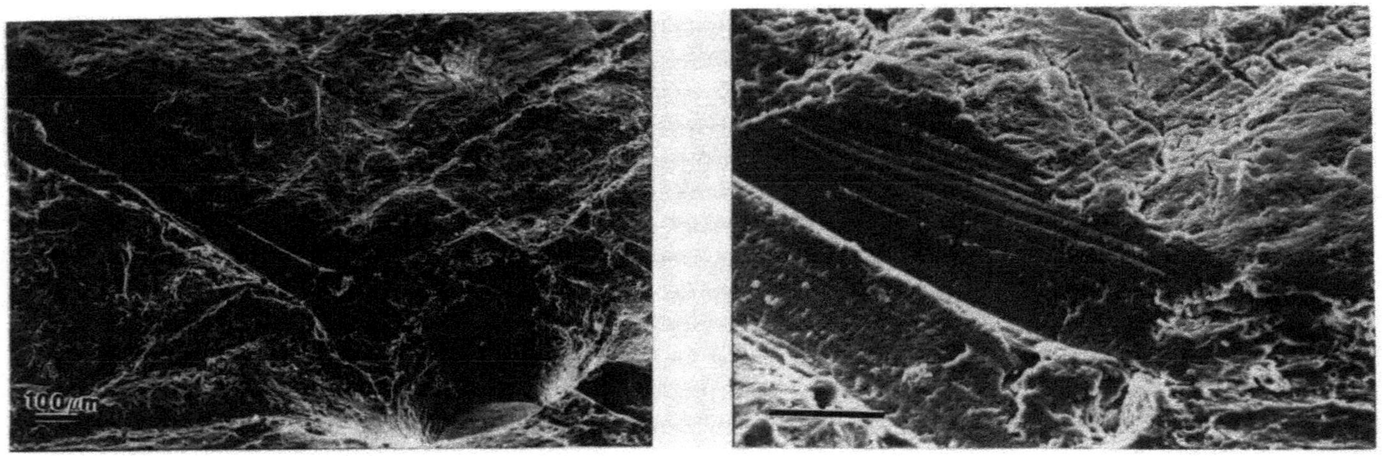

a b

<u>Fig. 3.</u> Comparison of fresh and old cutmarks on old bone. a) A newly-created slicing mark (upper left to vascular canal lower right) intersects and overlies an old slicing mark (upper right to lower left) made prior to the weathering of the bone. b) A close-up of the area of intersection reveals the lack of cracking and the greater detail in the fresh cutmark. The old cutmark, in contrast, is heavily cracked and has poorly visible fine striations.

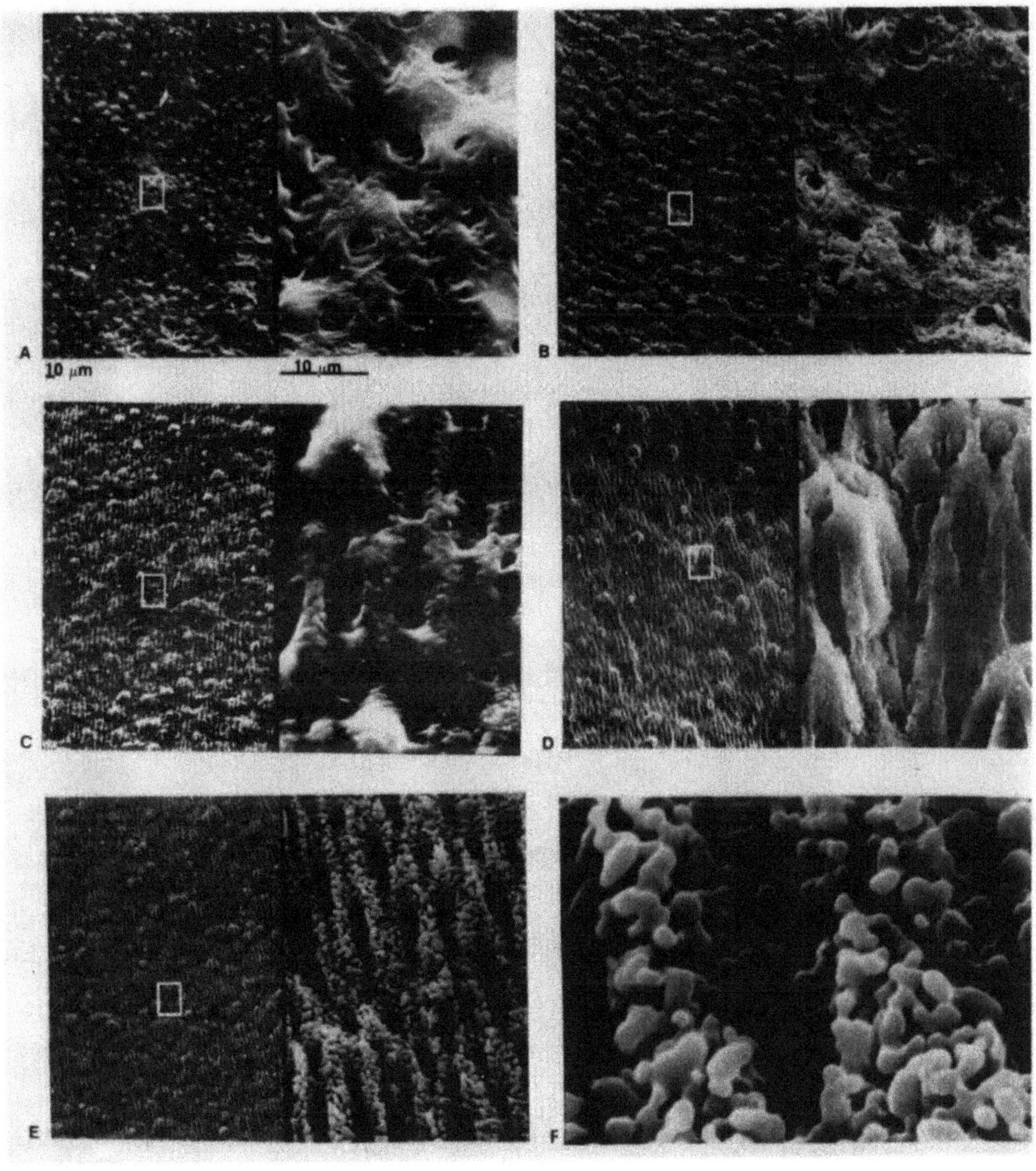

Fig. 4. Microscopic morphological stages produced by heating on a natural dentin surface of the pulp cavity. a) Stage I, normal dentin. b) Stage II. c) Stage III. d) Stage IV. e) Stage V. f) Stage V, shown at higher magnification. (Plate reproduced by permission from Shipman et al., 1984b, J. Arch. Sci., Academic Press.)

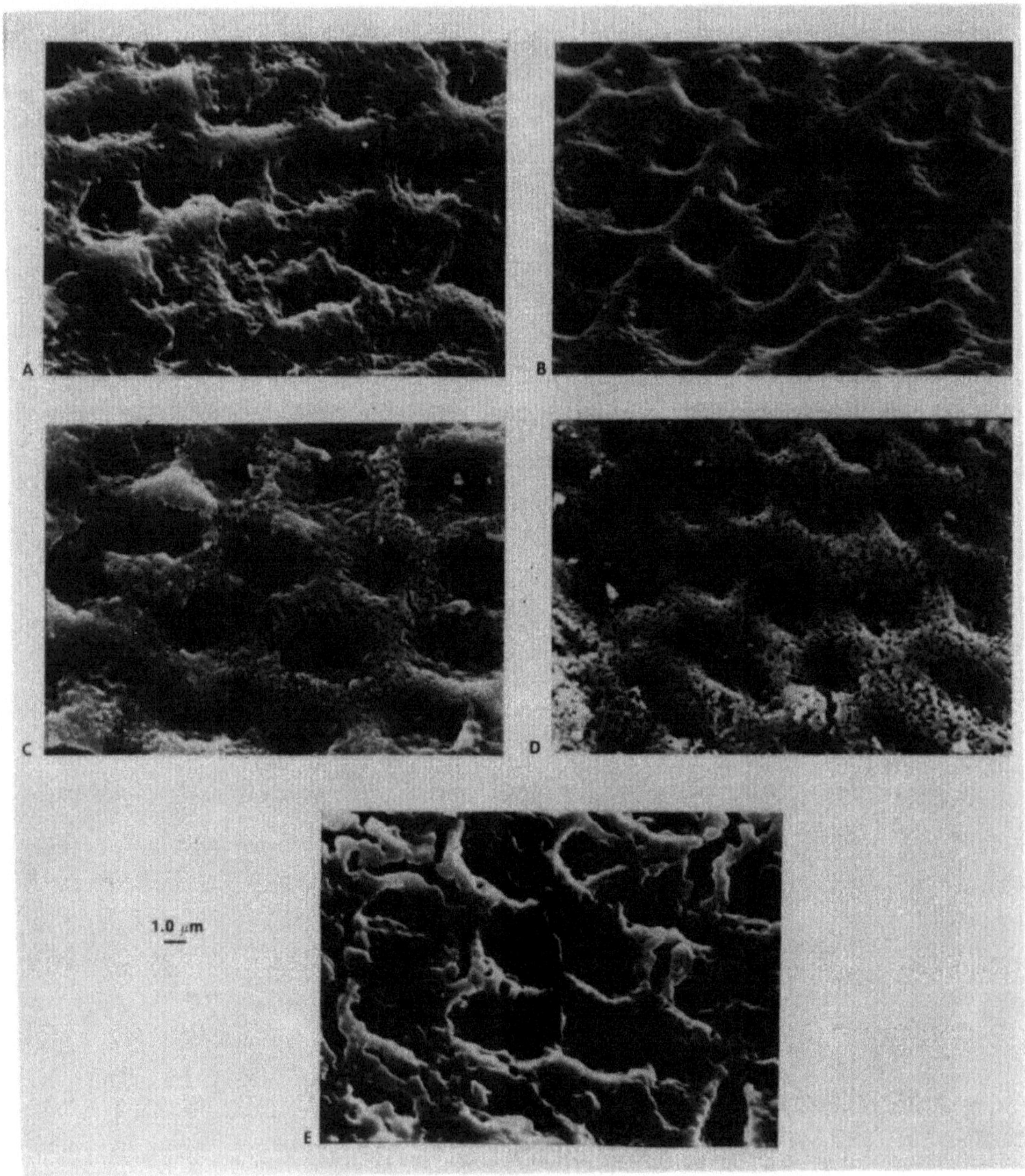

Fig. 5. Microscopic morphological stages produced by heating on natural, nonocclusal, enamel surfaces. a) Stage I, normal enamel. b) Stage II. c) Stage III. d) Stage IV. e) Stage V. (Plate reproduced by permission from Shipman et al., 1984b, J. Arch. Sci., Academic Press.)

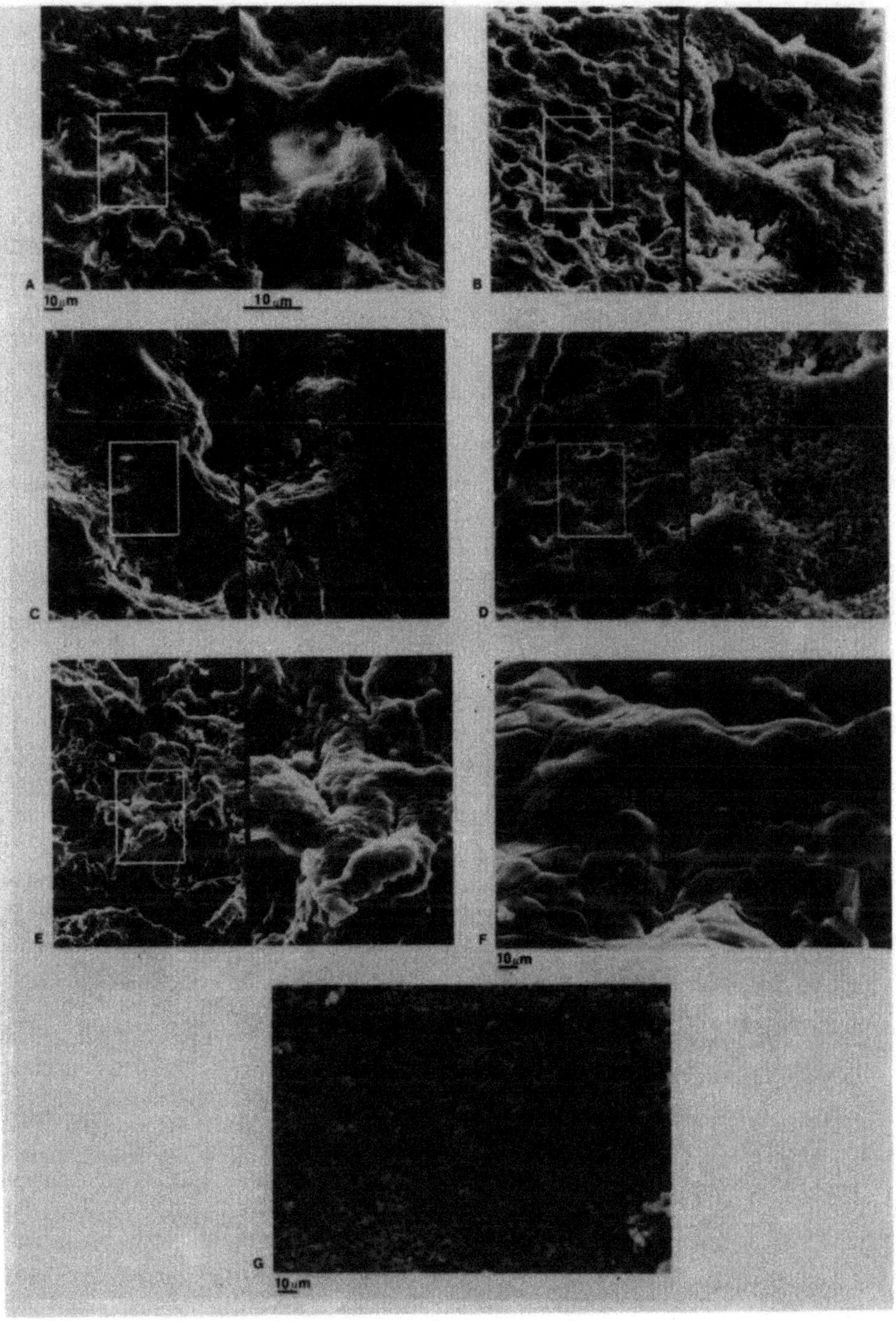

Fig. 6. Microscopic morphological stages produced by heating on intact bone surfaces. a) Stage I, normal subchondral bone. b) Stage II, subchondral bone. c) Stage III, subchondral bone. d) Stage IV, subchondral bone. e) Stage V, subchondral bone. f) Stage V, subchondral bone shown at higher magnification. g) Stage V, cortical bone. (Plate reproduced by permission from Shipman et al., 1984b, J. Arch. Sci., Academic Press.)

> ...in Stage I (20-<185°C), tissues are normal; in Stage II (185-<285°C), these tissues show increased surface roughness; in Stage III (285-<440°C), they become glassy and very smooth; in Stage IV (440-<800°C), they acquire a frothy or fleecy appearance; and in Stage V (800-940°C), these frothy protuberances coalesce into smooth-surfaces globules or nodules. (Shipman et al., 1984b:314-315).

Additionally, in bone or tooth heated to 645°C or more, which is in the morphological Stage IV or V, the crystal size of the hydroxyapatite has become larger, as judged from powder X-ray diffraction studies of the same samples that the microscopy was performed on. Although a similar change in crystal size has been observed in unburnt fossil bone, presumably as a result of some diagenetic changes, unburnt fossil material has not been observed to undergo the same morphological changes. The combination of morphology, crystal structure, and color can therefore be used to estimate the temperature to which a given skeletal element was heated. Although all three approaches should be used wherever possible, morphology is likely to be more reliable than crystal structure because of the possibility of diagenetic changes. Either morphology or crystal structure will be more reliable than color unless the specimen has been in a protected environment.

The primary benefit of knowing whether a bone or tooth has been heated lies in establishing the probability that the heating device was controlled by hominids. Figure 7 summarizes some of the literature about different types of fires or heating devices and the effects on skeletal elements. It is apparent that natural grass or brush fires do not produce sufficiently high temperatures for sufficiently long duration to heat skeletal elements to high temperatures. In contrast, natural forest fires may produce high temperatures of long duration. Thus, if other evidence suggests that a forest fire was highly unlikely and the skeletal elements show signs of being heated to high temperatures, manmade fires are a likely explanation. Boiling does not produce sufficiently high temperatures to be detected using this approach, nor would brief singeing of meat-covered bone. I have found no literature on such phenomena as guano fires in caves.

<u>Bone tools</u>. The topic of the utilization of bone as a raw material for making tools is a controversial one. On the one hand, there are Upper Palaeolithic bone tools and utilized bones that are so strikingly modified as to leave no doubt about the fact that these bones were altered and utilized, in some sense, by hominids. On the other hand, there are repeated claims for the existence of minimally modified but utilized Lower Palaeolithic bone tools - claims often received with a good deal of skepticism and vehement rebuttal. Among the best known claims are those of Dart (e.g., 1949, 1957) for the osteodontokeratic culture of the australopithecines. The acceptance, or even thoughtful evaluation, of his claims has been hindered by his flamboyant, evangelical style of writing. Now, many years after Dart's initial writings about the osteodontokeratic culture, the issue is finally receiving serious consideration (e.g., Schrenk, this volume). A more

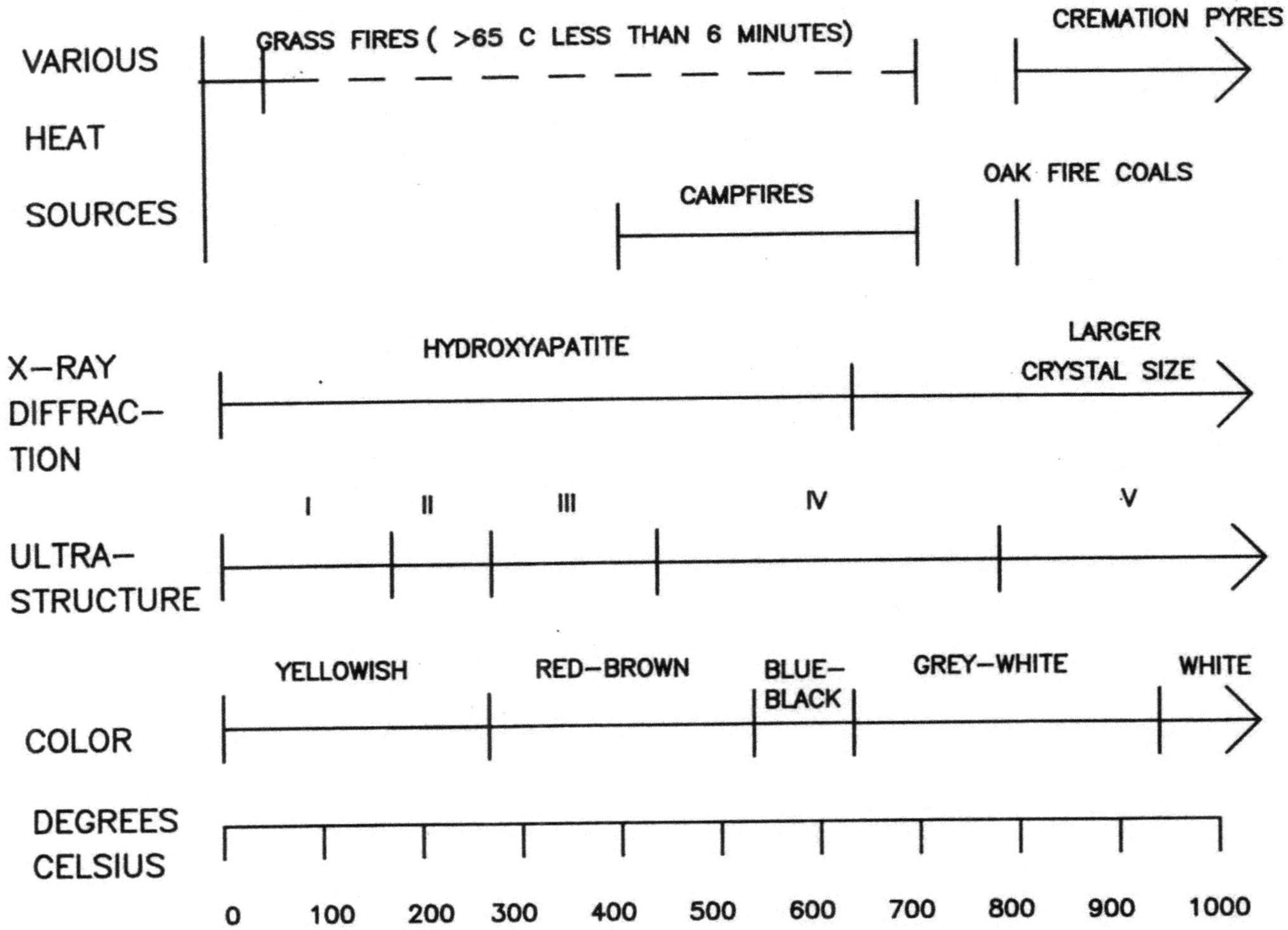

Fig. 7. Summary of temperatures produced by different heating sources and their effects (see Shipman et al., 1984b, for details).

cautious and conservative claim for utilized or modified bones has been made by Leakey (1971) about a small number of specimens from Beds I and II, Olduvai Gorge.

A set of experimental, actualistic studies were undertaken to determine whether utilized bones could be distinguished from unutilized ones on the basis of wear; some of the main conclusions from these studies have been referred to elsewhere (Shipman et al., 1984a; Shipman, in press), but the fullest description is in Chapter 21 of this volume (Shipman and Rose, this volume). Experiments were made using unmodified or simply fractured modern bones in processing soft substances, such as hides, meat, and soft vegetables, and in tasks where both hard and soft substances might be encountered, such as digging, butchery, and bark-working.

Ethnographic tools were also examined, to enlarge the sample to include heavily used and pre-ground bones; uses recorded for these implements included use as an ice pick, bark-scraper, hide-scraper, and tool for removing feathers from bird skins. Finally, a sample of eleven replicas of edges of fractured bones, purported to have been utilized, was prepared by Dan Fisher and sent to me without further information. I was not aware that some of the specimens were fractured but not used. All were identified correctly in general terms; diagnoses took the form of broad categories that referred to the duration of use (none, slight, moderate, extensive) and the type of substance (soft; mixed hard and soft; and hard). Again, the sample size is small but the probability of making these diagnoses correctly by chance are small. For duration of use, the probability is 4,194,304:1 of diagnosing all eleven specimens correctly by chance; for the substance worked, the probability is 177,147:1. It is unlikely that a larger sample size, including a wider range of uses, could be consistently diagnosed with 100% accuracy, but this blind test has considerably strengthened my confidence that utilized bones can be recognized based on their microscopic morphology. Figure 8a, b shows the same fracture surface, before and after utilization, to illustrate the general effects of utilization; Figure 8c, d contrast extensive utilization with extensive abrasion.

The weakest point to date in using the SEM to identify utilized bones has been the exploration of possible mimics. Thus far, I have examined the effects of sedimentary abrasion on broken bone surfaces, using both windborne and waterborne sediments varying in size from Mt. St. Helen's volcanic ash to pea-sized gravel. In most experiments, sedimentary abrasion produced rounding and smoothing of broken surfaces that could be distinguished from utilization by its overall pitted or pebbled texture (Fig. 8d). Under some circumstances, however, waterborne or windborne abrasion produces close mimics of utilization wear; these are discussed more fully in Chapter 21 (Shipman and Rose, this volume; see also Shipman, in press). In brief, it can be said here that extreme abrasion (200 hours in a rock tumbler) of broken bones with a slurry of water and fine sediment (loess) produced marked smoothing and rounding of the fracture surfaces and relatively little change on the intact bone surfaces (see Fig. 12 in Shipman and Rose, Chapter 21, this volume). Thus, such specimens show differential

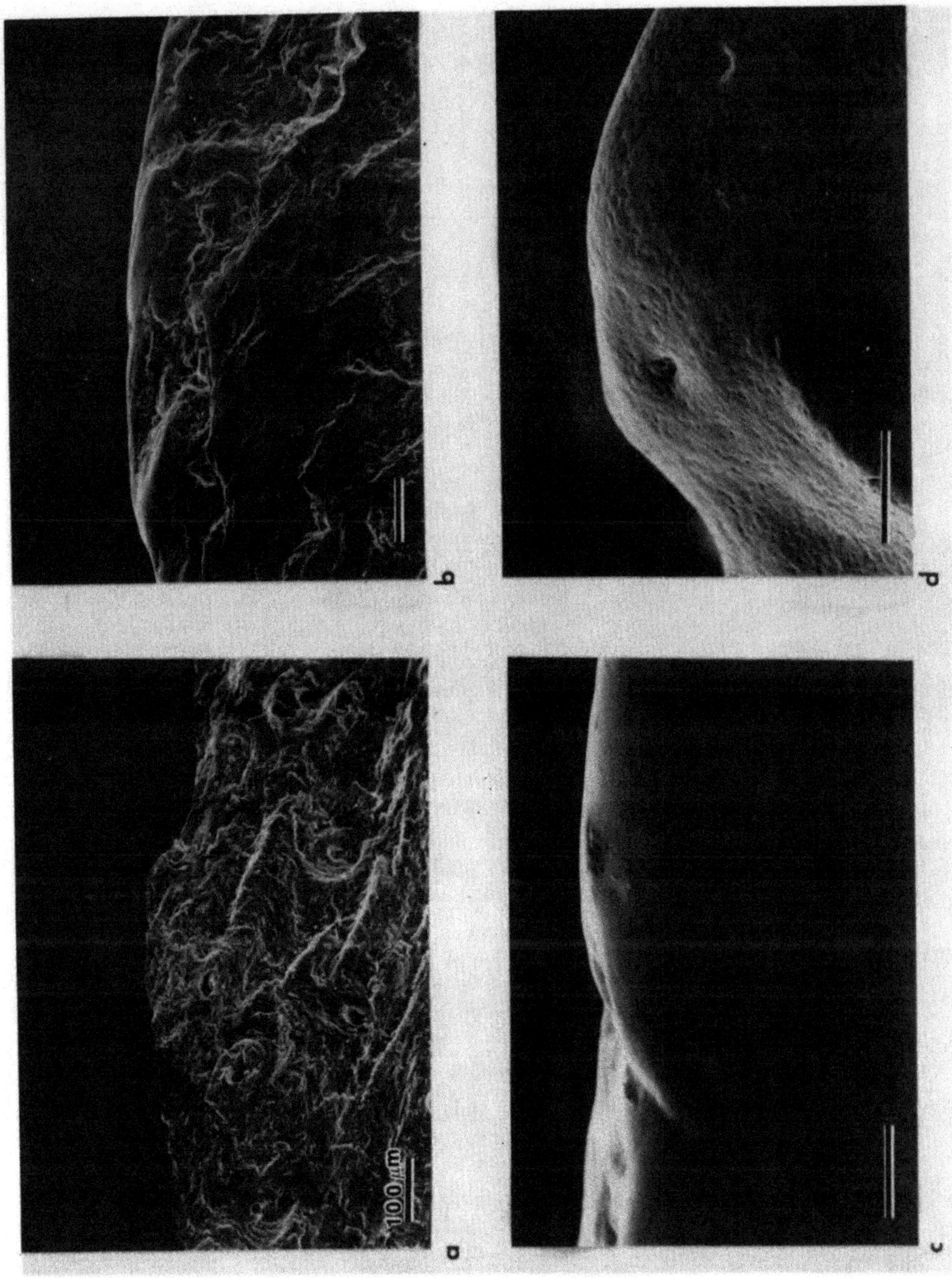

Fig. 8. Microscopic effects of different taphonomic events on fracture surfaces of bones. a) An unaltered fracture surface, showing typical rough texture and anatomical detail. b) The same surface shown in (a), after brief utilization on meat. Note the rounding and smoothing of the working edge and the loss of anatomical detail in the raised areas. c) An ethnographic bone tool, used extensively in hide-working, shows a fine, glassy polish that obliterates most anatomical detail. d) An experimentally-fractured elephant bone, abraded with air-driven loess for approximately 5 hours. Although the entire surface is rounded and smoothed, as in bone tools used extensively on soft substances, the overall pebbly texture is unlike the effects of extensive utilization shown in (c). This bone differs from briefly-utilized bone tools, as in (b), because abrasion alters the entire exposed surface rather than the immediate area of the working edge itself.

"wear" which mimics the effects of brief utilization on soft substances (compare Fig. 8b in this chapter with Fig. 12d in Chapter 21). Therefore, attention must be paid to the sedimentary context in which possible bone tools are found. Also, since the areas of bone that first show the abrasive effects of windborne or waterborne sediments are those where thin cortical bone overlies cancellous bone, special attention must be paid to such areas on possible bone tools showing what appears to be brief utilization on soft substances. Neither the abrasion experiments nor the actualistic studies of bone tools produced the extreme, glassy, rounding and polish observed on ethnographic bone tools.

Preliminary conclusions drawn from an SEM study of the purported bone tools from Olduvai are that many show microscopic features characteristic of utilization wear from either soft or mixed hard and soft substances (Shipman, in press). Further, there is a tendency for the bone tools that are independently judged to be larger, heavier objects to show mixed wear and for smaller, lighter objects to show soft substance wear. This is an intuitively satisfying observation, although it does not constitute proof that these objects were actually utilized. Analysis of the specimens identified as bone tools on microscopic criteria reveals a strongly biased sample of the total faunal assemblage; both large animals and skeletal elements with thick cortical bone are selected disproportionately often (Kolmogorov-Smirnov $Z = 3$, $p < .001$ on body size; chi-squared = 33.13, 6 d.f., $p < .001$ on skeletal element). These findings strongly suggest that specific bones of particular types of animals were being intentionally selected for use as tools. The question remains: what was being done with these tools? The actualistic studies performed to date have shown vividly what a poor raw material bone is for making a sharp cutting object; bone can neither be shaped to a very sharp edge nor does it hold its edge well through usage. Presumably, then, bone as a material would not be used in preference to stone or even in place of stone in such tasks unless no other alternative was available. Future experiments will focus on three points: 1) attempts to discover tasks for which bone is a good material; 2) attempts to refine further the categories of use that can be recognized; and 3) attempts to explore the range of possible mimics more fully.

Conclusions

Studies summarized here show a wide range of potential uses of the SEM in archaeological studies as well as a growing complexity in applying those uses appropriately. As I see it, we have grown out of the stage of simply describing observations made with this new (to archaeology) tool and into the more difficult stage of trying to define precise, useful, diagnostic criteria aimed at the recognition of important events. The transition from the infancy of this methodology into its childhood or even adolescence will be greatly eased by two factors. The first is resisting the temptation to throw the baby out with the bathwater. It is naive and even detrimental to insist that all of our first conclusions, drawn from our initial forays into the world of microscopy, either remain true or be abandoned as unprofitable lines of research. Science in general and archaeology in particular is full of awkward, peculiar truths and we must be prepared to wrestle with them. The SEM will provide no simple or magic answers. The second is stating and understanding exactly what our conclusions and applications mean, that is, which points are evidence, which hypothesis, and which deduction or elaboration. Without clarity in our writing and reporting, the greater complexity will give rise to greater confusion, not greater resolution.

Acknowledgments

Parts of this work were funded by the National Science Foundation (BNS 80-1397 and 80-2-1397), the Wenner-Gren Foundation, the Leakey Foundation, F.R.O.M., the Boise Fund, and the National Geographic Society. Fossils were studied by kind permission of the Government of Tanzania and the Office of the President of Kenya (permit #OP.13/001/6C79), with the help and assistance of the staff and director of the National Museums of Kenya. I am grateful to all.

References

Andrews, P. and Cook, J. 1985. Natural modifications to bones in a temperate setting. Man (N.S.) 20: 675-691.

Behrensmeyer, A.K. 1978. Taphonomic and ecologic information from bone weathering. Paleobiology 4: 2: 150-162.

Behrensmeyer, A.K., Gordon, K. and Yanagi, G. 1986. Trampling as a cause of bone surface damage and pseudo-cutmarks. Nature 319: 768-771.

Bunn, H. 1982. Meat-eating and human evolution. Ph.D. thesis, University of California at Berkeley.

Binford, L.R. 1981. Bones: Ancient Men and Modern Myths. San Francisco: Academic Press.

Bromage, T. and Boyde, A. 1984. Microscopic criteria for the determination of directionality in cutmarks on bone. American Journal of Physical Anthropology 65: 4: 359-366.

Brown, F., Harris, J., Leakey, R. and Walker, A. 1985. Early Homo erectus skeleton from west Lake Turkana, Kenya. Nature 316: 788-792.

Dart, R.A. 1949. The predatory implemental technique of Australopithecus. American Journal of Physical Anthropology 7: 1-38.

Dart, R.A. 1957. The Osteodontokeratic Culture of Australopithecus prometheus. Transvaal Museum Memoirs, No. 10.

Fiorillo, A.R. 1984. An introduction to the identification of trample [sic] marks. In First International Conference on Bone Modification: Abstracts, p. 11. Orono, Maine: Center for the Study of Early Man.

Gifford, D.P. 1977. Observations of modern human settlements as an aid to archaeological interpretation. Unpublished Ph.D. thesis, University of California at Berkeley.

Gifford, D.P., Isaac, G.L. and Nelson, C.M. 1981. Evidence for predation and pastoralism at Prolonged Drift: a pastoral Neolithic site in Kenya. Azania 15: 57-108.

Marshall, F. 1986a. Aspects of the advent of pastoral economies in East Africa. Unpublished Ph.D. thesis, University of California at Berkeley.

Marshall, F. 1986b. Implications of bone modification in a Neolithic faunal assemblage for the study of early hominid butchery and subsistence practices. Journal of Human Evolution 15(8):661-672.

Leakey, M.D. 1971. Olduvai Gorge, Volume III, Excavations in Beds I and II, 1960-1963. London: Cambridge University Press.

Oliver, J.S. 1984. Bone damage morphologies from Shield Trap Cave, Carbon Country, Montana. In First International Conference on Bone Modification: Abstracts, p. 27. Orono, Maine: Center for the Study of Early Man.

Olsen, S.L., and Shipman, P. In press. Surface modification on bone: Trampling versus butchery. Journal of Archaeological Science.

Potts, R. and Shipman, P. 1981. Cutmarks made by stone tools on bones from Olduvai Gorge, Tanzania. Nature 291: 577-580.

Russell, M.D., Shipman, P. and Villa, P. 1985. Cutmarks: immediate versus delayed carcass processing. American Journal of Physical Anthropology 66: 223.

Shipman, P. 1981. Applications of scanning electron microscopy to taphonomic problems. In Cantwell, A-M, Griffin, J.B. and Rothschild, N. (eds.), The Research Potential of Anthropological Museum Collections, Annals of the New York Academy of Sciences 276: 357-385.

Shipman, P. 1981b. Life History of a Fossil: An Introduction to Taphonomy and Paleoecology. Cambridge: Harvard University Press.

Shipman, P. 1983. Early hominid lifestyle: hunting and gathering or foraging and scavenging? In Clutton-Brock, J. and Grigson, C. (eds.), Animals and Archaeology, Volume 1, Hunters and their Prey, pp. 31-50. Oxford: British Archaeological Reports.

Shipman, P. 1986a. Scavenging or hunting in early hominids: Theoretical framework and tests. American Anthropologist 88: 1: 27-43.

Shipman, P. 1986b. Studies of faunal-hominid interactions at Olduvai Gorge. Journal of Human Evolution 15 (8):681-706.

Shipman, P. In press. Altered bones from Olduvai Gorge, Tanzania: Techniques, problems, and implications of their recognition. In Bonnichsen, R. and Sorg, M. (eds.), Bone Modifications. Orono, Maine: Center for the Study of Early Man.

Shipman, P., Fisher, D.C. and Rose, J.J. 1984. Mastodon butchery: microscopic evidence of carcass processing and bone tool use. Paleobiology 10 (3): 358-365.

Shipman, P., Foster, G. and Schoeninger, M. 1984. Burnt bones and teeth: An experimental study of color, morphology, crystal structure and shrinkage. Journal of Archaeological Science 11: 307-325.

Shipman, P. and Rose, J.J. 1983. Early hominid hunting, butchering and carcass-processing behaviors: Approaches to the fossil record. Journal of Anthropological Archaeology 2: 1: 57-98.

Shipman, P. and Rose, J.J. 1984. Cutmark mimics on modern and fossil bovid bones. Current Anthropology 25: 1: 116-117.

Shipman, P. and Rose, J.J. 1988. Bone tools: An experimental approach. In Olsen, S.L. (ed.) Scanning Electron Microscopy in Archaeology. Oxford: British Archaeological Reports, International Series.

Simpson, G.G. 1970. Uniformitarianism: an inquiry into principle, theory and method in geohistory and biohistory. In Hecht, M.K. and Steere, W.C. (eds.), Essays in Evolution and Genetics, pp. 43-96. New York: Appleton-Century Croft.

Villa, P., Bouville, C., Courtin, J., Helmer, D., Mahieu, E., Shipman, P., Belluomini, G. and Branca, M. 1986. Cannibalism in the Neolithic. Science 233:431-437.

Yellen, J.E. 1977. Archaeological Approaches to the Present: Models for Reconstructing the Past. New York: Academic Press.

ACTUALISTIC SEM STUDIES ON THE MAKAPANSGAT LIMEWORKS GREY BRECCIA BONE ASSEMBLAGE, TRANSVAAL, SOUTH AFRICA

Friedemann Schrenk and Judy M. Maguire

Summary

The Makapansgat Limeworks Hominid Site originally contained an enormous accumulation of bone fragments (mostly bovid) of Plio-Pleistocene age. Dart (1957) proposed that the hominid, *Australopithecus africanus*, practised an Osteodontokeratic Culture, interpreting many of the bone fragments as tools. However, a more holistic approach to taphonomy at the site seems to be required. Actualistic comparative investigations, including SEM work, are being undertaken to clarify the history of the bone breccia. A theoretical flow scheme includes stages and related processes in a bone's history from living animal to fossil museum specimen. Today we are left with the taphonomic effects, which have resulted from biological as well as non-biological agents. The potential kinds of microscopic damage to bone surfaces of the Makapansgat material are listed, all of which will receive further SEM studies. Examples of hyaena and porcupine damage and their bearing on the reconstruction of the taphonomic history of Makapansgat are discussed.

Introduction

The Makapansagat Limeworks site in Northern Transvaal, South Africa is well known for its immense wealth of fossil bones. It represents one of the most significant "bone caves" known, with an estimated total of at least two million bone fragments, of which several tens of thousands are still in situ (Dart, 1958). The collection at the Bernard Price Institute for Palaeontological Research, University of the Witwatersrand, Johannesburg, comprises some 250,000 specimens, all of which are available for taphonomic studies (J. Kitching, pers. comm. 1982).

Ten orders of mammals with 67 species are known from the Limeworks macrovertebrate assemblage. The relative abundance of species varies widely, with most being represented only by a few specimens. Members of the Bovidae, on the other hand, make up more than 90 per cent of the assemblage. The age of the deposits is assumed to be around 3.0 my (Brock *et al.*, 1977).

Makapansgat: An Archaeological Problem

Dart's (1957) controversial interpretation of the vertebrate remains from the Limeworks quarry was directly and indirectly responsible for the initiation of innovative research in such diverse fields as archaeology, taphonomy, ethnology and primate behaviour. The re-thinking began after Dart (1957) had published a statistical and descriptive analysis of 7159 bone fragments from the Limeworks site. He interpreted the differential survival of skeletal parts in the breccia as evidence that certain elements of animals were useful in an australopithecine tool economy. Certain bone modifications were thought to actually result from the deliberate fabrication and use of these parts as tools. Dart inferred that Australopithecus africanus was an omnivore, with a substantial meat component to its diet, that he killed for a living, sometimes even his own kind, and that he possessed a bone, tooth and horn culture (The "Osteodontokeratic Culture").

For Dart, the cave breccia material represented an australopithecine "midden deposit". While his archeological interpretation of the fossil material has not gone unchallenged (e.g. Washburn, 1957; Thenius, 1961; Heberer, 1966; Feustel, 1969) his hunter-killer-apeman theory has received worldwide attention through popularisation by Ardrey (e.g. 1961, 1976).

The controversies about the behaviour of early hominids still continue today. During the last thirty years a great deal of research has been undertaken on:

i) the formation of cave bone breccias (e.g. Brain, 1981),

ii) the taphonomic processes involved in the formation of vertebrate fossil deposits (e.g. Behrensmeyer and Hill, 1980),

iii) the behaviour of early hominids (e.g. Goodall and Hamburg, 1975, on primate behavior, and Lee and De Vore, 1976, with studies on the !Kung San Kalahari hunter-gatherers), and

iv) the appearance of diagnostic features of human activity in the fossil record (e.g. Potts and Shipman, 1981; and Binford, 1981).

It was shown in a historical review of research at Makapansgat (Schrenk, 1985) that the general ostodontokeratic concept was not the result, but rather the reason for undertaking investigations of bone breccias. The bone breccia story, as we know today, is extremely complex. If only a few aspects are considered, seemingly convincing arguments can be advanced to prove almost any predetermined explanation.

In order to answer the main question of how the bones got into Makapansgat cave, we have to first determine: a) what are the paleoecological implications of observed taphonomic effects on bone, b) what are the processes producing them, and c) how can these processes be reconstructed?

The only scientifically sound means of answering these questions is to conduct actualistic studies in order to establish controls, i.e. to observe or produce taphonomic effects on recent bones, with characteristics similar to those observed on fossil ones, under controlled conditions. Actualistic studies must be conducted in a way that allows causal relationship between observed processes and observed effects to be demonstrated (Binford, 1981).

In this paper we concentrate on a) clarifying the taphonomic processes in general and their effects on bone and b) initial actualistic SEM studies on modern bone damage. We would also like to provide ideas for further SEM studies on taphonomic effects by discussing the most important ones in the Makapansgat Limeworks grey breccia. They are macroscopically apparent but reveal their full story only after SEM investigations.

Taphonomic Processes and the Effects on Bone

A bone passes through many stages from the living animal to its incorporation as a fossil specimen in a museum collection. During this time, taphonomic processes are taking place, and these cause certain effects on the bone material and information involved (Maguire and Schrenk, 1985; Andrews and Cook, 1985; Cook, 1986).

Stages. In attempting to take into account possible events in the history of a bone a flow diagram (Table 1) was constructed, including defined stages through which bones can potentially pass. In reality this is a continuum and the stages listed merely represent potential stopping points. A bone does not necessarily pass through every stage listed in Table 1. For instance, an animal can die and become buried simultaneously in a collapsing burrow structure. The processes may stop here because the remains may never become fossilized. Animals may become preserved without passing through the stage of disarticulation. Any of the stages above can be the end point in the history of skeletal elements. It is also possible that bones may remain in one stage for a long period. A bone can pass through some of the stages more than once. For example, a buried bone may become exposed, transported and reburied in a different place.

The flow diagram in Table 1 is a very simplified scheme demonstrating the principle of taphonomy. In reality the situation is more complicated. Many complex alternative routes are possible, all of which involve transportation of the bones. Transportation may occur in several different ways. Runoff of rainwater on a hill slope in combination with gravity is commonly responsible, especially where caves are concerned. Animals or hominids also transport bones over some distance.

From stage to stage, both material and information are lost. This is due to taphonomic processes which are affecting the remains along the way and ultimately biasing the paleontological sample.

Processes. The passing of a bone from one stage to the next is controlled by a variety of biological, geological, physical and chemical processes. These are mediated by the ecological framework and the living community, as well as by the conditions for fossilization and even by the quality of research conducted at the site.

Effects. How do taphonomic processes alter bones and bone assemblages, and how diagnostic are these alterations? Can the alterations be used to identify the processes involved in the formation of a fossil assemblage? Many may indeed be potentially unequivocal indicators, but as yet there are far too little actualistic experimental and comparative data available. Several principal statements can be made however.

Table 1. Taphonomic stages and processes on bone from the living animal to incorporation as a fossil specimen in a museum collection.

STAGES		PROCESSES
LIVING ANIMAL ↓	— DEATH —	Senility, Dehydration, Starvation, Disease, Accident, Predation
DEAD ANIMAL ↓	— DESTRUCTION OF CARCASS —	Consumption, Butchering, Disarticulation, Attack by animals, Toolmaking, Tool use, Chemical Decomposition, Biological Decay
SKELETAL REMAINS ↓	— DESTRUCTION OF SKELETAL REMAINS —	Weathering, Trampling, Insect Attack, Biological Decay, Rodent Action, Root etching
SKELETAL REMAINS AVAILABLE FOR BURIAL ↓	— BURIAL —	Sediment Coverage, Sediment Infiltration, Compression, Depression, Decay
BURIED REMAINS ↓	— FOSSILIZATION —	Impregnation, Mineralization, Leaching, Cracking, Shearing
FOSSILIZED REMAINS ↓	— RECOVERY —	Erosion, Excavation, Blasting, Drilling, Wedging, Leavering
RECOVERED REMAINS ↓	— PREPARATION AND CONSERVATION —	Selection, Mechanical Preparation, Acid Preparation, Surface Coating
STORED AND DISPLAYED REMAINS	— STUDY —	Handling, Breakage, Destruction, Measuring, Disappearance

If bones are being damaged by a certain agent, both the physical properties of the bone, such as fragility and size, and the properties of the damaging agent affect the resulting alterations. In addition, a fossil bone assemblage is not only characterized by the individual bone morphology and surface damage, but also at the assemblage level. The relative abundance of species and of skeletal elements, among other factors, are then of interest to the taphonomist.

Macroscopic/Microscopic Damage

Numerous macroscopic effects on bone exist which are beyond the scope of this paper. Statistical comparison of macroscopic alterations might be useful in analyzing and comparing modern and fossil bone assemblages (e.g. Maguire et al., 1980). Microscopic features, however, may be crucial in separating surface damage which appears indistinguishable macroscopically.

Microscopic taphonomic effects potentially present on Makapansgat Limeworks specimens listed in Table 2 are grouped according to the processing causing them. Only the processes in Table 1 known to leave microscopic effects on bone surfaces are given in Table 2. Several transportation processes are included as well. It is important to differentiate between "preferential alterations" and "nonpreferential alterations". The former ones are effects which occur at a preferred spot on a bone surface, while the latter ones are randomly distributed. Some effects may look microscopically similar, but due to non-random distribution can still be differentiated.

The microscopic effects listed in Table 2 are very general because in many cases specific and unequivocal descriptions do not yet exist. "Abrasion", for instance, occurs several times, as the resulting effect of gravity transport, handling, measuring and tool use. There are, at present, insufficient comparative and experimental data to differentiate between different types of abrasion. The related taphonomical problem is that of assigning specific causes to observed damage. Difficulties may arise not only because one process may give rise to several different effects, but also different processes may give rise to very similar effects. It may also be the case that numerous sets of effects caused by a variety of processes are superimposed or that some processes occurred more than once (Maguire and Schrenk, 1985).

Our questions (what do microscopic details of surface damage look like and can they indicate the process or agent having produced the surface damage on the Makapansgat specimens?) led to the application of SEM in this study. Although the relatively low magnification required may easily be reached with a light microscope, the SEM technique is preferred because it provides a considerably greater depth of field. In the present study only original specimens have been used in order to eliminate potential errors introduced through application of replicating techniques. All micrographs were taken on a Cambridge S.4 scanning electron microscope at the Electron Microscope Unit, University of the Witwatersrand, Johannesburg.

Table 2. Potential micro-damage to bone surfaces of the Makapansgat Limeworks bone breccia material. Alterations may be preferential or non-preferential and are not mutually exclusive in the two columns.

PROCESSES		PREFERENTIAL ALTERATIONS	NON-PREFERENTIAL ALTERATIONS
Death	DISEASES PREDATION	pathological damage carnivore tooth marks	pathological damage carnivore tooth marks
Destruction of carcass	CONSUMPTION DISARTICULATION BUTCHERING TOOL MANUFACTURE AND USE	carnivore tooth marks cut & chop marks abrasion, polish wear microchipping	carnivore tooth marks trampling scratches soft part removal abrasion, polish wear
Destruction of skeletal remains	INSECT ATTACK CHEMICAL DECOMPOSITION BIOLOGICAL DECAY RODENT ACTIONS	gnawing marks	borings, runnels abrasion, etching soft part removal gnawing marks
Burial	SEDIMENT COVERAGE DECAY	depression & compression features	abrasion, polish root grooves insect borings
Recovery	EROSION DRILLING EXCAVATION		abrasion "fresh" scratches
Preparation & Conservation	MECHANICAL PREPARATION ACID PREPARATION SURFACE COATING		preparator marks etching, smoothing obliteration of surface
Transportation	HILL WASH WATER PREDATORS WIND	abrasion of extruding parts	abrasion abrasion, erosion stomach acid etching attrition, pitting

Some of the surface damage produced by hyaenas and porcupines are described in this report. The comparative samples are derived from recent collections for which the damaging agent is known. The hyaena sample (Crocuta crocuta) was collected at the Kruger National Park, South Africa; the porcupine sample (Hystrix africaeaustralis) comes from the Hartebeesthoek porcupine lair west of Pretoria, South Africa. A large number of specimens have been studied in order to establish the typical microscopic features by which certain taphonomic effects are characterized.

Hyaena Damage

Punctate depressions are caused by the robust conical bone-crushing premolars (P^3 and P_3 and P_4) of adult hyaenas. These depressions are frequently present on one surface of the bone only. They range in size from 2 mm to 10 mm. Small punctate marks probably produced by the teeth of juvenile animals have been called pitting marks (Maguire et al., 1980). The micrograph demonstrates the typical features of punctate depressions (Fig. 1). They have a rounded outline and are conical in cross-section, with particles of bone crushed inwards. Potts and Shipman (1981) discuss the differences between these marks and chopping marks or hard hammer percussion marks.

Lunate fracture scars (Fig. 2) are portions of a punctate mark left along a fracture margin when the strength of the bite is sufficient to split the bone. Typically, these marks, which are elongated, round at one end, and up to 10 mm wide, show fragments of bone material crushed slightly inwards. No fine striations, sometimes created by a tooth sliding over the bone surface, were seen on the lunate scars.

Striations are typically short, straight, and oriented subperpendicular to the long axis of the bone. Sutcliffe (1970) suggests that striations are made by juvenile animals. Macroscopically U-shaped elongated grooves up to 1 mm wide (Fig. 3), striations typically show very fine substriations within and parallel to the main groove (Fig. 4). This observation, apparent on every single example of hyaena striation studied, is in contradiction to the results of studies by Potts and Shipman (1981) who used the occurrence of fine striations within the main groove as an argument for cut marks ("slicing marks") and against tooth marks. The situation seems to be even more complex in that the depth of penetration into the bone tissue also possibly affects the production of microscopic traces.

Porcupine Damage

Porcupine gnawing marks are broad, contiguous and shallow scrape marks produced by the chisel-shaped upper and lower incisors. They are up to 5 mm wide and are characterized by V-shaped sup-grooves between superficially rounded ridges (Figs. 5 and 6). The slip of a tooth may produce steps perpendicular to the direction of tooth movement (Fig. 5). Compared to hyaena tooth striations, porcupine gnawing marks cut more deeply into the bone. Frequently, both articular ends of long bones such as humeri are prone to destruction. Hyaena tooth striations

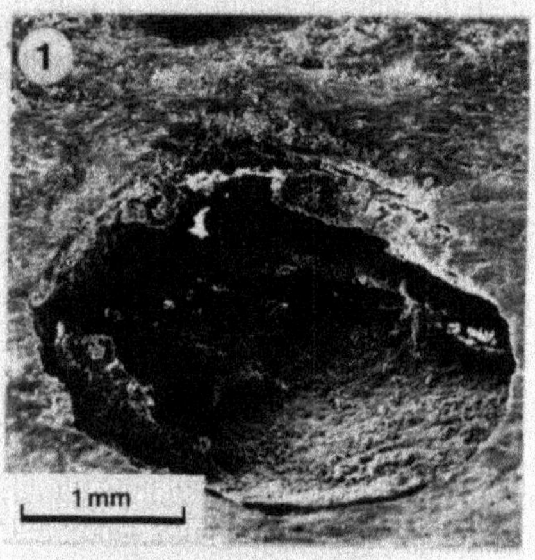

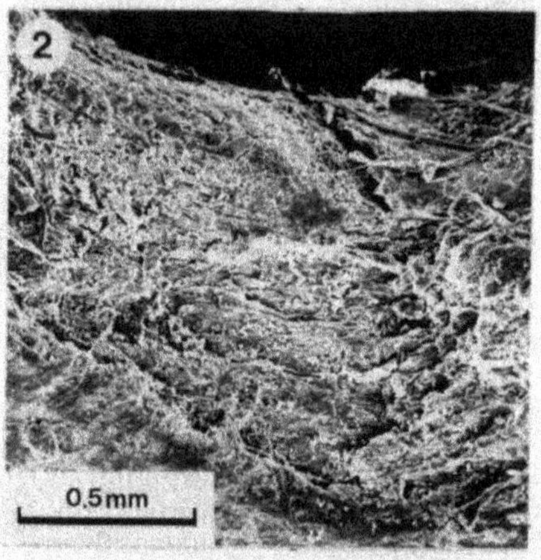

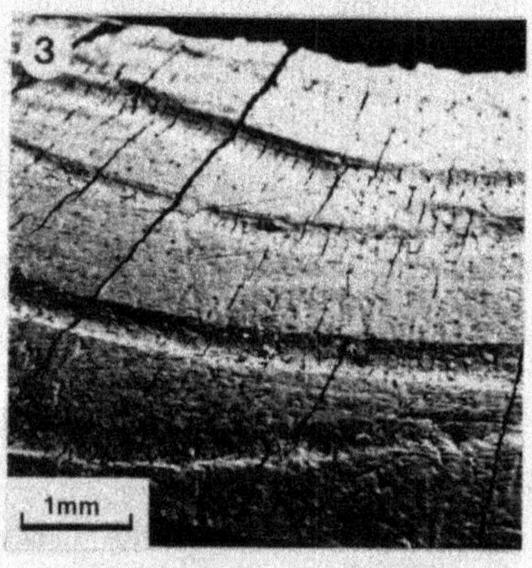

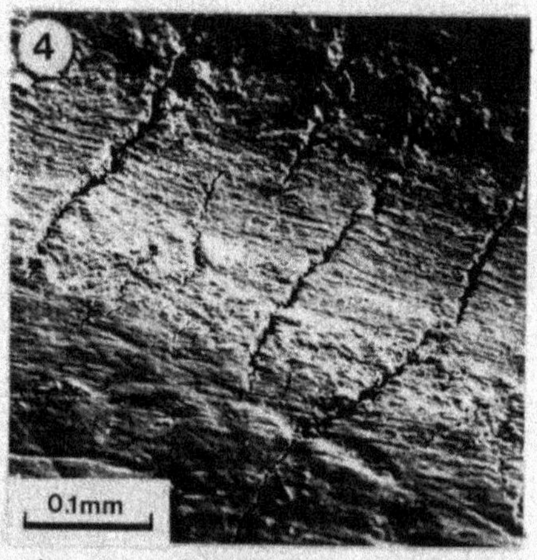

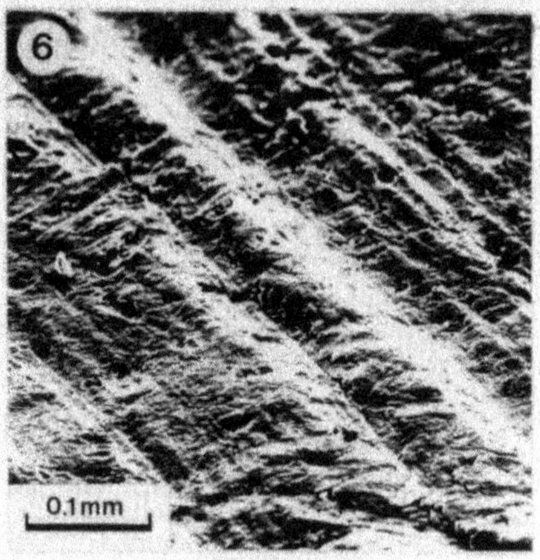

Fig. 1. Punctate mark from hyaena tooth on modern bone.
Fig. 2. Lunate fracture scar made by hyaena tooth on modern bone.
Fig. 3. Striations made by hyaena teeth on modern bone.
Fig. 4. Striations made by hyaena teeth on modern bone.
Fig. 5. Porcupine gnawing marks on modern bone.
Fig. 6. Porcupine gnawing marks on modern bone.

are usually much less frequent on a bone surface.

Possible Agents of Bone Damage and Bone Accumulation at Makapansgat

There are nine significant processes or agents through which bones may have accumulated inside ancient dolomitic caves, such as Makapansgat Limeworks, or within their immediate catchment areas:

1. "hillwash", or periodic sheetflooding (Wells, 1974; Partridge, 1979);

2. caves acting as natural death traps (Gow, 1980);

3. voluntary entry into caves by animals for refuge or to obtain water, and failure to emerge (Gow, 1973; Wells, 1973);

4. birds: owls which may use caverns as roosts and eagles and vultures which may use krantzes in the vicinity of cave entrances as nesting sites (Brain, 1981; Levinson, 1982);

5. leopards (Brain, 1968, 1969b, 1970);

6. sabre-toothed cats (Brain, 1981);

7. hyaenas: striped, brown, spotted and possibly the extinct hunting hyaenas (Ilani, 1975; Skinner et al., 1980; Mills and Mills, 1977; Bearder, 1977; Sutcliffe, 1970; Maguire et al., 1980; Brain, 1981);

8. porcupines (Hughes, 1958; Hendey and Singer, 1965; Maguire, 1976, 1978); and

9. hominids (Brain, 1969a, 1974).

It has not yet been established to what degree each of these agents was responsible for the formation of the bone assemblages at Makapansgat. Through the deposition of their pellets, owls were probably responsible for the accumulation of small mammal bones leading to the formation of rodent breccias in various parts of the cave. The remains of leopards, sabre-toothed cats, hyaenas, porcupines and australopithecines found in the grey breccia suggest that any of these animals could have been involved in the accumulating process at Makapansgat. The surface damage observed in the collection suggests that at least hyaenas and porcupines have played a minor role. In a sample of 2000 fossil metapodials and metapodial flakes of bovids, 25% bore clear evidence of hyaena damage. Only in 4% of the cases was porcupine damage clearly identifiable.

On the basis of comparative background studies we were thus able to determine at least two agents of bone damage at Makapansgat. Whether these agents were also involved in the accumulation of bones at the site is a different question. Although damage and accumulation of bones are functionally independent processes, experience shows that they often occur simultaneously.

Since Kruuk's (1972) study on the spotted hyaena, the fact that hyaenas are active predators is no longer in doubt. The question of whether or not hyaenas accumulate bones in their dens has also been answered. Contradictory to these results of studies by Hughes (1954a, b, 1958) in South Africa, many publications have reported that recent hyaenas collect bone (Buckland, 1822, 1824, 1828; Sutcliffe, 1970; Kruuk, 1972, 1976; Ilani, 1975; Skinner, 1976; Bearder, 1977; Mills and Mills, 1977; Mills, 1978; Hill, 1978; Owens and Owens, 1979; Henschel et al., 1979; Hill, 1980; Behrensmeyer and Dechant Boaz, 1980; Skinner et al., 1980; Maguire et al., 1980; Brain, 1981). Examples of bone collections made by fossil hyaenas have been described by Buckland (1822, 1824), Abel (1935), Ehrenberg (1959, 1962), Thenius (1960), Klima et al. (1962), Tratman et al. (1971), Malez (1975), Turner (1981) and others. As far as the porcupines are concerned, the bone collecting and gnawing habits of modern Hystrix are well known (Shortridge, 1934; Alexander, 1956; Maguire, 1976; Brain, 1980, 1981).

It should be noted that between 50 and 80% of all the bones collected by modern hyaenas bear traces of hyaena damage. In the case of accumulations in porcupine lairs the figure is even higher: around 90% of bone fragments collected by modern porcupines bear porcupine gnawing marks (Maguire et al., 1980). Even if this does not necessarily prove that fossil bones from Makapansgat bearing traces of porcupine or hyaena damage were also accumulated by these animals, it does render it highly likely. Moreover, it could in fact mean that more than 50% of the bone fragments at the Limeworks site were accumulated by hyaenas and porcupines even though they left their marks on only 29-34% of the bone material. Nevertheless, the problem remains that the accumulating agent is not definitely known for around half of the bone fragments found at Makapansgat.

The ultimate aim of the ongoing study is to re-analyze this Makapansgat collection, now numbering some 250,000 specimens, and to identify the taphonomic processes which have influenced the formation of this assemblage. It has become clear that presently it is virtually impossible to identify all the processes to which bones may have been subjected. The accumulation of experimental and comparative data will have to proceed a long way, before all resulting effects can be defined. There is yet too little information on bone damage caused by biological agents (e.g. leopards, hyaenas, porcupines, man) and even worse, by non-biological processes (e.g. chemical weathering, water transport, fossilization, etc.). Further actualistic background studies are required (c.f. Bromage and Boyde, 1984). Excavation of in situ bone breccias at the Limeworks site may solve further problems associated with this collection, such as selection and preparation biases. A microscopic analysis of surface damage on those specimens of the collection originally claimed to be tools by Dart (1957) is in progress.

Conclusions

The story of bone accumulation in the Makapansgat Limeworks cave is complex and not yet fully understood. However, it was shown that actualistic SEM studies can help reveal at least part of this story.

If good comparative samples are available, certain producers of tooth marks, for instance can be identified. A related problem is the demonstration that the damaging agent may also have been the accumulating agent. Hyaena and porcupine tooth marks on fossil bones from Makapansgat suggest involvement of these animals in the accumulation process. Detailed knowledge of as many possible types of bone surface damage and of their characteristic microstructure is indispensible in taphonomic studies. It would be extremely helpful to have a larger range of actualistic data available from SEM studies of modern bone damage before more of the questions in Makapansgat may be resolved.

Acknowlegements

We thank the staff of the SEM Unit at the University of the Witwatersrand, Johannesburg, for assistance. The Institute for the Study of Man in Africa (ISMA) funded parts of this research.

References

Abel, O. 1935. Vorzeitliche Lebensspuren Jena: G. Fischer.

Alexander, A.J. 1956. Bone carrying by a porcupine. South African Journal of Science 55: 257-258.

Andrews, P. and Cook, J. 1985. Natural modifications to bones in a temperate setting. Man 20: 675-691.

Ardrey, R. 1961. African Genesis. New York: Atheneum.

Ardrey, R. 1976. The Hunting Hypothesis. London: Collins.

Bearder, S.K. 1977. Feeding habits of spotted hyaenas in a woodland habitat. East African Wildlife Journal 15: 263-280.

Behrensmeyer, A.K. and Dechant Boaz, D.E. 1980. The recent bones of Amboseli National Park, Kenya in relation to East African paleoecology. In Behrensmeyer, A.K. and Hill, A. (eds.), Fossils in the Making, pp. 72-93. Chicago: University of Chicago Press.

Behrensmeyer, A.K. and Hill, A. (eds.), 1980 Fossils in the Making. Chicago: University of Chicago Press.

Binford, L.R. 1981. Bones: Ancient Men and Modern Myths. New York: Academic Press.

Brain, C.K. 1968. Who killed the Swartkrans Ape-man? South African Museum Association Bulletin 9: 127-139.

Brain, C.K. 1969a. Faunal remains from the Bushman rock shelter, Eastern Transvaal. South African Archaeological Bulletin 24: 170-171.

Brain, C.K. 1969b. The probable role of leopards as predators of the Swartkrans Australopithecines. South African Archaeological Bulletin 24: 170-171.

Brain, C.K. 1970. New finds at the Swartkrans Australopithecine site. Nature 225: 1112-1119.

Brain, C.K. 1974. Human food remains from the iron age at Zimbabwe. South African Journal of Science 70: 303-309.

Brain, C.K. 1980. Some criteria for the recognition of bone-collecting agencies in African caves. In Behrensmeyer, A.K. and Hill, A. (eds.) Fossils in the Making, pp. 107-130. Chicago: University of Chicago Press.

Brain, C.K. 1981. The Hunters or the Hunted? Chicago: University of Chicago Press.

Brock, A., McFadden, P.L. and Partridge, T.C. 1977. Preliminary palaeomagnetic results from Makapansgat and Swartkrans. Nature 266: 249-250.

Bromage, T.G. and Boyde, A. 1984. Microscopic criteria for the determination of directionality of cutmarks on bones. American Journal of Physical Anthropology 65: 359-366.

Buckland, W. 1822. Account of an assemblage of fossil teeth and bones of elephant, hippopotamus, bear, tiger and hyaena and sixteen other animals; discovered in a cave at Kirkdale, Yorkshire in the year 1821. Philosophical Transactions of the Royal Society London 1: 171-236.

Buckland, W. 1824. Reliquiae diluvianae, or observations on the organic remains contained in caves, fissures and diluvial gravel, and other geological phenomena, attesting the actions of a universal deluge. 2nd edition, London: J. Murray.

Buckland, W. 1828. Letter of Professor Buckland to Professor Jameson and of Captain Sykes to Professor Buckland on the interior of dens of living hyaenas. Edinburgh Natural Philosophical Journal (1826/27): 377-380.

Cook, J. 1986. The application of scanning electron microscopy to taphonomic and archaeological problems. In D.A. Roe (ed.), Studies in the Upper Palaeolithic of Britain and Northwest Europe. B.A.R. Int. Ser. 296: 143-163.

Dart, R.A. 1957. The osteodontokeratic culture of Australopithecus prometheus. Transvaal Museum Memoir 10.

Dart, R.A. 1958. The minimal boner breccia content of Makapansgat and the Australopithecine predatory habit. American Anthropologist 60: 923-931.

Ehrenberg, K. 1959. Die urzeitlichen Fundstellen und Funde in der Salzofenhöhle, Steiermark. Archaeologica Austriaca 32: 1-23.

Feustel, R. 1969. Zur Problematik der Protolithischen Knochenkultur und der Osteodontokeratischen Kultur. Alt-Thuringen 10: 7-67.

Goodall, J. and Hamburg, D.A. 1975. Chimpanzee behaviour as a model for the behaviour of early man. In Hamburg, D.A. and Brodie, H.K.H. (eds.) New Psychiatric Frontiers, pp. 14-43, New York: Basic Books.

Gow, C. 1973. Habitual sheltering in an extensive cave system by baboons near Breadasdorp, South Africa. South African Journal of Science 69: 182.

Gow, C. 1980. Faunal remains from Hot Pot Cave Bredasdorp. Palaeontolgia Africana 23: 105-108.

Heberer, G. 1966. Uber die Osteodontokeratische Kultur der Australopithecinae. Quartar 17: 21-50.

Hendey, Q.B. and Singer, R. 1965. The faunal assemblages from the Gantoos Valley shelters. South African Archaeological Bulletin 20: 206-213.

Henschel, J.R., Tilson, R. and Blottnitz, F. von 1979. Implications of a spotted hyaena bone assemblage in the Namib Desert. South African Archaeological Bulletin 34: 127-131.

Hill, A. 1978. Hyaenas, bones and fossil man. Kenya Past and Present 9: 8-14.

Hill, A. 1980. A modern hyaena den in Amboseli National Park, Kenya. Proceedings of the 8th Panafrican Congress on Prehistory Nairobi 1977: 137-138.

Hughes, A.R. 1954a. Habits of hyaenas. South African Journal of Science 51: 156-158.

Hughes, A.R. 1954b. Hyaenas vs. Australopithecines as agents of bone accumulation. American Journal of Physical Anthropology 12: 476-486.

Hughes, A.R. 1958. Some ancient and recent observations on hyaenas. Koede 1: 105-114.

Ilani, G. 1975. Hyaenas in Israel. Tera Va'raetz 16: 10-15.

Klima, B., Musil, R. and Jelinek, J. 1962. Die Erforschung der Hohle Sveduv Stul 1953-1955. Anthropos 13.

Kruuk, H. 1972. The Spotted Hyaena, a Study of Predation and Social Behaviour. Chicago: University of Chicago Press.

Kruuk, H. 1976. Feeding and social behaviour of the striped hyaena (Hyaena vulgaris Desmarest). East African Wildlife Journal 14: 91-111.

Lee, R.B. and DeVore, I. 1976. Kalahari Hunter-Gatherers: Studies of the !Kung San and their Neighbors. Cambridge: Harvard University Press.

Levinson, M. 1982. Taphonomy of microvertebrates - from owl pellets to cave breccia. Annals of the Transvaal Museum 33: 115-121.

Maguire, J.M. 1976. A Taxonomic and Ecological Study of the Living and Fossil Hystricidae with Particular Reference to Southern Africa. Unpublished PhD thesis, Faculty of Science, University of the Witwatersrand, Johannesburg.

Maguire, J.M. 1978. Southern African fossil porcupines. South African Journal of Science 74: 144.

Maguire, J.M., Pemberton, D. and Collet, M.H. 1980. The Makapansgat Limeworks grey breccia: hominids, hyaenas, hystricids or hillwash? Palaeontologia Africana 23: 75-98.

Maguire, J.M. and Schrenk, F. 1985. From living animal to museum specimen: taphonomic processes and their effects on bone. Annals of the Geological Survey, South Africa 19: 69-71.

Malez, M. 1975. Hijenske Pecine u Doba Pleistocena na Podrucju Hrvatske. RAD Jugoslawenske Academije Znanosti i Umjetnosti 371: 307-316.

Mills, M.G.F. 1978. Foraging behaviour of the brown hyaena (Hyaena brunnea Thunberg, 1820) in the southern Kalahari. Zeitschrift fur Tierpsychologie 48: 113-141.

Mills, M.F.G. and Mills, M.E.J. 1977. An analysis of bones collected at hyaena breeding dens in the Gemsbok National Parks (Mammalia, Carnivora). Annals of the Transvaal Museum 30: 145-155.

Owens, D. and Owens, M.J. 1979. Communal denning and clan association in brown hyaenas (Hyaena brunnea Thunberg) of the Central Kalahari Desert. African Journal of Ecology 17: 35-44.

Partridge, T.C. 1979. Re-appraisal of lithstratigraphy of the Makapansgat Limeworks hominid site. Nature 279: 484-488.

Potts, R. and Shipman, P. 1981. Cutmarks made by stone tools on bones from Olduvai Gorge, Tanzania. Nature 291: 577-580.

Schrenk, F. 1985. Alte Höhle in neuem Licht: die Hominiden-Fundstelle Makapansgat Limeworks in Südafrika. Natur und Museum 115: 157-173.

Shortridge, E.C. 1934. The Mammals of South West Africa. London: W. Heinemann.

Skinner, J.D. 1976. Ecology of brown hyaena Hyaena brunnea in the Transvaal with a distribution map for Southern Africa. South African Journal of Science 72: 262-269.

Skinner, J.D., Davis, S. and Ilani, G. 1980. Bone collecting by striped hyaena (Hyaena hyaena) in Israel. Palaeontologia Africana 23: 99-104.

Sutcliffe, A. 1970. Spotted hyaena: crusher, gnawer, digestor and collector of bones. Nature 227: 1110-1113.

Thenius, E. 1960. Die pleistozänen und holozänen Wirbeltierreste der Griffener Höhle, Kärnten. Carinthia II 70: 26-62.

Thenius, E. 1961. Über die Bedeutung der Palökologie für die Anthropologie und Urgeschichte. 2. Österreichisches Symposium Burg Wartenstein 1959: 80-103.

Tratman, E.K., Donavan, D.T. and Campbell, J.B. 1971. The Hyaena Den (Wookey Hole), Mendip Hills, Sommerset. Proceedings of the University of Bristol Speleological Society 12: 245-279.

Turner, A. 1981. Ipswichian mammal faunas, cave deposits and hyaena activity. Quaternary Newletter 33: 17-22.

Washburn, S.L. 1957. Australopithecus: the hunters or the hunted? American Antiquity 59: 612-614.

Wells, L.H. 1973. Baboons sheltering in caves. South African Journal of Science 69: 279.

Wells, R. 1974. Reconstructing the past: excavations in fossil caves. Australian Natural History 18: 20.

Bone Tools: An Experimental Approach

Pat Shipman and Jennie J. Rose

Introduction

Recognizing bone tools that may have been minimally modified prior to use has been a problem of many years' duration. Although the identification of simple bone tools is difficult, it is nonetheless potentially important.

First, bone is a resource in addition to meat or fat that might have been derived from carcasses in the past; thus, evidence that bones were used as a raw material for tools would expand the catalogue of hominid-faunal interactions at that point in time.

Second, the need for bone as a raw material may have caused significant changes in patterns of exploitation of the local fauna. For example, my own and others' work (i.e., Stanford et al., 1981; Bonnichsen, 1979; Morlan, 1980; Shipman, in press a, b) suggest that bones with thick cortices are most suitable for flaking. Therefore, the use of flaked bone tools might have shifted the focus of exploitation from small- and medium-sized species to very large ones.

Third, although extensively modified bone tools, such as harpoons, needles, awls, and so on, are readily recognizable in the archaeological record, it is not clear when bone began being used as a raw material and for what tasks bone has been used.

Fourth, although bone and stone have different physical properties related to fracturing (Cotterell and Kamminga, 1987; Currey, 1984; see also Shipman, in press b), some purported bone tools appear to have been used for tasks similar to those performed with stone tools, raising questions of when and why bone was used rather than stone.

Finally, similar controversy has surrounded claims for early bone tools in North America and those for much earlier bone tools in the Old World (e.g., Binford, 1981; Binford & Bertram, 1977; Bonnichsen, 1979; Brain, 1967, 1981; Breuil, 1938; Clark, 1977; Dart, 1949, 1957; Frison, 1978, 1982; Freeman, 1978; Haynes, 1981, 1983, 1986; Hannus, 1985, in press a,b; Johnson, 1982; Morlan, 1980; Pei, 1938; Shipman & Phillips-Conroy, 1977; Stanford, 1979; Wolburg, 1970). Thus, the technological and experimental approach to bone tools which is undertaken here may be widely pertinent.

The problem has been one of recognizing bone expediency tools, that is, bones which may have been intentionally but simply fractured and then used briefly and discarded (Frison, 1974; Johnson, 1982, 1985). The sequence of events in the making of bone tools (Lyman, 1984) includes three crucial stages:

1) the primary decision about the appropriateness of the particular bone as a raw material;
2) the fracturing/manufacturing process that results in a suitably modified bone;
3) the use of the bone that results in wear or polish.

The source of the difficulties in recognizing bone expediency tools is clear: the bones are not sufficiently altered prior to use (stages 1 and 2) to be clearly recognizable as humanly-modified from gross inspection (Johnson, 1985). Therefore, as Frison maintains, it is the utilization and wear that are most diagnostic:

> ...bone tools used in most butchering contexts were of an expedient nature and their manufacture involved much the same breakage as occurred in some ordinary butchering processes. As a result, most of the identification as a tool is from <u>evidence of use</u> rather than manufacture. (1978:318-319, emphasis added).

Some or perhaps many of the types of breakage observed on possible bone tools can be created by a variety of taphonomic agents (Haynes, 1986); breakage patterns and shape <u>per se</u> do not unambiguously identify these bones as tools. However, an increasing number of studies indicate that utilization wear may be distinctive and diagnostic of bone tools (e.g., Arndt and Newcomer, 1986; Brain, in press; Clark, 1977; Fisher, 1984; Johnson and Shipman, 1986; Le Moine, 1985; Newcomer, 1974; Olsen, 1984, 1988; Semenov, 1964; Shipman, Chapter 18 this volume and in press a, b; Shipman et al., 1984). However, rigorous, verifiable, and replicable methods of identification based on sound experiments and clear characterization of results are needed (Johnson, 1985).

Many or most of the proposed bone expediency tools in both the Old and New Worlds are derived from large or very large mammals, usually proboscideans (e.g., Bonnichsen, 1979; Fisher, 1984; Frison, 1978; Frison and Todd, 1985; Hannus, 1985, in press a, b; Johnson, 1985; Miller, 1983; Morlan, 1980; Shipman in press a, b; Stanford et al., 1981; Steele and Carlson, in press). The general shape and form of these purported tools, and the skeletal elements from which they are derived, are often very similar across great expanses of time and space.

For example, possible bone tools from broken mammoth bones dated to 11,140 +/- 140 B.P. (accelerator date from associated organic material: AA905) from the Lange/Ferguson site in South Dakota (39SH33:Hannus, 1985, in press a, b) closely resemble broken proboscidean bones identified as tools dated to over 1 million years at Olduvai Gorge (Leakey, 1971; Shipman, in press b). Further, the microscopic modifications - possibly attributable to utilization wear- on the edges of these purported bone tools are also very similar. Close resemblances have also been noted (Shipman and Rose, 1983b) in the gross and microscopic morphology of gouges made on mastodon bones from Michigan (Fisher, 1984; Shipman et al., 1984) and on straight-tusked elephant bones from Torralba, Spain, although the former site is dated to between 10,000 and 12,000 b.p. and the latter is Acheulian (roughly 300,000 to 500,000 years b.p.). When these resemblances in form and modification occur on fossilized bones with negligible amounts of weathering, they can be attributed to close similarities in immediately postmortem taphonomic history. In short, these possible bone tools are similar in details of shape and microscopic morphology either because they have been subjected to the same, nonhuman agents of modification shortly after death or because the constraints of workingproboscidean bones dictate similar modes of treatment <u>even though the species of proboscidean and hominid involved are different</u> (Shipman, in press b).

Materials and Methods

An actualistic, experimental approach was taken to the problem of finding ways to identify bone expediency tools. A theoretical statement on the merits and limitations of this approach can be found in Shipman (Chapter 18, this volume). We argue that two criteria must be met for an explanation based on actualistic studies to be accepted as valid. First, a detailed similarity must be established between the effects of the events documented through experiments and those observed on the specimen of unknown taphonomic history. Second, the effects of the experiments must also be shown to be distinctive or recognizably different from those of other known sources.

In our view, it is important that possible bone expediency tools be judged on their own properties rather than on their association or lack of association with readily recognizable stone tools. Since the association of fossil bones with stone tools is no longer considered good evidence that the bones were acted upon by hominids (see Binford, 1981; Potts and Shipman, 1981; and Shipman, 1981 for a discussion of this point), we would argue that the <u>lack</u> of association of bones with stone tools cannot be taken as evidence that hominids had nothing to do with the bones.

In this instance, we explored four possible taphonomic histories that might explain the features of specimens proposed to be bone expediency tools. Only the last of these possible taphonomic histories involves human or hominid activities; the others all focus on natural, nonhominid agents of bone modification. These hypotheses do not exhaust the possible universe of taphonomic histories, but they do include some of the more probable and significant alternatives.

1) The bones were broken, chewed, and polished via partial digestion by carnivores;
2) The bones were broken by various causes and polished by waterborne sedimentary particles;
3) The bones were broken by various causes and polished by windborne sedimentary particles;
4) The bones were broken by hominids and polished during their use as tools for skinning, hide-working, and butchering carcasses or for digging plants or processing plant remains.

In order to evaluate the four hypotheses given above, experiments were undertaken to expand our understanding of: 1) the effects of bone tool use both on the working edge of the tool and on the bone surface damaged; 2) the effects of sedimentary abrasion on bones in fluvial environments; and 3) the effects of aeolian abrasion on bones. The effects of carnivore damage and digestion (hypothesis 1) are already well-known from previous work on a sample of about 200 tooth marks made by feral and captive predators (Shipman, 1981; Potts and Shipman, 1981; Shipman and Rose, 1983a) and about 100 bone fragments partially digested by various mammalian and avian predators. Other researchers have also reported on the effects of carnivore damage (e.g., Brain, 1981; Bunn, 1982; Haynes, 1981; Hill, 1980), although these studies have focused more on the gross morphology and breakage and less on the microscopic effects.

In addition to experimental studies, the working edges of six Eskimo bone tools, used and collected in the late 1800's, were replicated. These tools were selected to represent a diversity of tasks and were identified as: 1) a beaming tool/hide scraper; 2) a chisel used to remove birchbark from trees; 3) a graining tool or hide scraper; 4) a fat scraper; 5) a scraper or knife; and 6) an ice or root pick (Nelson, 1889). This is a small sample, so the results reported here must be viewed as preliminary. However, a larger sample of approximately 100 Plains Indian bone tools, which is still under study, appears to confirm the results reported here. Experiments conducted for the work reported here are described below.

In our experiments, the edges of bone tools were examined and replicated before and after use (see Rose, 1983, for the replication technique). Our bone tools were either spirally-fractured long bones or sharp-edged flakes derived from long bone shafts. They were manufactured by striking large cow long bones with a hammerstone. Many of the features of the utilization wear were found to be general, rather than specific to a particular experiment or a particular substance being worked; these common features are described following the description of the various experiments.

Experiment One: Bone Tools. Experiment One assessed the relative effectiveness of bone and obsidian flakes in butchering sheep forelimbs. Two fresh sheep joints, each consisting of an articulated right distal humerus and proximal two-thirds of an ulna and radius, were defleshed, disarticulated, and cleaned of soft tissues. One joint was processed using only bone flake tools manufactured from a bovid humerus; the other joint was processed only with a single, sidestruck obsidian flake. None of the tools was resharpened during the process. Tool edges were replicated before and after use. The joints were boiled gently to remove all remaining tissues; the bone surfaces were then inspected, replicated, and examined using a scanning electron microscope (SEM). Since the bones were covered in soft tissue at the outset of the experiment, all marks are presumed to have resulted from the experiment itself. Our previous work has shown that boiling does not alter the gross or microscopic features of marks on bones (Fig. 1a,b).

a

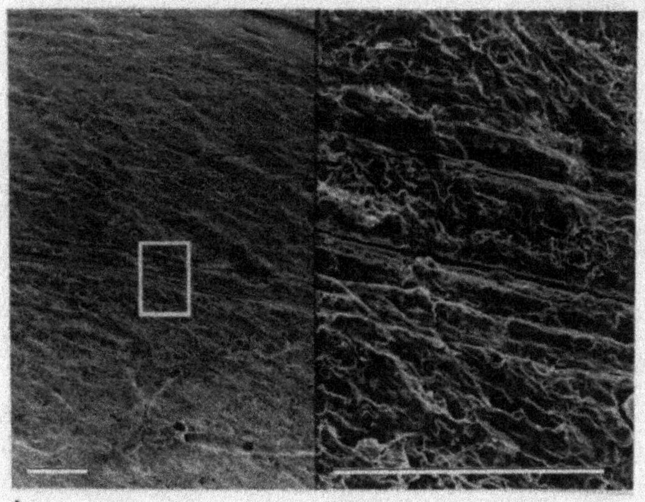

b

Fig. 1. The microscopic morphology of this experimentally-made slicing mark is comparable before (a) and after (b) boiling to remove the periosteum. The box indicates the area of the low-power micrograph that is enlarged in the righthand image in both (a) and (b). Anatomical features also remain unaltered, as can be seen by comparing the vascular grooves in the lower right corner of the low-power segment of each photograph.

There were several conclusions from this experiment. The obsidian flake was decidedly more efficient in cleaning and disarticulating a joint than were the bone flakes. Our subjective assessment was that using the obsidian tool required less energy expenditure and less time (about 20 minutes versus 90) to accomplish the task and resulted in a "cleaner" (more fully utilized) joint. There are several obvious reasons for this difference in efficacy. First, the obsidian flake has a more acute edge angle (less than 20 degrees) than any of the bone flakes (41 to 70 degrees). In fact, it is technically difficult to produce such an acute angle in bone. Thus, the obsidian edge was both objectively and subjectively sharper than the bone flakes. Second, the bone flakes tended to absorb grease and moisture from the flesh, dulling the edge and making it impossible to continue to use the same flake without considerable sacrifice in efficiency. This was not the case with the obsidian tool, which showed no obvious decrease in usefulness during the experiment. Third, the general shape of the bone flakes made them more awkward to use when applying force than the obsidian flake.

Both stone and bone tools produce slicing marks with common microscopic features (Fig. 2a,b): these are elongate grooves with multiple, fine, parallel striations lying within a main groove. (For a definition of slicing as used here, see Potts and Shipman, 1981). In marks made by either type of tool, these striations result from the sinuous edge of the tool and its microscopic irregularities. Scraping and chopping marks produced by bone tools are also similar to those described for stone tools (Shipman and Rose, 1983a).

Two differences were observed between stone tool marks and bone tool marks. First, during this experiment many more stone tool marks were made than bone tools marks, even though both joints were in similar condition at the onset of the experiment (Shipman and Rose, 1983a). This is not an unexpected result, given that the obsidian flake was sharper, and therefore more capable of producing distinctive marks, than the bone flakes. The implications of this finding for bones from pre-Clovis sites are interesting. Since it is well-documented (Guilday et al., 1962; Crader, 1983) that only a small percentage of bones from butchered animals show stone tool marks, this experiment suggests that the bones of animals butchered with bone tools may have very few, if any, cutmarks. Thus, a dearth of cutmarks on bones from sites lacking lithic industries cannot be used to refute the hypothesis that the bones represent animals processed by hominids. It would, of course, be equally absurd to consider that the lack of marks on such bones confirms the hypothesis, either.

Second, bone tool marks appear to be somewhat shallower and broader than stone tool marks, although the differences are difficult to measure accurately. Despite the fact that these distinctions are real and important, they are insufficient to distinguish bone tool cutmarks from stone tool cutmarks reliably. Both types of marks are readily distinguished from tooth marks of various types upon microscopic inspection in most cases. Criteria for distinguishing cutmarks from trampling marks are suggested by Andrews and Cook (1985) and Olsen and Shipman (in press).

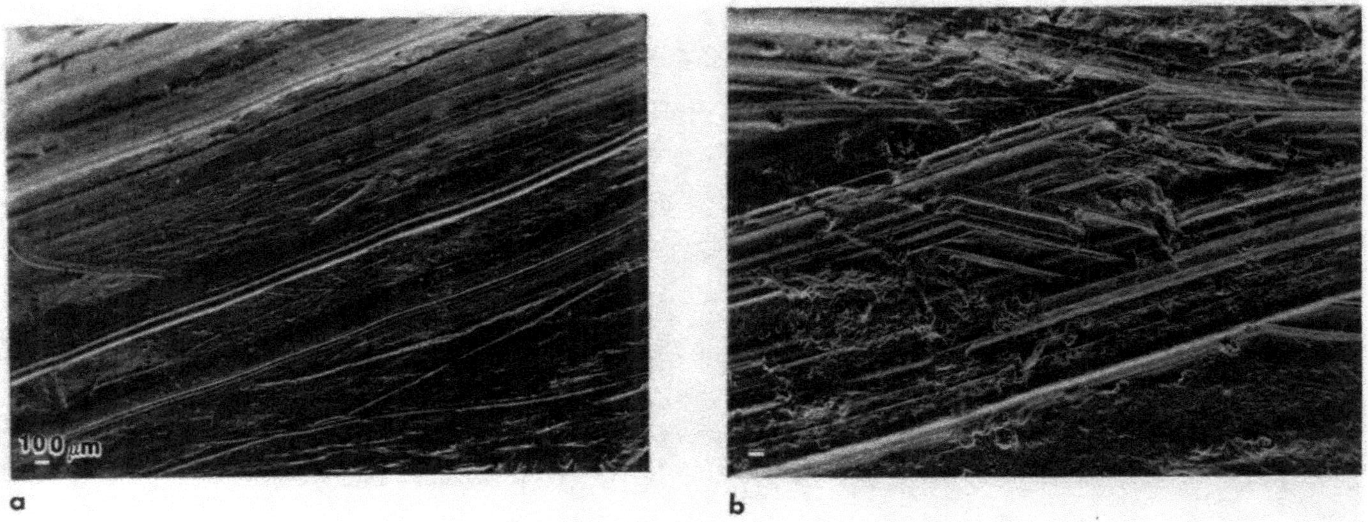

Fig. 2. Cutmarks made with bone and stone tools share the same diagnostic features. (a) Experimentally-made scraping marks produced with an obsidian tool. (b) Experimentally-made scraping marks produced with a bone tool.

Finally, utilization produced a distinctive pattern of wear or polish that readily identified the specimen as used rather than simply fractured. Under the SEM, it is easy to distinguish utilized edges from unutilized ones (see Fig. 8 a, b in Shipman, Chapter 18, this volume).

Experiment Two: Bone Tools. The second experiment of the bone tool series involved the use of two bone flakes from a cow humerus as pestles for grinding. We ground leaves, small twigs, and grass blades in shallow, natural mortars on the medial surface of clean cow ilia. One pestle was preshaped by grinding it on a piece of granite to produce a smooth edge prior to use; the other flake was used without any alteration to the broken edge. Both pestles and mortars were replicated before and after use.

Gross inspection of both mortars and pestles after use revealed surface polish. Microscopically, both mortars were smooth planes traversed by many scratches (Fig. 3a). The unground pestle showed islands of smooth, almost glassy bone - representing areas that were initially raised - surrounded by rough-bottomed pits or recessed areas. The pre-ground pestle showed numerous scratches and sectioned vascular grooves caused by the manufacturing process (Fig. 3b). The extent of use of the preground pestle was insufficient to alter its microscopic morphology appreciably. Even after use, the surface morphology was dominated by scratches from the pre-use grinding.

Experiment Three: bone tools. The Ginsberg experiment was conceived and carried out by Dennis Stanford, Robson Bonnichsen, and Richard Morlan as a test of the effectiveness of bone tools in butchering proboscideans (see Stanford et al., 1981, for details). A large, female elephant (Ginsberg) that died of natural causes at the Franklin Park Zoo was butchered using only a few large, stone bifaces and a variety of bone tools manufactured from her own long bones; thus, the bone tools were exactly the type of objects referred to in the archaeological record as bone expediency tools. Both the manufacture of the tools and their use was documented in detail. We replicated both the edges of the broken bones that had been used as tools and the surfaces damaged by those tools. All replicas were then inspected under the SEM. The results of this study confirm those of Experiment One. The microscopic appearance of the marks and of the utilized edges is strikingly similar to those resulting from our experiments. Further, Stanford et al. (1981) subjectively evaluated the relative efficacy of their stone and bone tools in terms that parallel our own, although our notes were written before their paper was published or known to us.

Experiment Four: bone tools. Additional replicas of utilized bone tool edges and marked surfaces were made from bones used in butchery experiments by Gary Haynes and Dan Fisher. These expanded our sample of the effects of utilization on bone tools, but did not produce any new results. The advantage of replicating and inspecting bone tools made and used by others is that we can be confident that our prejudices about bone tools and their use have not affected the results. Further,

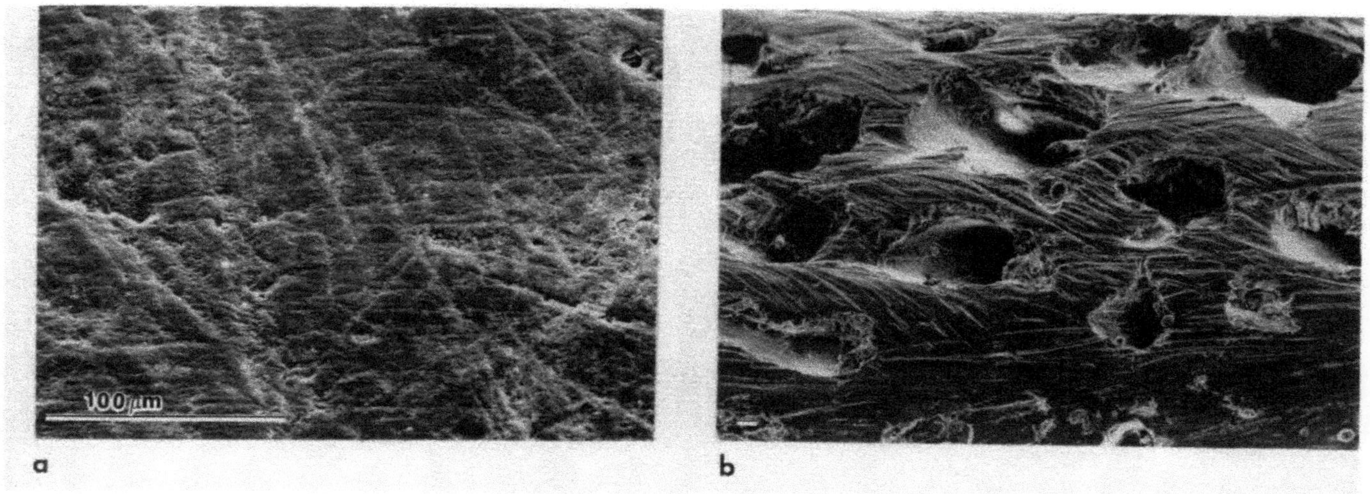

Fig. 3. (a) This bovid ilium was used as a mortar in which vegetable matter was ground, producing numerous scratches. (b) A pre-ground bone pestle from the same experiment showed numerous scratches and sectioned vascular canals, both produced by the initial shaping with a piece of granite. This pestle was not used sufficiently long for utilization wear to erase or substantially alter the coarser scratches produced by pre-grinding.

Dan Fisher included in his sample of replicas (N = 11) two replicas of bone tool edges that he had manufactured and never used. We were not informed that the sample he sent to us involved a unused specimens. In this blind test, we were able not only to correctly distinguish between used and unused edges, we were also able to distinguish accurately between briefly-use and heavily-used edges. In addition, we correctly identified an edge that had been ground prior to use. Thus, the collaborative part of this study greatly increased our confidence that utilization wear per se was genuinely recognizable and clearly distinct from unused, but fractured, bone surfaces.

Ethnographic bone tools. SEM inspection of our limited sample of Eskimo bone tools revealed many features seen in the bone tools used in plant processing and butchering experiments. A major difference in the ethnographic bone tools was that all except the root/ice pick had working surfaces with an extremely fine, glassy polish (Fig. 4) that exceeded the degree of polish on any of the experimental tools. All of the ethnographic tools were used apparently more extensively than our experimental ones. In addition, the five hide-working and bark-working tools were not routinely, or perhaps ever, worked against another bone or other abrasive surface. These differences in extent and type of use may account for the finer polish on the ethnographic tools than on the experimental ones.

Small, shallow pits or dimples, taken to be structural features of the bone tissue rather than an effect of use, were consistently observed on working edges. These pits did not resemble replication artifacts observed in other contexts. Fine, shallow scratches were present but not common on bone tools used exclusively on soft tissues. The bark-working tool showed multiple striations that were large enough to be visible macroscopically; these were probably the remnants of pre-use grinding scratches (Fig. 5). Where these scratches intruded into areas that had been extensively used, they showed a general softening in appearance, a rounding of edges, and a loss of fine detail at the microscopic level. This same tool also showed scattered, large, rough-edged pits and more frequent fine scratches than the other bone tools; both types of feature may have resulted from use. Unfortunately, the collectors of these ethnographic tools did not provide detailed accounts of their manufacture and use, nor are photographs of Eskimos using these particular tools available.

In summary, these bone tool experiments and studies of ethnographic bone tools yielded a set of microscopic features that characterize utilized edges and the marks caused by such edges on unmodified bone during butchering; our work also demonstrates that these features can be used to identify bones of unknown taphonomic history with some confidence. Actualistic studies of various types of bone tools by Olsen (1984, 1988), Brain (in press), and LeMoine (1985) have yielded generally similar results that confirm our findings.

Regardless of the substance upon which the edges were used, all bone tools made and used experimentally in this study showed common features. The utilized edges developed a gloss or polish usually

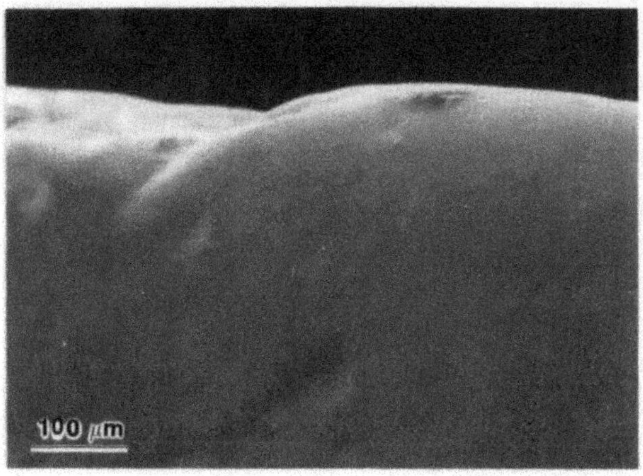

Fig. 4. Ethnographic bone tools used on soft substances, such as this Eskimo tool used as a fat scraper, show extremely fine, glassy polish with few or no scratches.

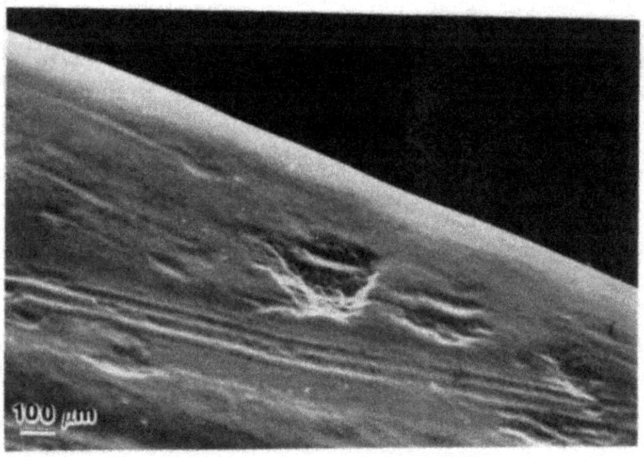

Fig. 5. This Eskimo woodworking tool, used to remove birchbark from living trees, shows large, parallel scratches with soft edges and poor detail. These probably represent remnants of pre-grinding scratches partially obliterated by later use. The large, rough-edged pits may also have been caused by utilization, but their origin is less clear.

visible without magnification, but the changes were sometimes subtle and required close inspection. At the microscopic level, wear is apparently restricted to raised areas of the edge or rugosities that actually come into contact with the substance being cut; these initially raised areas become rounded, glassy, and smooth and lose surface detail progressively with continued utilization (Fig. 6). Such areas may even appear to be melted, although it is not known whether the heat produced by the friction of utilization is capable of generating sufficiently high temperature to melt the hydroxyapatite. Anatomical features such as osteons, vascular grooves, or bone laminae retain definition only if they lie in recesses or if they are protected from wear by soft tissues.

Limited utilization of an edge or surface may leave raised islands of glassy bone surrounded by low-lying areas with pronounced surface roughening that is the result of the fracturing process (Fig. 7). More extensive use may wear down these raised areas until nearly the entire surface is uniformly polished; this can be seen most clearly on the ethnographic tools, which presumably were used far more extensively than our experimental bone tools. Bone tools used to butcher animals or to grind vegetable matter frequently come into contact with hard particles of bone, dirt, or other substances. Such tools show many coarse and fine scratches and frequent pits on their polished surfaces. Bone tools used only on soft tissues, judging from ethnographic tools used on fat (see below), show very few or no scratches or pits. It is important to note that the utilization features of a bone tool thus reflect not only the substance that was processed but also the foreign matter adhering to the substance. Therefore, tools used only on soft substances free of particulate matter can be distinguished from those used on harder substances or on those that had abrasive particles clinging to them. However, at this point we cannot make any more precise identification of the substance on which a given bone tool was use based on the appearance of the wear _per se_.

So far, it is possible to distinguish only among unused edges, edges used on soft substances (such as fat or skin) and those used on more abrasive substances (such as vegetable matter or mixed bone and flesh). We would emphasize that, on briefly-used bone tools, the polish or utilization wear occurs differentially, being most closely concentrated on the working edge. This differential distribution of polish or wear is true both on the gross level and on the microscopic level, where polish and rounding is observed to occur first on the raised areas of the working edge itself. A few millimeters away from the working edge, polish can be expected to be more poorly developed or absent. This pattern of polish occurs because the details of the shape of the bone tool and the angle at which it is held greatly influence its usefulness in a particular task and its contact with other materials. In contrast, any bone that has acquired a more general, widespread polish or patina may well have been subjected to other taphonomic events, such as trampling or sedimentary abrasion, in which the shape and orientation of the object are less relevant than in use.

Although our experiments produced microscopic features that undoubtedly represent the early stages of utilization wear and polish,

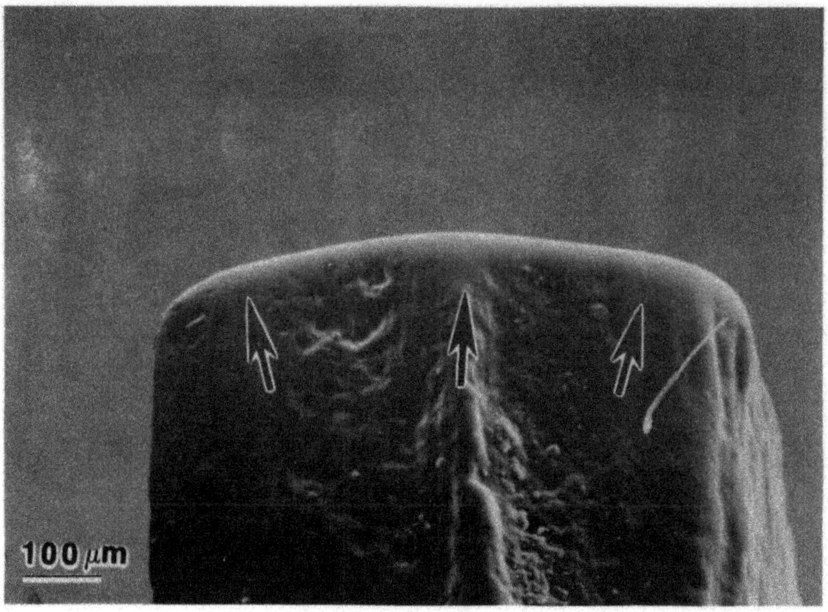

Fig. 6. Rounding and glassy polish are visible on the working edge (arrows) of this Eskimo hide scraper made from a scapula.

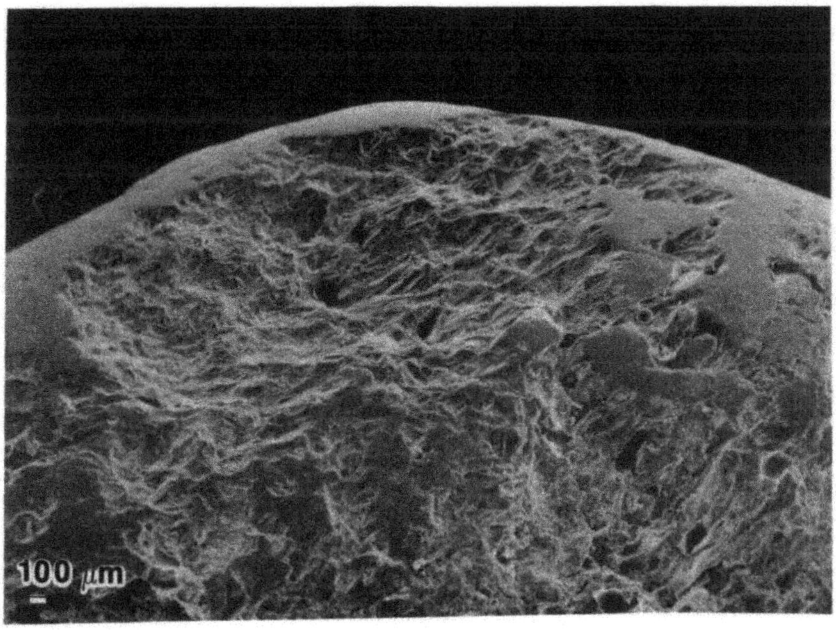

Fig. 7. Islands of glassy bone, polished by use as a pestle, are surrounded by lowlying areas showing unworn, irregular surfaces. The bone is a flake of bovid humerus that was broken and used experimentally.

none of the experiments produced gross rounding and smoothing comparable to that on the ethnographic tools. We believe this difference is due to the fact that our experimental tools, unlike the ethnographic tools, were neither used for long duration nor pre-ground, the exception being the pre-ground pestle which was used too briefly even to obliterate the grinding scratches. Although tool marks made on bones during butchery are readily identifiable as a general category by their microscopic features - and can be distinguished from pseudo-cutmarks produced by trampling by a range of criteria (Olsen and Shipman, in press) - we do not at present known of any features that serve as signatures for bone tool cutmarks as distinct from cutmarks made by tools made of stone or other raw materials.

Abrasion studies: hydraulic abrasion. A second major hypothesis accounting for the rounding and apparent wear of proposed bone tools is that they have been altered by hydraulic abrasion. At some sites, proposed bone expediency tools have been found in channel or riverine sediments. Even in low-energy sedimentary deposits, pedoturbation such as trampling, solifluction, and other processes may cause abrasion, polishing and rounding of fracture surfaces (see review in Olsen and Shipman, in press).

Sedimentologists have long studied the abrasion of sedimentary particles during hydraulic transport, but few workers (Korth, 1979, excepted) have addressed the pertinent issue here: what are the effects of sedimentary abrasion during hydraulic transport of bones? While some aspects of the process of abrasion of rock particles are analogous to those of bone abrasion, there are important differences. Sedimentologists routinely deal with homogeneous quartz particles of reasonably constant size, shape and density; bones are neither homogeneous in composition nor constant in size, shape or density. Because sedimentary particles are often similar when derived from a single source, quantification can proceed along the lines of size reduction, percentage of material lost, and rate of wear based on the attributes of particles at different distances downstream from the source. Individual particles are not commonly followed through the abrasion process, although an individual group of particles may be experimentally abraded in a tumbling barrel and examined at various points during the procedure.

Despite these fundamental differences in approach, some ideas, principles, and procedures can be borrowed from sedimentology for use on bone abrasion. Pettijohn (1975), in a review of abrasion studies, concludes that the size reduction of sedimentary particles can be attributed to five factors:

1) the size of the material at onset;
2) the nature and durability of the material;
3) the nature and violence of the action (rigor);
4) the size and proportion of the associated material (other particles); and
5) the duration or distance involved in transport.

Many of these factors also influence the abrasion of bones or bone fragments.

Our experimental studies, described below, identify the following influences on the rate and nature of abrasion of bone by sedimentary particles:

1) the grain size of the sedimentary particles with which the bones are transported;
2) the composition of the sedimentary particles;
3) the presence or absence of soft tissue on the bone;
4) the condition of the bone at the onset (fresh or weathered; broken or whole; mineralized or unmineralized);
5) the presence or absence of water in the sedimentary system; and
6) the duration or distance involved in transport.

Other factors not investigated here may also prove to be important.

Different experiments were undertaken to clarify the role of these factors and to document the sequential effects of abrasion on individual bones. A set of 14 experiments was undertaken abrading bones with sediments in a tumbling barrel. In these experiments, three circumstances were varied: the state of the bones at the onset (fresh vs. weathered, whole vs. broken, and modern vs. fossil); the sediment used as an abrasive (loess, volcanic ash, fine sand, mixed sand, coarse sand, and gravel); and the inclusion of water in the tumbling barrel (none vs. 700 ml). Bones were replicated before being subjected to abrasion and at intervals thereafter. The effects of abrasion at different stages were determined and documented using the SEM.

We investigated the effects of abrasion on bones under experimental rather than natural conditions. 300 ml. of sediment, 700 ml. of water (when water was used), and clean mammal bones were placed in a small tumbling barrel (16 cm. in diameter) that is capable of 1800 revolutions per hour. The barrel was roughly three-quarters full of bones and sediment, to allow ample room for movement of the contents. In most of the abrasion experiments, small mammal (rat, cat, and dog) bones were used; only one experiment in this series and in the aeolian abrasion experiments involved fragments of elephant bone.

There were several reasons we decided upon this approach to the problem. First, one of the only systematic studies of bone abrasion to date (Korth, 1979) also used a tumbling barrel. Using a similar methodology would therefore increase the value of our results by making them comparable to Korth's. Second, although the tumbling barrel has only a single velocity and subjects the bones to continuous contact with the sedimentary particles (aspects of the experimental situations which are not paralleled under natural conditions), we felt that this method would provide an adequate and fairly simple first approximation of the effects of natural abrasion. Third, attempts to monitor fluvial transport and abrasion under field conditions have suffered from the extreme difficulty of relocating bones after transport has occurred (Behrensmeyer, pers. comm.). For example, Hanson (1980) reports that only 42 (35%) of 120 bones placed in the shallow East Fork River were relocated after one year. Use of a tumbling barrel eliminates this problem and makes it possible to be certain that a missing bone has been destroyed (broken into tiny pieces) rather than buried or transported farther downstream. Fourth, Hanson (1980:176) reports that cow "...elements carried nearly 2 kilometers in the East Fork River showed virtually no abrasion or other transport-related damage..." as determined with the naked eye by comparing transported bones with detailed, but macroscopic, photos of those bones taken at the outset. His finding suggests that long-distance or long-duration transport might be necessary to produce visible abrasion. Although the duration or distance of transport cannot be controlled in natural settings, it is readily controlled in our experimental situation. Therefore, although we are well aware that a tumbling barrel full of bones, sediment, and sometimes water does not accurately reproduce all aspects of hydraulic transport and sedimentary abrasion that might occur under natural conditions, it does offer substantial advantages in terms of simplicity, replicability, control over variables, and a good return in knowledge for effort expended.

We recognize at least three significant aspects in which these hydraulic abrasion experiments fail to resemble actual fluvial abrasion:

1) the rotation of the tumbling barrel and its content eliminates any semblance of directionality in the abrasion, while large bones undergoing natural hydraulic abrasion may sometimes be stationary with a directional current transporting sediment-laden water over them;
2) the rotation of the barrel subjects the bones to sediment impact more continuously than is probably realistic, thus abrading the bones more rapidly than would occur in many natural settings (Behrensmeyer, pers. comm.); and
3) the force with which particles strike the bones is roughly constant, whereas actual current velocities are highly variable (Potter and Pettijohn, 1963).

All bones in these experiments were tumbled for at least 80 hours; some experiments ran as long as 200 hours. It is apparent from observing the movement of bones and sediments in the tumbling barrel that the bones are not transported around the circumference of the tumbler with each revolution. Instead, they move through slightly more than one-third of the circumference and then fall to the opposite side (Fig. 8). With a tumbler of about 50 cm. in circumference, a single revolution thus represents perhaps 40 cm. of travel for the bone.

The experiments were most conveniently conducted in terms of duration in hours; the results are given in these terms because it is difficult to determine what distance under natural transport conditions is the equivalent of one hour of transport in this tumbler. For readers who are interested in such equivalencies, however, a distance travelled (in kilometers) in the tumbler can be calculated as follows:

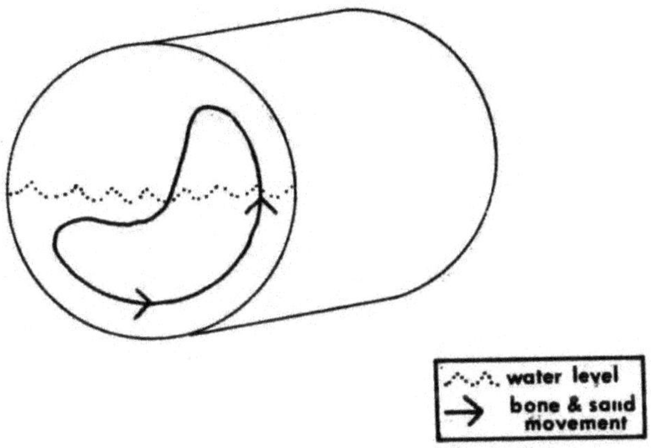

Fig. 8. This diagram indicates the movement of bones and sediment in the experimental tumbling barrel, which rotates in the direction shown by the arrow. The contents of the barrel move around part of the circumference of the barrel before gravity overcomes the centrifugal force and causes them to fall to the opposite side.

$D_k = T \quad R \quad D_r$

where D_k = total distance travelled in kilometers
T = time in hours (duration)
R = revolutions per hour (1800)
D_r = distance travelled per revolution in kilometers (.0004 for our tumbler)

Thus in terms of actual distance travelled, 80 hours is approximately 57.6 kilometers and 200 hours, the maximum duration of any of our experiments, is approximately 144 kilometers. To our knowledge, this distance far exceeds the documented distance of transport of bones under natural conditions.

As indicated above, Behrensmeyer (pers. comm.) suggests that the tumbling action may expose the bones to faster abrasion than would occur under natural circumstances because of the continuous impact of sedimentary particles in the tumbler. Therefore, our rates of abrasion cannot be translated with confidence into rates that would occur under more realistic conditions. In fact, our minimum distance of actual transport may be more nearly a maximum distance of transport under natural conditions. We do not offer the distance equivalency formula above with any confidence that distance travelled in the tumbler is the same as distance travelled under natural conditions; we caution readers strongly against making this leap of faith.

Although we plan to report the results of these studies more fully in future publications, a few comments pertinent to the bone tool controversy can be made here. The first is that grossly visible abrasion occurs extremely slowly. A few bones have lost sufficient material from their surface to expose cancelli after 20 hours of tumbling, but many have not. The majority of bones, regardless of their initial condition or of the sediment size, show grossly visible abrasion after 35 hours. The degree of abrasion seen on such bones ranges from slight to moderate. Shape changes occur throughout the abrasion process, but these are frequently undetectable unless a replica showing the original, unmodified bone shape is available. Gross shape changes in the early stages of abrasion are extremely subtle. Microscopic changes occur fairly rapidly. Cutmarks on bones lose all of their diagnostic, microscopic features after as little as 5 hours of abrasion although the indentations are still visible after 80 hours (Figs. 9a, b). This means that hydraulically transported bones cannot be expected to show cutmarks that can be identified on the basis of SEM inspection. In our experiments, we were unable to produce a profound abrasion or polish, such as is seen on many purported bone tools.

The second important point is that none of our tumbling experiments produced large numbers of scratches or linear grooves regardless of the particle size of the sediment. Not a single specimen showed more than a few scratches, even upon microscopic inspection. In addition, none of these scratches mimicked cutmarks in microscopic appearance (Fig. 10). Nonetheless, cutmark mimics and other linear grooves and scratches have been reported to occur frequently on trampled bones (e.g., Behrensmeyer et al., 1986; Oliver, 1984, 1986;

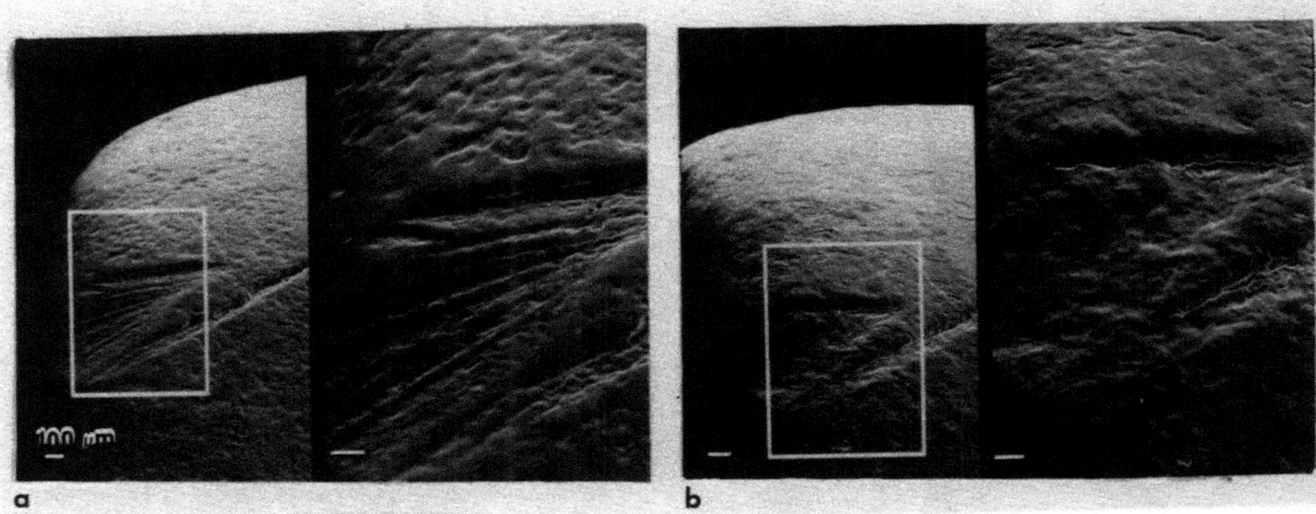

Fig. 9. (a) The experimental cutmark on this bone shows all the expected microscopic features prior to abrasion. (b) After 5 hours of abrasion in a tumbling barrel, the fine striations have disappeared, leaving only a broad indentation.

Fig. 10. Experimental abrasion rarely produced scratches on bones; the few scratches that are produced (arrow) do not mimic cutmarks. These scratches appeared on this femoral condyle after 35 hours of tumbling with dry gravel.

Fig. 11. The Mount Saint Helen's ash used in our abrasion experiments was comprised of very small, angular particles of volcanic glass; this ash was a more effective abrasive than various sands or gravels we used.

Fiorillo, 1984; but see also Andrews and Cook, 1985, and Olsen and Shipman, in press). Since abrasion in a tumbling barrel without water is a reasonable approximation of trampling, the absence of linear grooves in our experiments requires explanation. One possible explanation is that our tumbling experiments far exceeded, in duration and hence in potential abrasion, the trampling experiments or conditions documented in the literature. Thus, while relatively brief trampling may create grooves and scratches, prolonged trampling or dry sedimentary abrasion may produce a more general polish and loss of surface detail that eventually obliterates individual scratches. Alternatively, the free movement of both bones and sediment in the tumbling barrel may prevent sedimentary particles from traversing bone surfaces for any distance with sufficient force to leave a distinct scratch.

Third, our experiments suggest that abrasion, the removal of bone material caused by the impact of sedimentary particles, is a surface area phenomenon. That is, the smaller the grain size, the faster the rate of abrasion, all other things being equal. Thus, in our experiments, the volcanic ash and the loess were the most abrasive sediments.

Fourth, the composition of the sedimentary particles was important. One of our experiments was run using Mt. St. Helen's ash. The ash itself was comprised primarily of very small particles of sharp-edged volcanic glass (Fig. 11). The ash abraded bone considerably faster than did fine sand, suggesting that the attributes of the ash (sharpness, brittleness) may have played a major role in its effectiveness as an abrasive substance.

Fifth, although the velocity generated by our tumbler (20 cm/sec) was similar to that of a stream competent to move fine-grained sediments, it was not great enough to break any of the bones used in any of the experiments, with the exception of the thin vault bones of a mouse skull. Given that the cortical thickness of bones increases allometrically with body size (Currey, 1984; McMahon, 1973), it seems unlikely that faster streams would be able to break larger bones unless they became wedged against rocks or other obstacles or unless the bones were in a dry, friable state at the outset.

Sixth, broken bone surfaces appear to abrade much more rapidly or more visibly than intact surfaces. Extreme abrasion (200 hours) of fresh, broken bones in loess produced marked smoothing and polishing of the fracture surfaces and relatively little visible change on the intact surfaces. Figure 12 shows the sequence of abrasion on a broken dog ulna in this experiment. Such long-term abrasion with loess smooths and polishes the fracture surface to a point that its features might be confused with the effects of brief utilization on soft substances (compare Fig. 12d with Fig. 8b in Shipman, Chapter 18, this volume). However, it is important to note that extreme abrasion in our experiments did not scratch the bones. Therefore, although extreme abrasion mimics brief utilization on soft substances, it does not mimic the effects of extensive utilization on either soft substances or those of mixed hardness, such as occur either in hide-working or butchery.

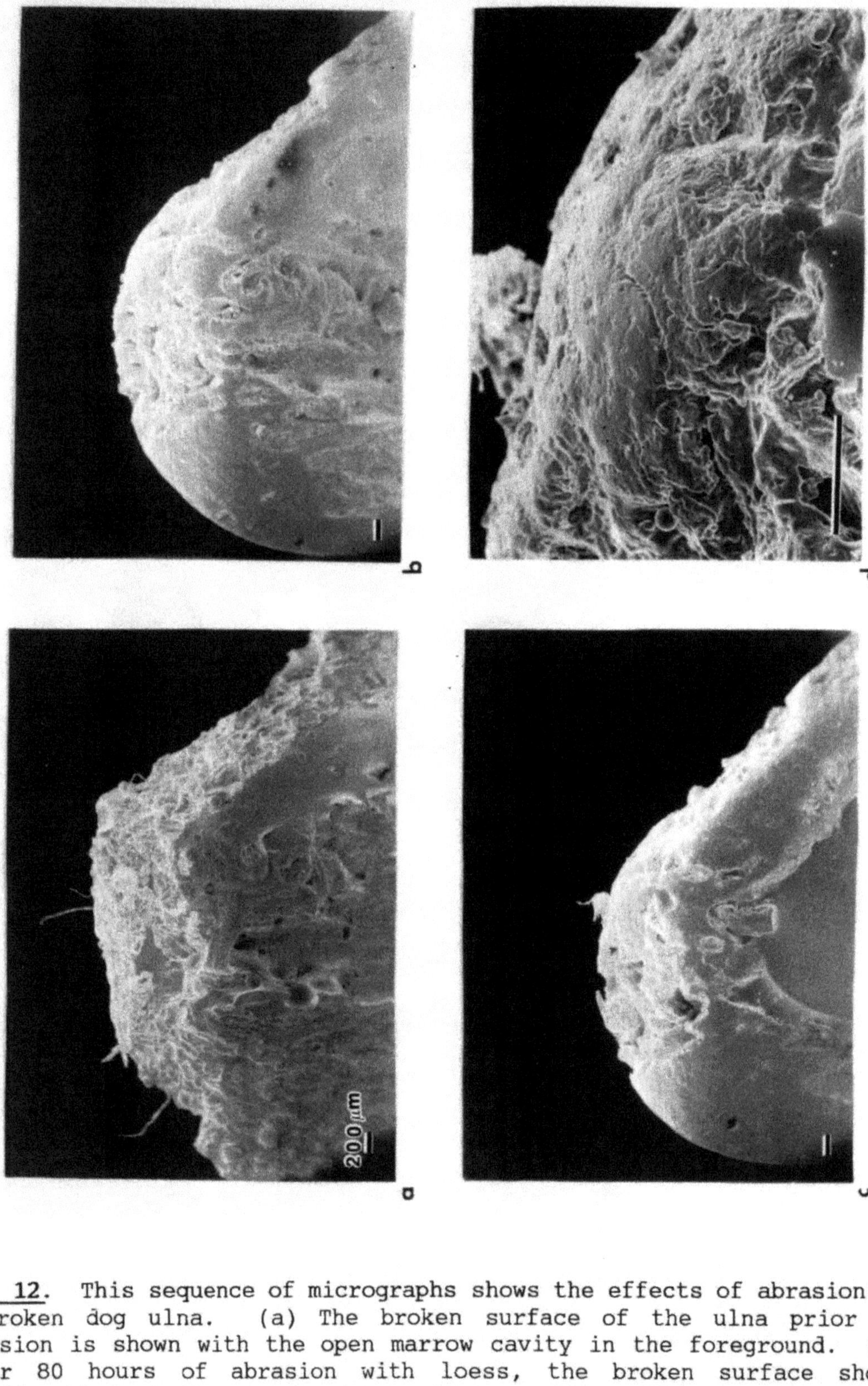

Fig. 12. This sequence of micrographs shows the effects of abrasion on a broken dog ulna. (a) The broken surface of the ulna prior to abrasion is shown with the open marrow cavity in the foreground. (b) After 80 hours of abrasion with loess, the broken surface shows considerable rounding and some polish. (c) After 164 hours of abrasion with loess, the bone shows little additional rounding or polish. (d) A close-up of the surface after 164 hours of abrasion reveals a similarity to bones used briefly on soft substances. Compare this image with that in Fig. 8b in Shipman (Chapter 18, this volume).

Seventh, comparison of longitudinal sections of bones with the location of earliest detectable abrasion show that, as expected, areas with thin cortical bone overlying cancellous bone show the earliest signs of abrasion. These areas include anatomical projections and crests, such as the greater trochanter of the femur and the greater tuberosity of the humerus. Areas where abrasion apparently occurs more slowly, or is only visible after greater duration, are gently concave subchondral areas or the thick cortical bone of diaphyses. Our work suggests that the initial stage of abrasion on subchondral bone merely removes asperities and irregularities, smoothing the surface but not yet exposing bone of sufficiently different texture to be grossly visible (Fig. 13).

Eighth, prolonged abrasion of fractured elephant bone (over 200 hours) with loess was inadequate to produce the degree of gross rounding and smoothing or of glassy, microscopic polish seen on ethnographic tools.

Abrasion studies: aeolian abrasion. Another major hypothesis is that the purported bone expediency tools have been modified by windborne sedimentary particles. This hypothesis receives circumstantial support at many pre-Clovis sites in North America, where bones are found in Peorian loess, which is classically a windborne sediment (Graham, 1981; Stanford, 1979). Although we perceived no a priori reason why windborne loess would produce different abrasion than waterborne loess, we devised an experiment to explore the effects of aeolian abrasion on bone.

A modern elephant femur was cleaned of flesh and fractured; its fracture surface was replicated to document its original microscopic morphology. The fracture surface was then abraded with loess using a Paasche abrasion gun driven by pressurized air (25 lbs/in^2). The nozzle was directed at the tip of the broken elephant femur in order to maximize the extent of abrasion. The experiment was continued intermittently for about 5 hours total, at which time alteration of the bone was grossly apparent and our tolerance for further inhalation of loess was minimal.

Figure 14a,b compares the fracture surface of the elephant bone before and after abrasion. When viewed under SEM, the fracture surface of the elephant bone is dramatically altered by aeolian abrasion. There is a noticeable loss of bony material and a pronounced rounding and smoothing of formerly irregular surfaces and edges. Normal vascular openings and other holes in the bone's surface are enlarged. However, even at high magnification, the bone's surface is strikingly pitted and irregular in texture. We were unable to find a single scratch, coarse or fine, on this specimen.

Despite the dramatic microscopic alteration of the elephant bone, it is important to note that the extent of grossly visible edge rounding produced by this experiment was orders of magnitude less than that seen on the ethnographic tools or on putative bone tools from various sites. Further, our intuitive assessment was that, if the

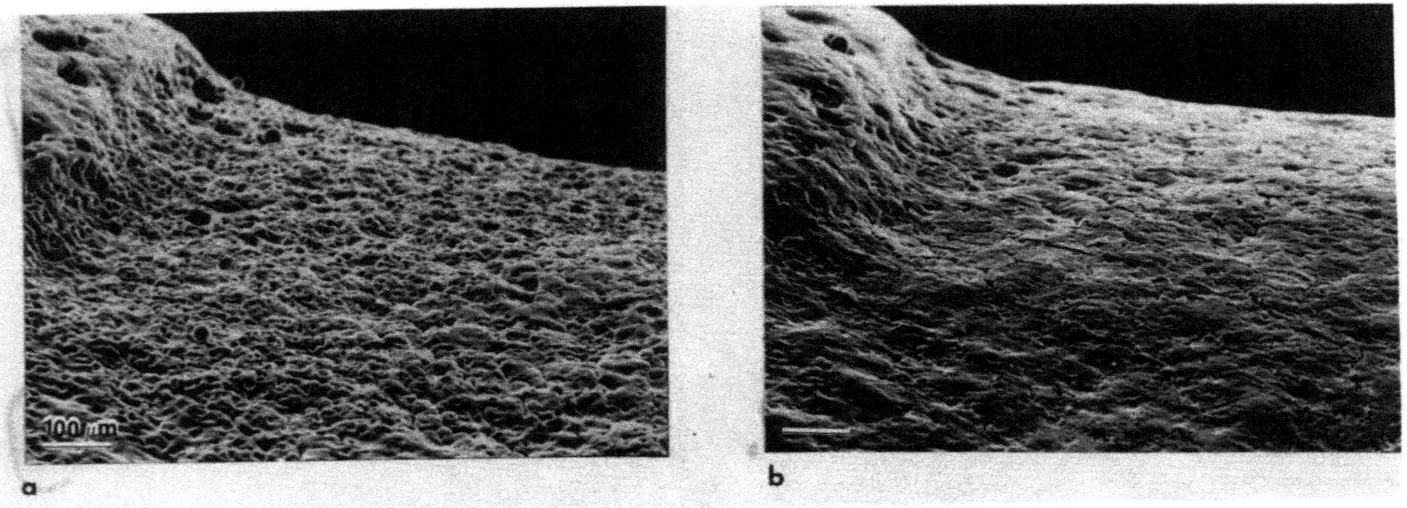

Fig. 13. (a) Before abrasion, the subchondral bone of the semilunar notch of an ulna shows a typical, highly vascularized surface. (b) After 35 hours of abrasion in sand and water, the microscopic morphology indicates that some of the bone surface has been lost. However, this change was so subtle as to be undetectable from gross visual inspection.

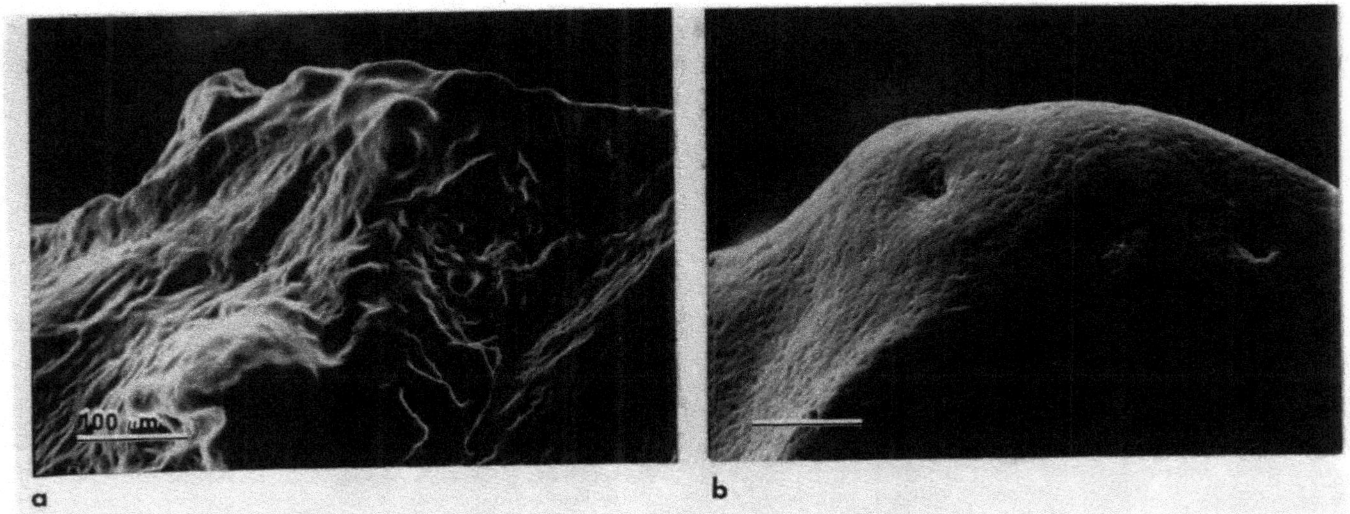

Fig. 14. These micrographs compare the same small point on a fractured elephant femur before (a) and after (b) aeolian abrasion with loess. Although the degree of rounding and smoothing is impressive, the bone surface in (b) remains pitted and irregular in texture. At the microscopic level, aeolian abrasion did not produce a fine, glassy polish typical of utilization.

elephant bone were fixed in position with only the tip exposed (thus permitting differential distribution of the polish to be created), and if we were not recycling the loess, the bone tip would have been buried in sediment by the end of the experiment, if not sooner.

The most significant conclusions of this experiment are, first, that windborne abrasion of large mammal bone with loess, to the extent achieved here, does not create either a fine polish or scratches like those resulting from extensive utilization. It is possible that much more prolonged abrasion could produce such features, but the microscopic morphology of the abraded elephant bone is much less similar to that of the ethnographic tool than are the briefly-used experimental bone tools. Second, aeolian abrasion in this experiment failed to produce grossly-apparent rounding of the elephant bone comparable to that observed on ethnographic or proposed bone expediency tools. In order for aeolian abrasion to create such gross rounding, the windborne abrasion would have had to continue for a great deal longer than the duration of our experiment, without burying the bone in question. While this is theoretically possible, we judge it unlikely.

Summary and Conclusions

Although the results of these abrasion and utilization studies are promising, they also indicate the need for caution in interpreting the microscopic features of possible bone tools. One of the most important findings was that hydraulic abrasion of long duration produces effects on previously-fractured surfaces that are strikingly similar to those produce by brief utilization on soft substances. Therefore, if such features are observed on a bone of unknown history, only the first of our two criteria for a positive identification, that is, similarity of features to those of a known, is met. Distinctiveness of those features from ones caused by other taphonomic events is not demonstrated. Thus, in such cases, a positive identification based on the microscopic features along cannot be made.

However, in some cases a positive identification of brief utilization on soft substances can be made. Such utilization is insufficient to expose cancelli on prominent anatomical features, such as the olecranon, the greater trochanter, the lip of the glenoid fossa, or the greater tuberosity of the humerus, to name a few, especially if these are some distance from the working edge. In contrast, abrasion of sufficient duration to produce features like those of brief utilization (in our experiments, 200 hours of abrasion) will invariably expose cancelli on such anatomical features. Visible abrasion on such areas, if they are isolated from the proposed working edge, indicates that sedimentary abrasion of some sort was probably one of the taphonomic agents at work. Therefore, if abrasion is lacking on such features, natural abrasion can be eliminated as unlikely, leaving three alternatives to be evaluated:

1) hominid utilization caused the alteration of the fracture surface;
2) the bone was abraded while the articular end was protected from abrasion by implantation in sediment; or

3) the bone was abraded while remaining soft tissues protected most surfaces from damage.

It may be difficult to choose among these three alternatives in cases of bones altered in ways that microscopically resemble brief utilization on soft tissues. The distribution of apparent wear or abrasion may enable the analyst to decide if hypothesis #1 or #2 is more likely. If sedimentary abrasion occurred while the bone was implanted in sediments (hypothesis #2), then the abrasion ought to be unifacially distributed. That is, the abrasion ought to occur predominantly and perhaps exclusively on the upstream side of the bone and not on the downstream side. In contrast, a utilized bone may shown a preferred axis of movement, but this ought to be restricted to a specific functional area, not to an entire face or side of a specimen's end. Therefore, we propose that hypothesis #1 should be favored over #2 if the following rigorous criteria are met:

1) there are appropriate microscopic features concentrated primarily on the functional working edge or tip;
2) the microscopic polish is not unifacial;
3) there is a demonstrably unabraded articular end or other area possessing prominent anatomical tuberosities or crests where underlying cancelli are not exposed.

If these criteria are met, the task then becomes one of evaluating the relative merits of hypotheses #1 and #3. Hypothesis #3 is less likely if bones show pre-fossilization weathering, which can only occur after soft tissues are removed, especially if the fracture appears to have been made while the bone was fresh (i.e., prior to the weathering). The presence of rodent gnawing, which also generally occurs after soft tissues are removed, also suggests that hypothesis #3 is less likely. Finally, hypothesis #3 is less likely if there is a pattern of possibly utilized bones that are the same or functionally similar portions of skeletal elements. That is, it is possible that the proximal portion of the shaft of one broken humerus, complete to its unabraded distal end, might have been abraded while the more distal end is covered in soft tissues but this explanation becomes increasingly less likely as additional specimens of this type that also possess strongly differential polish are found.

It is inevitable that many cases will be found that cannot be resolved on the basis of our current knowledge of the effects of abrasion and utilization. It is also probable that, if bone expediency tools do exist, many real tools will be "explained away" in analysts's zeal to evaluate the specimens critically. This is an acceptable alternative to accepting every oddly-shaped or oddly-worn bone as utilized without applying rigorous tests. As our knowledge expands, new criteria can be expected to emerge and already analyzed specimens re-assessed.

It is important to note that neither type of sedimentary abrasion, as studied here, produces alterations that approach the extent of modification seen in ethnographic bone tools or in some proposed bone expediency tools from Old or New World sites. By far the easiest and fastest means to produce gross rounding and fine microscopic polish, differentially distributed, is to utilize the broken edges in tasks involving at least some working of soft tissues. Clearly, additional actualistic studies and detailed inspections of purported bone tools are needed. This study indicates that assessments of the taphonomic history of proposed bone expediency tools can be aided greatly by a careful comparison of the microscopic morphology of the possible working edges with those of known bone tools.

Acknowledgments

We thank Dennis Stanford, Gary Haynes, Dan Fisher, and Rick Potts for providing specimens of utilized and unutilized bones; other altered bones were provided by or collected under the sponsorship of the National Museums of Kenya (on permit #OP.13/001/6C79) with the assistance of R.E. Leakey and his staff. Simon Pincus and Amanda Alexander assisted in the abrasion studies, using, in part, volcanic ash provided by Sally Shamp. Ginsberg and Margie, two elephants, provided their personal osteological specimens without which this project could not have proceeded. Simon Walker rescued one of the elephant bones from untimely destruction. Funding was provided by the National Science Foundation (BNS 80-1397 and 80-2-1397), the Boise Fund, and a National Institutes of Health Biomedical Research Grant (5 S07RR07041-13).

References

Andrews, P., and Cook, J. 1985. Natural modifications to bones in a temperate setting. Man (N.S.) 20: 675-691.

Arndt, S.L., and Newcomer, M.H. 1986. Breakage patterns on prehistoric bone points: An experimental study. In Roe, D., (ed.), Studies in the Upper Paleolithic of Britain and Northwest Europe, pp.165-173. Oxford: British Archaeological Reports, International Series 296.

Behrensmeyer, A.K. 1978. Taphonomic and ecologic information from bone weathering. Paleobiology 4:150-162.

Behrensmeyer, A.K., Gordon, K. and Yanagi, G. 1986. Trampling as a cause of bone surface damage and pseudo-cutmarks. Nature 319:768-771.

Binford, L.R. 1981. Bones: Ancient Men and Modern Myths. San Francisco: Academic Press.

Binford, L.R., and Bertram, J. 1977. Bone frequencies and attritional processes. In Binford, L.R., (ed.), For Theory-Building in Archeology, pp. 77-156. New York: Academic Press.

Bonnichsen, R. 1979. Pleistocene Bone Technology in the Beringian Refugium. Archaeological Survey of Canada Paper No. 89, Mercury Series. National Museum of Man, Ottawa.

Bonnichsen, R., and Young, D. 1980. Early technological repertoires: Bone to stone. Canadian Journal of Anthropology 1(1):123-129.

Brain, C.K. 1967. Bone weathering and the problem of bone pseudo-tools. South African Journal of Science 63:97-99.

Brain, C.K. 1981. The Hunters or the Hunted? An Introduction to African Cave Taphonomy. Chicago: University of Chicago Press.

Brain, C.K. in press. Recent excavations at Swartkrans: The southern African 'robust' australopithecines in temporal, archaeological, and paleoecological context. In Grine, F.E. (ed.), The Evolutionary History of the Robust Australopithecines. Chicago: Aldine.

Breuil, H. 1938. The use of bone implements in the old Paleolithic period. Antiquity 12:56-67.

Bryan, A.L. 1965. Paleo-American prehistory. Idaho State University Museum Occasional Papers 16.

Bunn, H.T. 1982. Meat-eating and Human Evolution. Ph.D. dissertation. University of California, Berkeley.

Clark, J.D. 1977. Bone tools of the earlier Pleistocene. Eretz-Israel 13: 23-37.

Cotterell, B., and J. Kamminga, 1987. The formation of flakes. American Antiquity 52:675-708.

Currey, J. 1984. The Mechanical Adaptations of Bones. Princeton: Princeton University Press.

Crader, D.C. 1983. Recent single-carcass bone scatters and the problem of 'butchery' sites in the archaeological record. In Clutton-Brock, J., and Grigson, C. (eds.) Animals and Archaeology, Volume I, Hunters and their Prey, pp. 107-141. Oxford: British Archaeological Reports.

Currey, J. 1984 The Mechanical Adaptations of Bones. Princeton: Princeton University Press.

Dart, Raymond A. 1949. The predatory implemental technique of Australopithecus. American Journal of Physical Anthropology 7:1-38.

Dart, Raymond A. 1957. The Osteodontokeratic Culture of Australopithecus prometheus. Transvaal Museum Memoirs No. 10, Pretoria.

Fiorillo, A. 1984. An introduction to the identification of trample [sic] marks. In First International Conference on Bone Modification Abstracts, p.11. Orono: Center for the Study of Early Man.

Fisher, D.C. 1984. Taphonomic analysis of late Pleistocene mastodon occurrences: Evidence of butchery by North American Paleo-Indians. Paleobiology 10:338-357.

Freeman, L.G. 1978. Mousterian worked bone from Cueva Morin (Santander, Spain). In Freeman, L.G. (ed.), _Views of the Past_, pp. 29-51. The Hague: Mouton.

Frison, G.C. 1974. _The Casper site: A Hell Gap Bison Kill on the High Plains_. New York: Academic Press.

Frison, G.C. 1978. _Prehistoric Hunters of the High Plains_. New York: Academic Press.

Frison, G.C. 1982. Bone butchering tools in archaeological sites. _Canadian Journal of Anthropology_ 2:159-167.

Frison, G.C., and Todd, L. 1985. _The Colby Mammoth Site: Taphonomy and Archaeology of a Clovis Kill in Northern Wyoming_. Albuquerque: University of New Mexico.

Graham, R.W. 1981. Preliminary report on late Pleistocene vertebrates from the Selby and Dutton archaeological/paleontological sites, Yuma County, Colorado. _Contributions to Geology_ 20(1):33-56.

Guilday, J.E., Parmalee, P.W., and Tanner, D.P. 1962. Aboriginal butchering techniques at the Eschelman site (36LA12), Lancaster County, Pennsylvania. _Pennsylvania Archaeologist_ 32:59-83.

Hannus, L.A. 1985. The Lange/Ferguson site - An event of Clovis mammoth butchery with the associated bone tool technology. Ph.D. thesis, University of Utah.

Hannus, L.A. in press a. Flaked mammoth bone from the Lange/Ferguson site, White River Badlands area, South Dakota. In Bonnichsen, R., and Sorg, M., (eds.), _Bone Modifications_. Orono: Center for the Study of Early Man.

Hannus, L.A. in press b. 6,000 pounds of meat to fillet: That's mammoth! _Proceedings of the World Archaeological Congress_ Southampton, U.K.

Hanson, C.R. 1980. Fluvial taphonomic processes: Models and experiments. In Behrensmeyer, A.K., and Hill, A.P. (eds.), _Fossils in the Making_, pp. 156-181. Chicago: University of Chicago Press.

Haynes, C.V., Jr. 1978. The Clovis culture. _American Quaternary Association Fifth Biennial Meeting Abstracts_: 129-132.

Haynes, G. 1981. Bone modifications and skeletal disturbances by natural agencies: Studies in North America. Ph.D. thesis, Catholic University of America.

Haynes, G. 1982. Utilization and skeletal disturbance of North American prey carcasses. _Arctic_ 35:266-281.

Haynes, G. 1986. Spiral fractures and cut mark-mimics in noncultural elephant bone assemblages. _Current Research in the Pleistocene_ 3:45-46.

Hill, A. 1980. Early postmortem damage to the remains of some contemporary East African mammals. In Behrensmeyer, A.K., and Hill, A.P. (eds.) Fossils in the Making, pp.131-155. Chicago: University of Chicago Press.

Johnson, E. 1982. Paleo-indian bone expediency tools: Lubbock Lake and Bonfire Shelter. Canadian Journal of Anthropology 2:145-157.

Johnson, E. 1985. Current developments in bone technology. In Schiffer, M. (ed.), Advances in Archaeological Method and Theory. 8:157-235.

Johnson, E., and Shipman, P. 1986. Scanning electron microscope studies of bone modifications. Current Research in the Pleistocene 3:47-48.

Korth, W.W. 1979. Taphonomy of microvertebrate fossil assemblages. Annals of the Carnegie Museum 48:235-285.

Leakey, M.D. 1971 Olduvai Gorge, Volume III, Excavations in Beds I and II, 1960-1963. London: Cambridge University Press.

LeMoine, G. 1985. Experimental use wear analysis of bone tools. M.A. thesis, University of Calgary.

Lyman, R.L. 1984. Broken bones, bone expediency tools, and bone pseudotools: Lessons from the blast zone around Mt. St. Helen's, Washington. American Antiquity 49:315-333.

McMahon, T. 1973. Size and shape in biology. Science 179:1201-1204.

Miller, S.J., 1983. Osteo-archaeology of the mammoth-bison assemblage at Owl Cave, the Wasden site, Idaho. In LeMoine, G., and MacEachern, A.S. (eds.), A Question of Bone Technology, pp. 39-54. Calgary: University of Calgary.

Morlan, R.E. 1980. Taphonomy and Archaeology in the Upper Pleistocene of the Northern Yukon Territory: A Glimpse of the Peopling of the New World. National Museum of Canada, Archaeological Survey of Canada, Paper no. 94.

Nelson, W.H. 1889. Report of the United States National Museum Archives. Washington, D.C.

Newcomer, M.H. 1974. Study and replication of bone tools from Ksar Akil (Lebanon). World Archaeology 6:138-153.

Oliver, James S. 1984. Bone damage morphologies from Shield Trap Cave, Carbon County, Montana. In First International Conference on Bone Modifications: Abstracts, pp.27. Orono: Center for the Study of Early Man.

Oliver, James S. 1986. The Taphonomy and Paleoecology of Shield Trap cave (24CB91), Carbon County, Montana. M.Sc. thesis, University of Maine.

Olsen, S.L 1984. Analytical approaches to the manufacture of bone artifacts in prehistory. Ph.D. thesis, London University.

Olsen, S.L. 1988. Applications of scanning electron microscopy in archaeology. In Hawkes, P., (ed.), Advances in Electronics and Electron Physics 71: 357-380. New York: Academic Press.

Olsen, S.L, and Shipman, P. In press. Surface modification on bone: Trampling versus butchery. Journal of Archaeological Science.

Pei, W.C. 1938. Le Role des Animaux et des Causes Naturelles dans la Cassure des Os. Palaeontologica Sinica n.s. D, 7.

Pettijohn, F.J. 1975. Sedimentary Rocks. Third Edition. New York: Harper and Row.

Potter, P.E., and Pettijohn, F.J. 1963. Paleocurrents and Basin Analysis. New York: Academic Press.

Potts, R., and Shipman, P. 1981. Cutmarks made by stone tools on bones from Olduvai Gorge, Tanzania. Nature 291:577-580.

Rose, J.J. 1983. A replication technique for scanning electron microscopy: Applications for anthropologists. American Journal of Physical Anthropology 62:255-263.

Semenov, S.A. Prehistoric Technology: An Experimental Study of the Oldest Tools and Artefacts from Traces of Manufacture and Wear. 1964. Tr. by M.W. Thompson. Chatham, Kent: Barnes & Noble, Inc.

Shipman, P. 1981. Applications of Scanning Electron Microscopy to Taphonomic Problems." In Cantwell, A.-M., Griffin, J.B., and Rothschild, N., (eds)., The Research Potential of Anthropological Museum Collections, pp 357-385, vol 276, Annals of the New York Academy of Sciences.

Shipman, P. 1988. Actualistic studies of animal resources and hominid activities. In Olsen, S. (ed.), Scanning Electron Microscopy in Archaeology. Oxford: British Archaeological Reports, International Series.

Shipman, P. In press a. Altered bones from Olduvai Gorge, Tanzania: Techniques, problems, and implications of their recognition. In Bonnichsen, R., and Sorg. M. (eds.) Bone Modifications. Orono: Center for the Study of Early Man.

Shipman, P. In press b. Body size and broken bones: Preliminary interpretations of proboscidean remains. In Fox, J., and Smith, C., (eds.) Mammoths, Mastodons, and Human Interactions. Waco: Baylor University Press.

Shipman, P., and Phillips-Conroy, J. 1977. Hominid tool-making versus carnivore scavenging. American Journal of Physical Anthropology 46:77-86.

Shipman P., and Rose, J.J. 1983a. Early hominid hunting, butchering and carcass-processing behaviors: Approaches to the fossil record. Journal of Anthropological Archaeology 2:57-98.

Shipman, P., and Rose, J.J. 1983b. Evidence of butchery and hominid activities at Torralba and Ambrona; An evaluation using microscopic techniques. Journal of Archaeological Science 10:465-474.

Shipman, P., Fisher, D.C., and Rose, J.J. 1984. Mastodon butchery: Microscopic evidence of carcass processing and bone tool use. Paleobiology 10:358-365.

Stanford, D. 1979. The Selby and Dutton sites: Evidence for a possible pre-Clovis occupation of the High Plains. In Humphrey, R.L., and Stanford, D. (eds.), Pre-Llano Cultures of the Americas, pp. 101-123. Anthropological Society of Washington, D.C.

Stanford, D., Bonnichsen, R., and Morlan, R.E. 1981. The Ginsberg experiment: Modern and prehistoric evidence of a bone-flaking technology. Science 212:438-440.

Steele, D.G., and Carlson, D. In press. Excavation and taphonomy of mammoth remains from the Duewall-Newberry site, Brazos County, Texas. In Bonnichsen,R., and Sorg, M. (eds.), Bone Modifications. Orono: Center for the Study of Early Man.

Wolburg, D.L. 1970. The hypothesized osteodontokeratic culture of the australopithecinae: A look at the evidence and the opinions. Current Anthropology 11:23-30.

THE IDENTIFICATION OF STONE AND METAL TOOL MARKS ON BONE ARTIFACTS

Sandra L. Olsen

Introduction

The study described here is a small part of a larger project to identify a wide range of manufacturing marks visible on prehistoric bone artifacts. This research focuses on distinguishing between marks made by stone and those made by metal tools. In studying small collections of bone artifacts from Bronze Age and Iron Age sites in Britain it was noted that, not surprisingly, there was an increase in the use of metal tools in working bone and antler through time.

The 73 Early Bronze Age artifacts from West Row Fen (Mildenhall 165), in Suffolk, dating to around 1400 bc, showed only a few examples of marks attributable to metal tools, while the rest appeared to have been made with stone tools. In contrast, the 56 artifacts from the Iron Age site of Fiskerton (ca. 510 bc), in Lincolnshire, were made predominantly with metal tools. In order to better see the transition, material from another site of Late Bronze Age date (900-700 bc) was examined. These artifacts from Egham Runnymede Bridge, in Surrey, (Longley 1980) were intermediate in terms of their manufacture, showing slightly more use of metal tools than at West Row.

This study, which necessitated the development of distinguishing criteria for separating traces from stone and metal tools, contributes to our understanding of ancient metallurgy, as well as bone technology. The number of Bronze Age metal tools known from Britain is quite small and almost certainly does not reflect the full range then available. Their original rarity, recycling of worn tools, and decomposition prevent us from discovering all of the types of tools then in existence.

By looking at manufacturing marks on a wide variety of materials, including bone artifacts, it should be possible to expand our understanding of the types of metal tools present in the Bronze Age and the kinds of materials upon which they were used. The increase in the frequency of their use through time in the manufacture of bone artifacts may suggest increased availability of metal and a greater dependence on metal for utilitarian purposes.

Descriptions of the Collections of Bone Artifacts

The Early Bronze Age site of West Row is located on the edge of a fen 5 km west-northwest of Mildenhall, Suffolk. The area occupied was on a sand ridge and an adjacent hollow that was subsequently overlain with peat. It was excavated from 1982 to 1986 by the Suffolk Archaeological Unit, headed by E.A. Martin.

A total of 73 bone and antler artifacts were recovered during excavation. Table 1 lists and enumerates the artifacts by type. Figures 1 and 2 illustrate a few of the more distinct examples. The identifiable species from which the raw material was derived include

Table 1
West Row (MNL 165) Bone Artifacts

Artifact	Count
Antler Flaker	
(Rod)	2
(Tine)	2
Worked Antler Tine	2
Antler Mace-head	1
Dagger	1
Awl	57
Rib Knife or Awl Base	1
Spatula	1
Scoop	1
Tubular Bead	3
Miscellaneous Fragments	2
TOTAL:	73

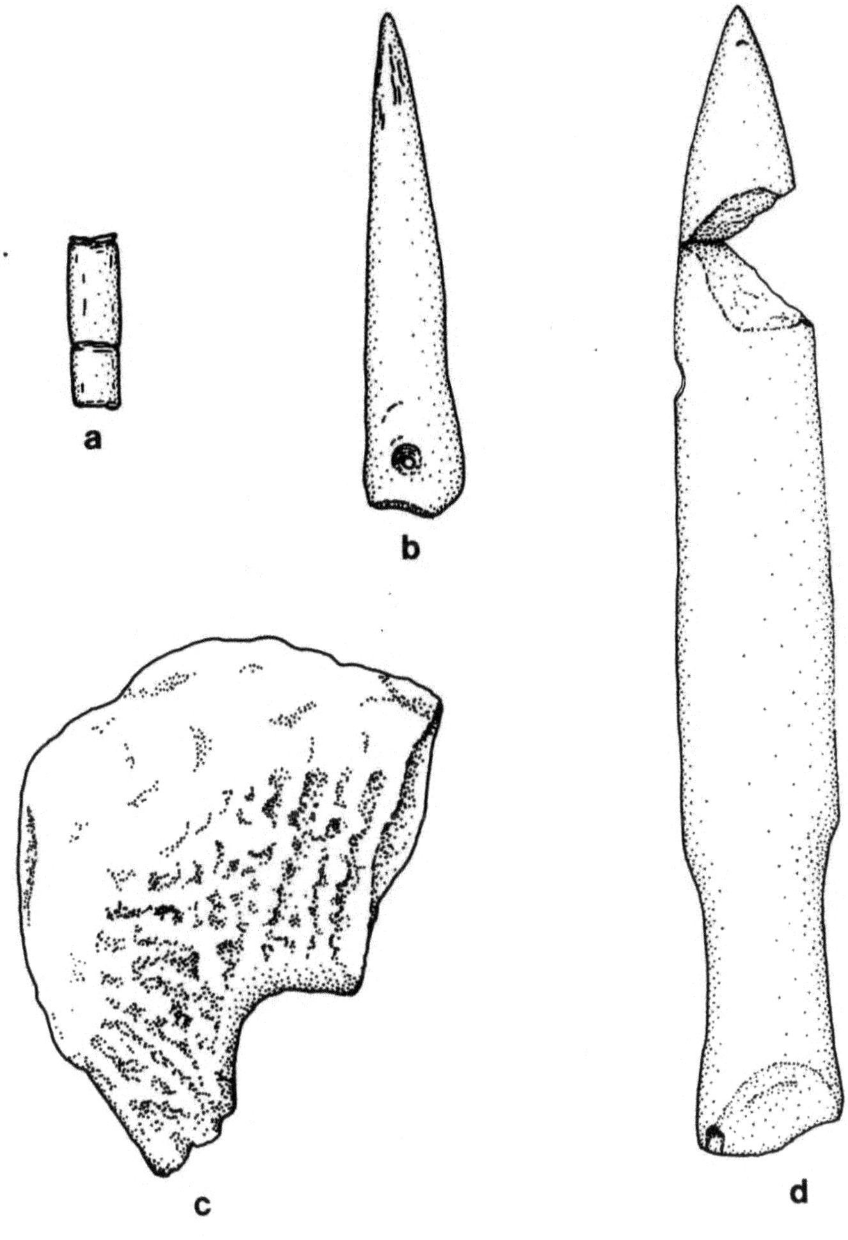

<u>Fig. 1</u>. Bone and antler artifacts from the Early Bronze Age site of West Row, Suffolk. a. tubular bead made by grooving and snapping both ends of a bird bone. Note the incomplete groove around the middle, probably meant to section the bead into two pieces. b. eyed awl made from a sheep or goat metapodial. c. fragmentary mace-head made on a shed red deer antler. Oblong hole (lower right) would have received the wooden handle. d. bone "dagger" or lance, probably originally fit into a socketed handle.

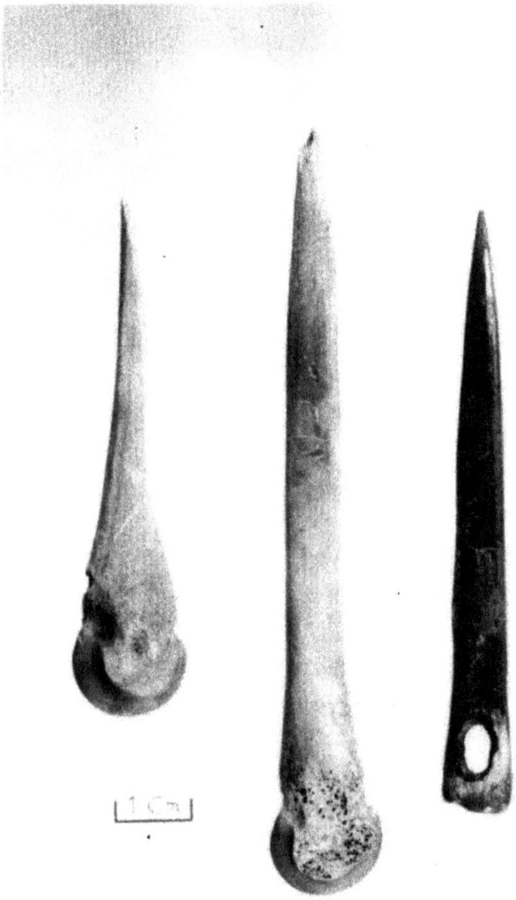

Fig. 2. Typical awls from West Row.

red deer (Cervus elaphus), roe deer (Capreolus), sheep or goat (Ovis aries or Capra hircus), and cattle (Bos taurus).

Fiskerton is an Iron Age site in Lincolnshire which was sealed by a later Romano-British flood deposit. The Iron Age levels produced 56 bone and antler artifacts. Of these, 54 were of a type commonly found in Iron Age sites on Britain, first described in detail by Cunnington (1923: 82-8) and classified as "gouges" (Fig. 3). Forty of the 54 Fiskerton "gouges" were made from the tibiae of sheep and goats. The remainder include seven from other sheep/goat bones, three pig (Sus scrofa) tibiae, two roe deer metatarsals, one cattle radius, and one too fragmentary to identify. The other two objects consisted of a bipointed needle and a worked red deer antler tine.

The artifacts from Runnymede which were examined for comparison were antler tine off-cuts or manufacturing debitage, cheekpieces, and a toggle, as well as notched bones, bone awls, bipoints, and spatulae.

Methods

Several methods were employed in the examination and analysis of manufacturing marks on the bone artifacts from West Row and Fiskerton. Firstly, the objects were studied for macroscopic traces. Manufacturing marks are generally visible to the unaided eye and are relatively easy to identify with confidence. To bring out the more subtle characteristics of the marks and some of the finer detail, a hand lens (10X) and a stereomicroscope (18X) were used to study each piece.

After examining all of the specimens and locating a variety of distinct manufacturing traces, samples were selected to be studied through the SEM. The SEM offered high resolution images, with a great depth of field and a wide range of magnifications. The curvature of the bone and the depth of many of the manufacturing marks inhibits high quality photomicrographs with an optical microscope, but with the SEM most or all of the surface of the object could be brought into focus at once (Olsen, 1988).

Since, in this case, most of the pieces were whole, or nearly so, they were too large to be studied directly in the SEM chamber at my disposal. Instead, small silicone rubber molds (Dow Corning Silastic 9161) of the surfaces bearing traces were made and prepared for viewing. After curing, the mold was peeled off, attached to an aluminum stub with Durofix adhesive, and sputter-coated with gold-palladium. Low magnifications in the range of 30 to 100 times were generally used for viewing these specimens.

Making silicone rubber molds yielded a second advantage, since identification of particular kinds of manufacturing marks is often facilitated by seeing the profile of the marks. To do this, a thin strip of the silicone rubber mold was cut with a sharp scalpel across the striations, pits or perforations. The strip showing the cross-section of the manufacturing marks was then laid on its side, mounted on a stub, coated, and viewed in the SEM. The profile of the mold is the reverse of the actual specimen's profile, so after photographing

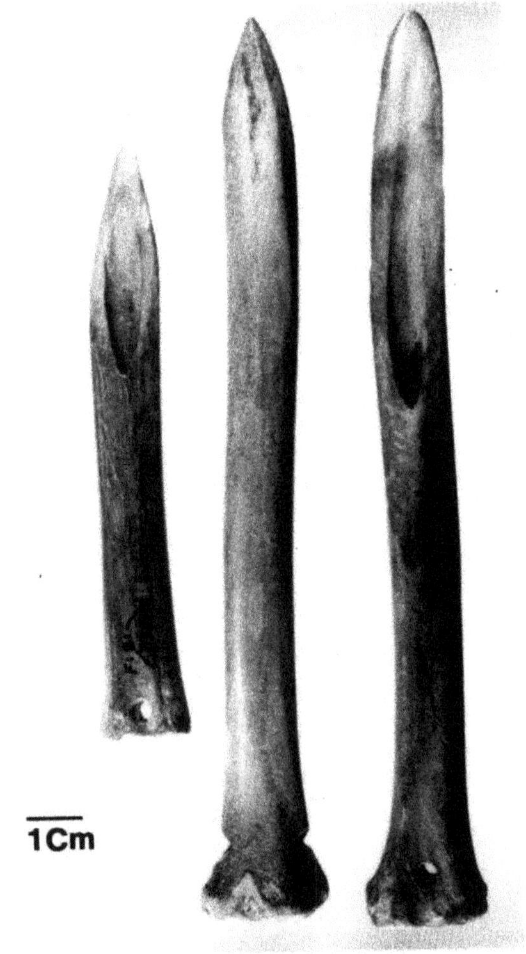

Fig. 3. Examples of 'gouges' from the Iron Age site of Fiskerton.

the negative replica in the SEM, the outline of the profile was then transferred to tracing paper and oriented so that it could be seen in correct perspective. This method may not be accurate enough for precise measuring of the dimensions of asperities or recessions in the surface, but it is very useful for interpreting the morphology of manufacturing marks (see Figs. 7 and 20).

To better comprehend how manufacturing marks are formed and indeed which kinds of tools would most likely produce marks like those on the ancient pieces, experimental replication was conducted on modern bone and antler, using both stone and metal tools.

Results

The major kinds of manufacturing techniques identified on the bone artifacts from West Row and Fiskerton include grooving and snapping, sawing, scraping, chiseling, abrading, filing, chopping and drilling (Table 2). In general, metal tools can be said to make more uniform patterns on the bone, often removing material more effectively, leaving behind either no striations or striations of a more uniform depth and spacing than when stone tools were used. When a metal tool is used on bone, it is often possible to match a series of striae from one mark (Fig. 4) to those in another because the irregularities on the metal tool's edge that bite into the bone do not wear away as quickly as those on a stone tool.

One of the most common methods of reducing bone in a predetermined way during the prehistoric manufacture of artifacts was the groove-and-snap technique (Olsen 1984). This is accomplished by incising a deep groove in the bone or antler through bidirectional scoring with a piercer, a blade, or other fine, sharp stone tool (Fig. 5). When the groove is sufficiently deep, the blank is "snapped", or broken free from the remaining bone. The edges of the tool blank and the offcuts or debitage bear very diagnostic traces: a series of fine, parallel striations on the groove wall and a rough, irregular surface below the groove where the bone was snapped in two.

When metal tools replaced stone ones, sawing was used as a substitute for the groove-and-snap technique for sectioning bone. Fragments of saws do occur in the Late Bronze Age, but metal saw marks were not visible on either the West Row or Runnymede bone artifacts. At the Iron Age site of Fiskerton, one saw fragment was recovered and the basal ends of many of the bone tools and an antler tine bore traces of sawing (Fig. 6). If well-preserved, a groove made by a saw is distinct from cut marks made by stone tools or metal knives because it has a square profile (Fig. 7) instead of a V-shaped section. There are parallel striae running along the bottom and sides, while the top edges may have some minute chipping or crumbling due to the back and forth motion of the saw (Fig. 8).

Scraping is one of the most common techniques for removing unwanted material on bone artifacts dating back to the early Upper Palaeolithic (Newcomer 1974). Longitudinal scraping with the strong edge of a stone tool, such as the facet of a burin, was frequently used to sharpen awls and projectile points, for example. In the Early and

Table 2. Major Manufacturing Techniques for Bone and Antler Artifacts

Action	Purpose	Stone tools	Metal tools
Sawing/ Grooving	to remove articular ends and cut out blanks	blade, flake, piercer	saw
Scraping	to remove excess material sharpen tool, smooth surface	burin, scraper, flake	draw-knife, spoke-shave, cabinet scraper
Chiseling	to remove excess material	--	chisel
Abrading/ Filing	to remove excess material, sharpen tool, smooth surface	granular stone abrader	file
Chopping	to sever, usually transversely	chopper, axe	axe
Drilling	to perforate bone for needle eye, thong hole, rivet hole, or decoration	drill, piercer	drill

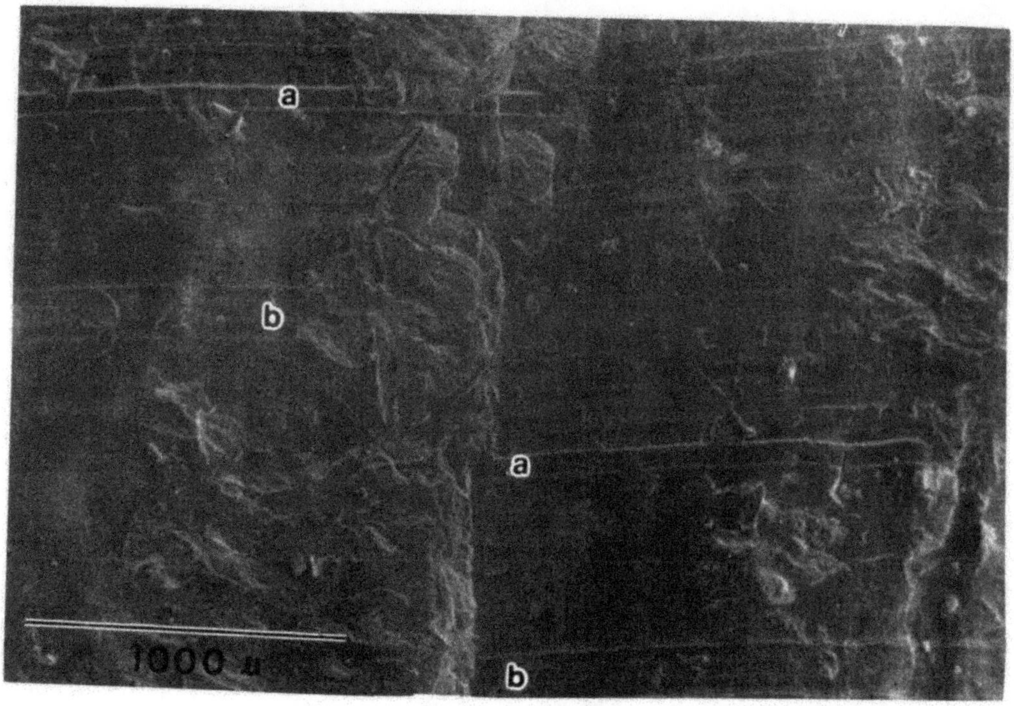

Fig. 4. Chisel marks on a 'gouge' from Fiskerton. Note how series of striations (a-b) made by asperities on the chisel's edge are repeated in adjacent facets.

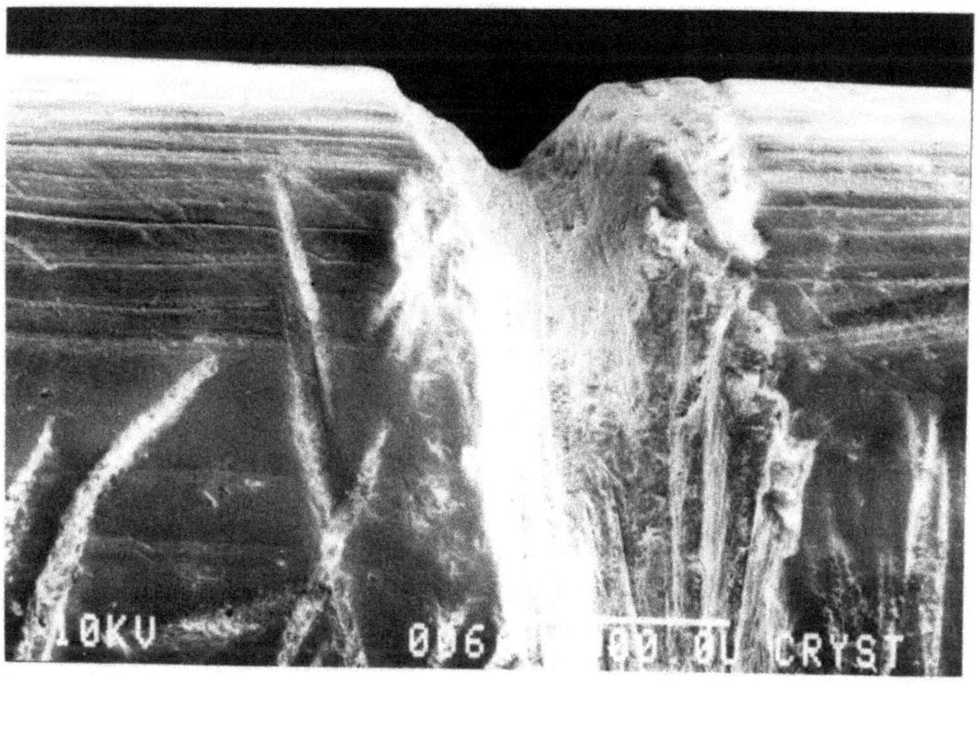

Fig. 5. Groove made with a stone tool in the West Row bead illustrated in Fig. 1a.

<u>Fig. 6</u>. Antler tine from Fiskerton with a sawn base.

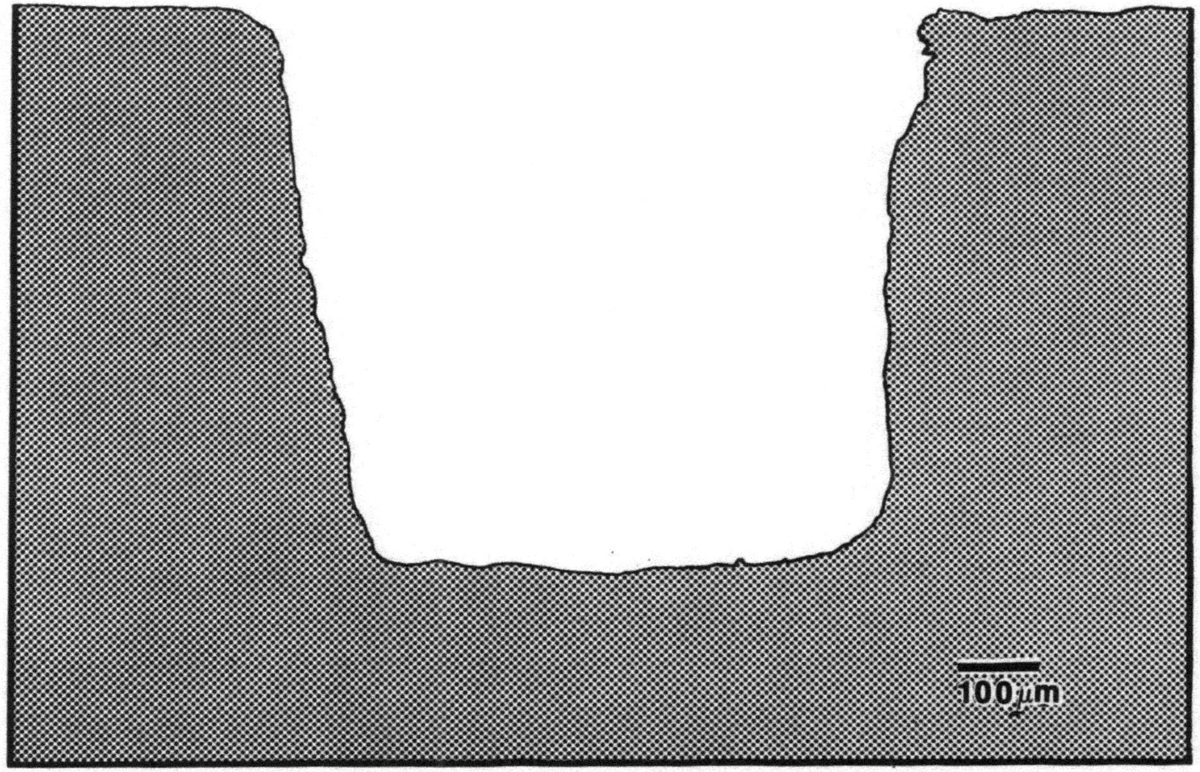

Fig. 7. Profile of experimental (metal) saw mark in bone.

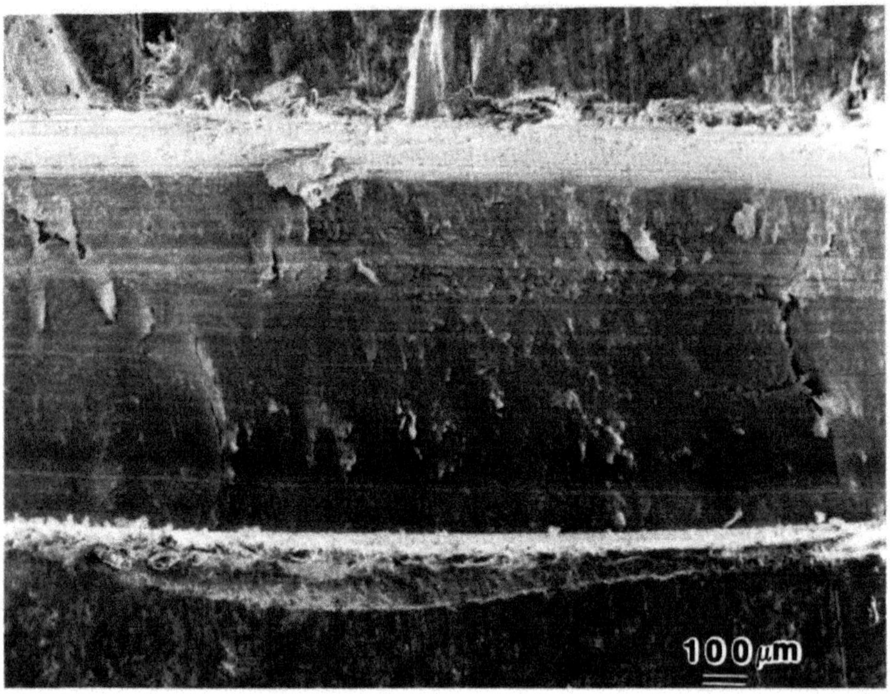

Fig. 8. Details of experimental saw mark. Note vertical sides, flat bottom, longitudinal striations within groove, and chipping of groove lips.

Fig. 9. Experimental scrape marks made with a flint burin. The direction of movement was across the photo. Note the fine, irregularly spaced, wavy longitudinal striations and the transverse ridges, or 'chattermarks'.

Late Bronze Age this technique continued, but by the Iron Age scraping with a metal tool began to replace it. When a stone tool was used the longitudinal striations created on the bone were usually somewhat wavy and of variable depth. Transverse ripples known as "chattermarks" (Newcomer 1974: 149) were produced when too much pressure was applied, causing the stone tool to bounce over the bone's surface (Fig. 9).

Scraping with a metal tool would have been performed with a long blade, such as a drawknife, a spokeshave, or a cabinet scraper. Quite wide longitudinal facets are formed by such tools. The striae running along inside these facets are straight and of shallow, uniform depth. Chattermarks may be formed with a metal scraping tool, if sufficient force is applied to cause the tool to skip over the bone's surface (Figs. 10 and 11). Scraping marks made by a metal implement were frequently observed on the Iron Age Fiskerton artifacts.

If a narrow tool, such as a small chisel, is used, then the facets are narrower and slippage is more likely to occur. As a result of slippage, the striations frequently angle off, following the contours of the bone. The marks are often shorter than those made with a broad blade and may terminate suddenly as when the chisel slips or bites too deeply into the bone's surface.

If the chisel is allowed to dig into the antler or bone, then thicker pieces of material may be removed instead of fine shavings (Fig. 10: a). This is done with shorter strokes and results in an undulating surface made up of a series of concave hollows, each containing striations running parallel to the direction of movement of the metal tool.

It is sometimes possible to match up the sets of parallel striations in one chisel mark with those in another made by the same tool. This has been demonstrated in one of the Fiskerton "gouges" (Fig. 4), for example. It should, then, in some cases, be possible to determine if two tools were made about the same time with the same chisel, so long as wear or resharpening had not significantly altered the chisel's working edge during the interval between their manufacture.

Chisel marks were found on two antler tines from West Row (Fig. 12: a) and were more common than metal tool scraping marks in the Fiskerton collection. Chisels are known from the Late Bronze Age and the Iron Age. No analogous marks made with stone tools have been observed in these collections.

Metal axes were used for chopping antler and bone in the Early Bronze Age and thereafter. Deep, narrow, V-shaped chopmarks have been observed on the same antler tines from West Row that were chisel-marked (Fig. 12:a). Chopmarks made with a stone tool tend to have a broad V-shaped profile, sometimes with parallel striations running vertically down into the cut, while a chopmark made with a metal tool may be quite deep and narrow, with smooth, or lightly striated walls (Figs. 12:c and 13). At Fiskerton, rivet holes were made in the handle of one of the gouges with a metal axe (Fig. 14:a).

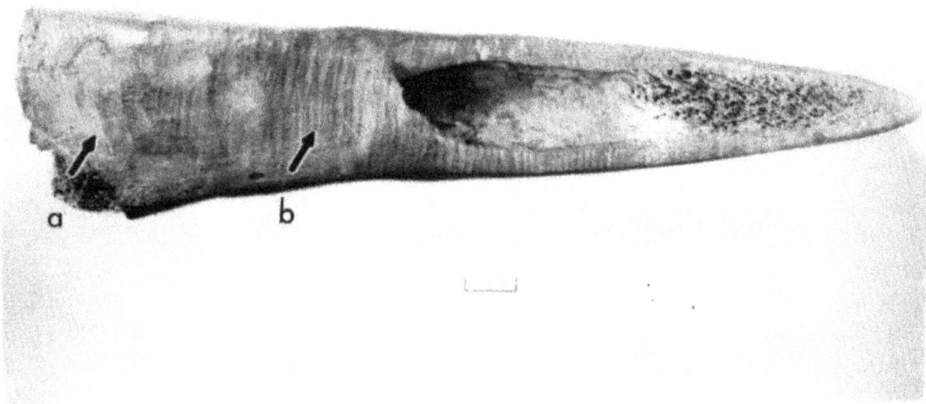

Fig. 10. Fiskerton 'gouge' with a. short, scooped out chisel marks and b. pronounced chattermarks made by a metal scraping tool.

Fig. 11. Close-up of the 'gouge' in Figure 10, showing broad facet with chattermarks and fine striations.

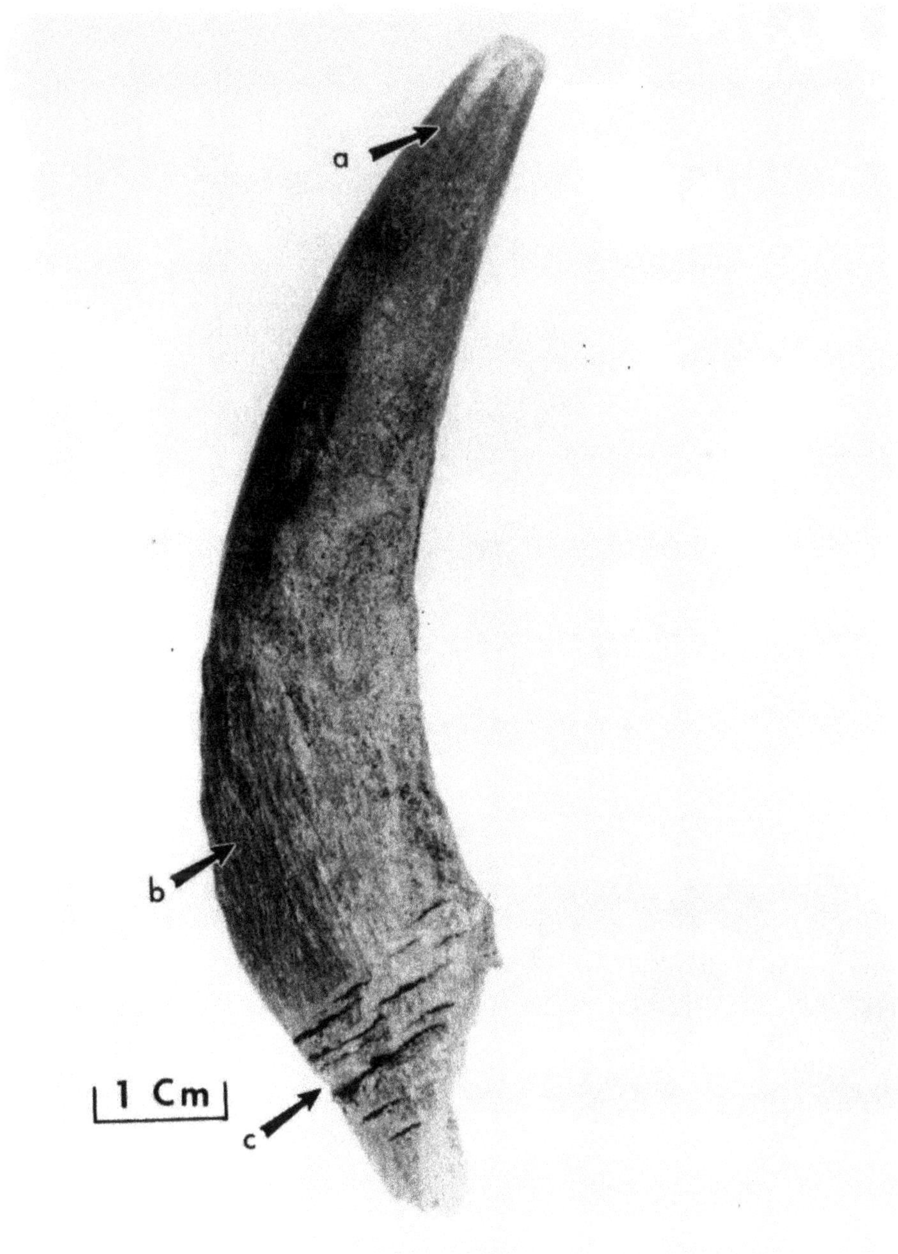

Fig. 12. Worked antler tine from West Row. a. chisel marks at the tip. b. metal file marks. c. chopmarks made with a metal axe.

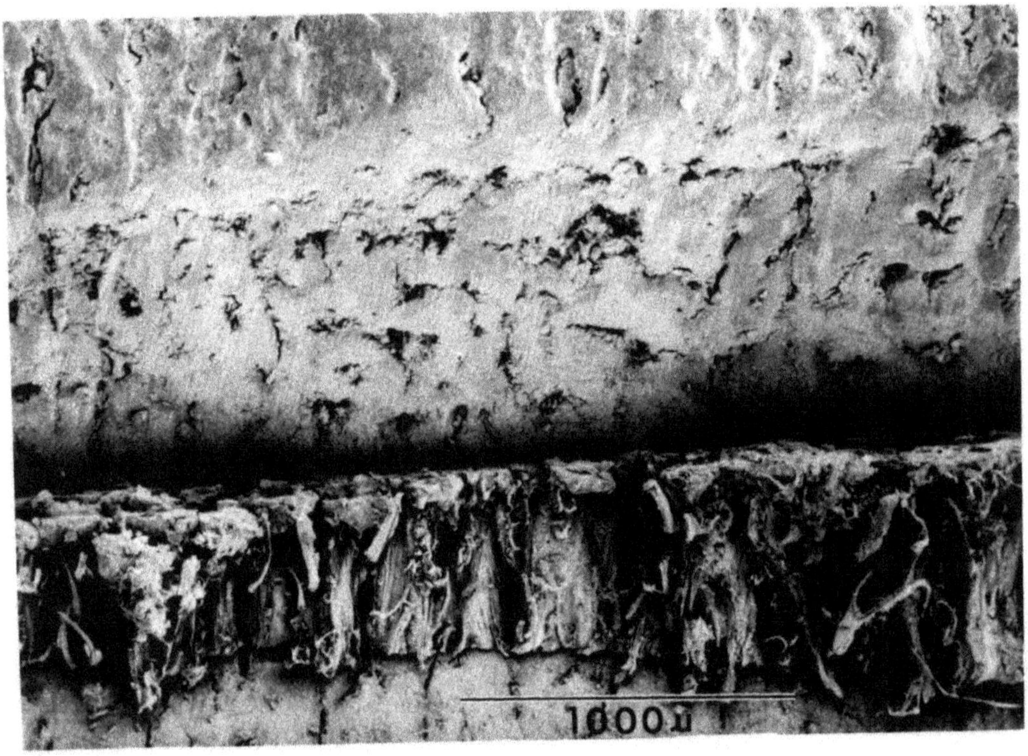

Fig. 13. Experimental chopmark made in antler with a metal axe.

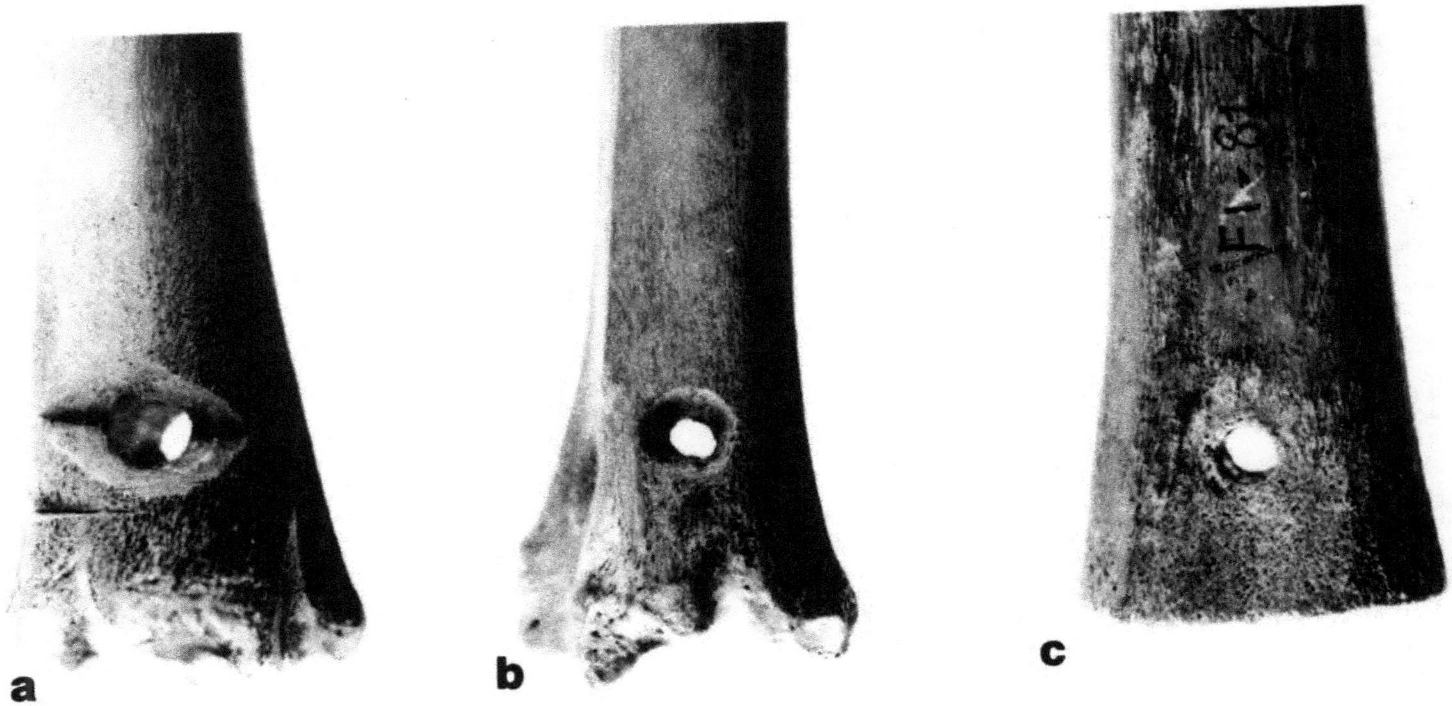

<u>Fig. 14</u>. Fiskerton 'gouge' handles with rivet holes to secure the tang of a wooden handle in the socketed base. a. hole made with a metal axe. b. countersunk rivet hole. c. hole made with a parallel-sided metal drill.

The Bronze Age collections of worked bone and antler from West Row and Runnymede contained many examples of pieces modified by a granular stone abrader. Abraders were used to shape and smooth bone implements by grinding away excess material. The manufacturing traces left by abrading are broad areas of fine, closely spaced, parallel striations (Fig. 15). The profile of the abraded area has a low relief consisting of few asperities and numerous shallow, ragged-bottomed valleys. Because the abrader is usually rubbed over the bone's surface in at least two directions and there is not restriction on directionality, sets of striae may cross over, intersect, and form zig-zagged patterns. When an area of bone is rubbed repeatedly in one plane, a facet will be formed.

In the Iron Age, metal files seem to have replaced granular stone abraders to remove other manufacturing traces and unwanted rugosities and to sharpen bone tools. Well-preserved examples of files from Fiskerton show that they had long, flat, course, single-cut files with the teeth cut transversely rather than diagonally as they are today.

Many of the bone artifacts from Fiskerton retain file marks on the basal end where a socket was made to receive a wooden handle (Fig. 16). It is usually possible to see how the bone tool was turned between filing strokes, so that the grooves made by the file form a radial pattern on the bone implement's base. Some of the gouges also have file marks on the shaft and tip where this method replaced or supplemented scraping.

File marks differ from those made by scraping in that their striations run diagonally or transversely within each longitudinal facet on the bone shaft (Fig. 17). With scraping, the striations always run parallel to the facet. Filing striations are more uniformly spaced than those made by a scraper or a stone abrader, since they are made by cut teeth. The early files were generally courser than sandstone, so the size of the striations created by Iron Age files may be considerably larger than those made by a stone abrader.

Seven of the awls from West Row were drilled with a stone drill. One awl bears the light traces of the initial stage of drilling on both sides, but the perforation was never completed. Stone drills generally produce a conical hole if drilled from one side or a biconical hole if drilled from both sides. Rarely are both sides parallel unless the hole has been reamed after drilling. The concentric striations inside and sometimes around the top are deep and irregularly spaced (Fig. 18). In a perforation made by a stone drill, the lip is generally sloping and any bone debris has usually been scraped off by the shoulder of the drill.

The use of metal drills seems to come in by the Late Bronze Age in Britain. Antler cheek-pieces from Runnymede and other sites appear to have been drilled with a metal drill. A tool identified as a bradawl from Runnymede, having a narrow, round shaft, chisel-like tip, and a square tang, would have been perfect for drilling bone. The artifacts from Fiskerton were all drilled with metal rather than stone drills.

With a metal drill, concentric striae are either absent altogether, or are very shallow and uniform (Fig. 19). The sides,

Fig. 15. Experimental abrading marks on bone made with a flat piece of sandstone.

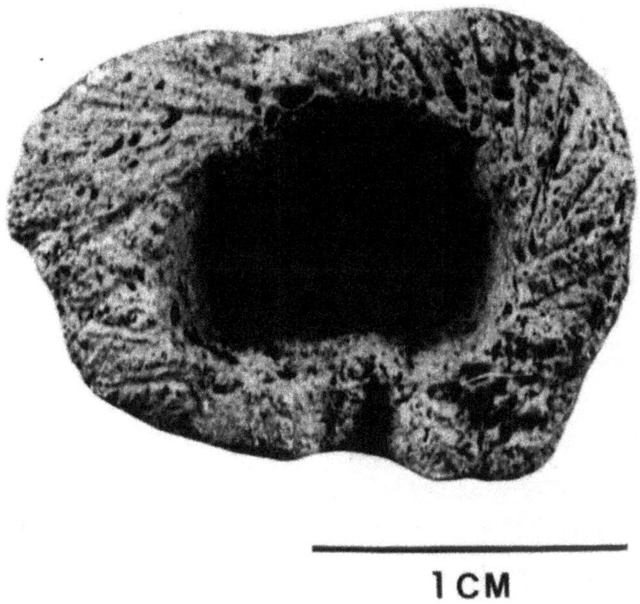

Fig. 16. Base of a 'gouge' from Fiskerton showing socket hole and a radiating pattern of file marks made by turning the bone occasionally between filing strokes.

Fig. 17. File marks on experimental bone made with a flat, course-cut, steel file. Note the evenly-spaced diagonal striations formed in the longitudinal facet. Direction of movement was across the image.

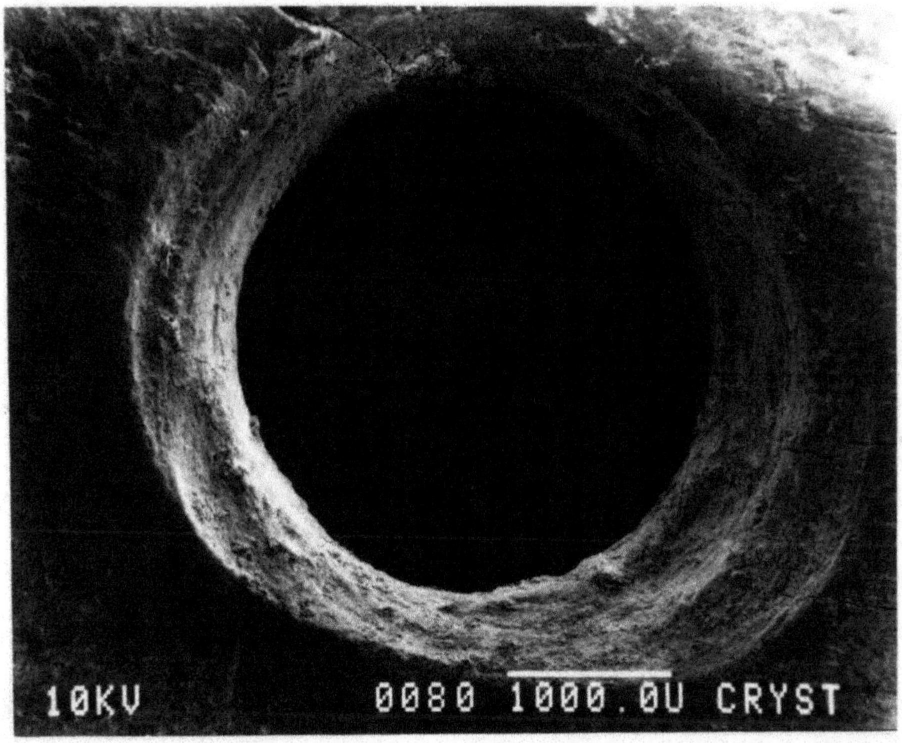

Fig. 18. Eye in an awl from West Row made with a stone drill. Note uneven striations on the walls and lip of the perforation.

Fig. 19. Hole drilled in experimental bone, using a spade-tipped metal drill bit. Note the straight, tapered walls with uniformly spaced striations of even depth.

whether conical or parallel are usually straighter when a metal drill is used, and the lip of the perforation is often very crisp (Fig. 19) or may even show a fringe of pushed-up bone debris (Fig. 14: c).

At Fiskerton there is evidence to suggest that in some cases the rivet holes on the bone tools were countersunk with a spade-like drill tip after being drilled with a parallel-sided drill. The profiles in Figure 20 show the difference in the outline of holes made with a parallel-sided drill bit (archaeological), a spade-tipped bit (experimental), and one that has been countersunk (archaeological). Countersinking of the rivet holes on the Iron Age gouges allowed the rivet to be set flush with the handle's surface. Pins or rivets of wood, iron, and bone have been found in place in holes on gouges from Grimthorpe (Mortimer 1905: 151), Wookey Hole, Hanging Langford Camp, and All Cannings Cross (Cunnington 1923: 86-87). No rivets, however, were found in the Fiskerton assemblage.

Conclusions

This is only a cursory coverage of small samples of artifacts, but it introduces the possible scope for further research into manufacturing marks on bone artifacts and shows the potential for learning more about the use of early bronze and iron tools. The SEM provided a means of seeing the manufacturing traces at low magnifications with high resolution and great depth of field. The use of silicone rubber molds in replicating the profiles of marks on archaeological specimens was extremely useful in establishing criteria for distinguishing between manufacturing traces caused by metal as opposed to stone tools.

The information learned from analysis of an early Bronze Age and an Iron Age assemblage of bone artifacts sheds light on the gradual replacement of stone tools by metal ones in the manufacture of other objects. This research demonstrates that it is generally possible to distinguish between the manufacturing traces left by stone and metal tools and to identify the type of metal tool employed in various stages of manufacture. Studies of manufacturing traces on bone, as well as other materials, can expand our understanding of the range of metal tools available and their specific uses through time.

Acknowledgements

I would like to extend my gratitude to M.H. Newcomer, Institute of Archaeology, University of London, for his advice on this project, E. Martin, Suffolk Archaeological Unit, for the opportunity to study the worked bone from West Row, Naomi Field, Trust for Lincolnshire Archaeology, for allowing me to analyze the bone artifacts from Fiskerton, and Stuart Needham, British Museum, for access to the material from Runneymede. The SEM work was conducted in the Crystallography Department, Birkbeck College, University of London.

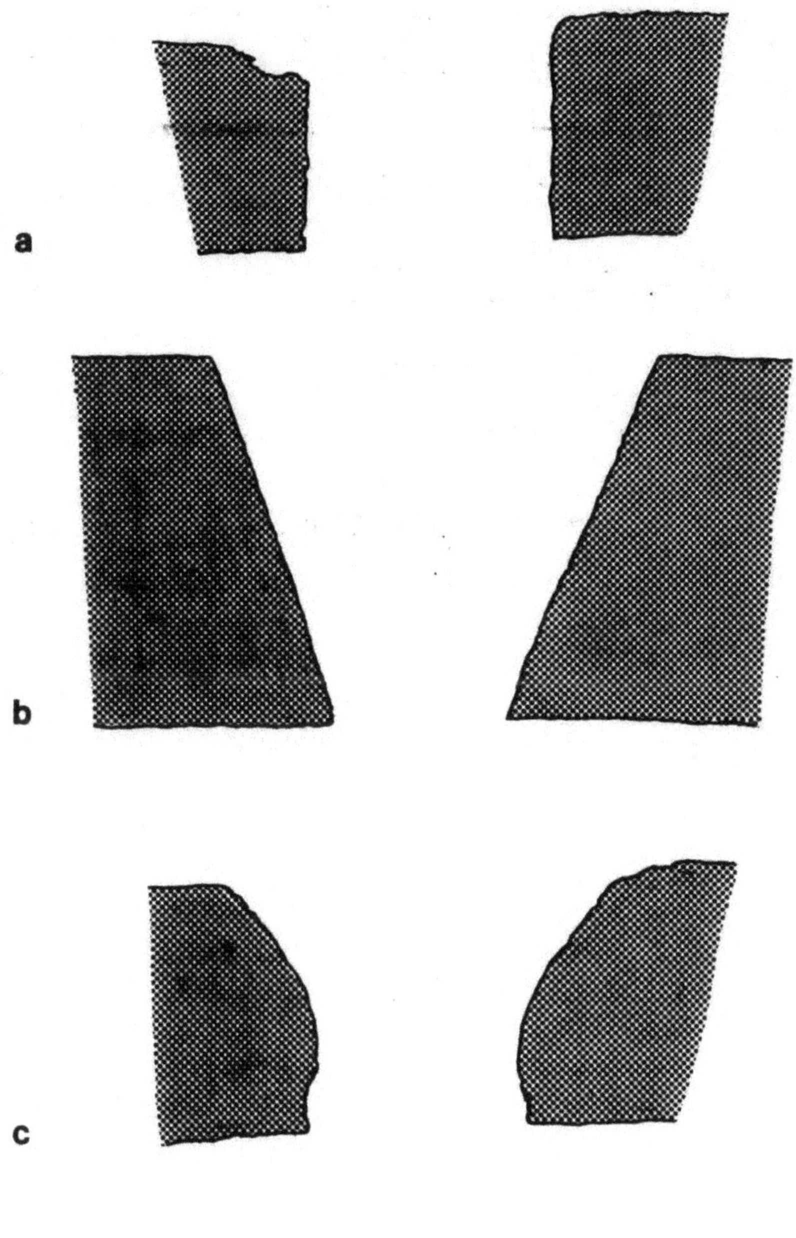

Fig. 20. Profiles of three different holes drilled in bone, using a. a parallel-sided metal drill bit (Fiskerton 'gouge' handle). b. a spade-tipped metal drill bit (experimental bone). and c. first a parallel-sided metal drill bit, followed by a spade-tipped bit to countersink the hole (Fiskerton 'gouge' handle).

References

Cunnington, M.E. (1923). The Early Iron Age Inhabited Site at All Cannings Cross Farm, Wiltshire. George Simpson and Co., Ltd., Devices, England.

Longley, D. (1980). "Runnymede Bridge 1976: Excavations on the Site of a Late Bronze Age Settlement," Research Volume of the Surrey Archaeological Society, No. 6. Surrey Archaeological Society, Castle Arch, England.

Mortimer, J.R. (1905). Forty Years' Researches in British and Saxon Burial Mounds of East Yorkshire. A. Brown and Sons, Ltd., London.

Newcomer, M.H. (1974). "Study and replication of bone tools from Ksar Akil (Lebanon)." World Archaeology 6(2): 138-153.

Olsen, S.L. (1984). Analytical Approaches to the Manufacture and Use of Bone Artifacts in Prehistory. Unpublished Ph.D. thesis, Institute of Archaeology, University of London.

Olsen, S.L. (1988). "Applications of Scanning Electron Microscopy in Archaeology." In P.W. Hawkes (ed.), Advances in Electronics and Electron Physics, 71: 357-380.

PART VII:

HOMINID REMAINS

FLUORIDE CONTENT AND MICROWEAR OF TEETH IN THE PENINJ AUSTRALOPITHECUS

P.F. Puech, F. Cianfarani, A. Markitziu and I. Gedalia

Summary

A new methodology for the evaluation of the effect of fluoride upon enamel wear was developed and its applicability to fossil study presented. A different rate of wear occurring between fluoride-enriched enamel and normal controls was demonstrated "in vivo" and "in vitro". A possible cause for the thick enamel in East African Australopithecus may have been attributed to life in the vicinity of volcanoes where high fluoride levels are present.

Introduction

The molar crowns of early hominids are covered by a thick enamel layer which has been interpreted by Simons (1964) as an adaptive phenomenon. The variability of enamel thickness has been connected to diet by specific dental wear (Gantt, 1981; 1985; and Kay, 1981). The enlargement of the masticatory surface is also considered an adaptation against continuous wear. The enamel of *Australopithecus* throughout its successive forms: *A. afarensis*, *A. africanus*, *A. robustus* and *A. boisei*, strengthens this assumption. The size of the smallest molar and premolar of *A. boisei* is superior to that of the largest ones of *A. africanus* (Wood, 1981).

In East Africa the fossil remnants of *Australopithecus*, as well as those of early *Homo*, were found in volcanic regions. Large quantities of fluoride compounds are produced during volcanic eruptions, increasing the fluoride content of plants and animals growing up in the close vicinity to an abnormally high level (Le Moan, 1984). The aim of the present study was to find out whether a relationship exists between wear and fluoride content of tooth enamel and its relevance to *Australopithecus* dentition.

Materials and Methods

Ten, 21-day-old, Sabra rats (Hebrew University strain) were maintained on a standard diet until the age of 40 days. Half of them had received 25 ppm of fluoride in the drinking water "ad libitum". The animals were killed with CO_2 and their mandibles dissected out and cleaned of soft tissue by successive washing with 0.1 M ammonium carbonate. The lingual surfaces of the molars were scratched with a calibrated tungsten burr after removing the pellicle (Puech, 1985). The specimens were attached to SEM stubs, gold plated and photographed.

Ten human teeth obtained from young adults living in an endemic fluorosis area (Khouribga, Morocco) with a tooth surface index of fluorosis (TSIF) value of 4 (Horowitz et al., 1984) and the same number of teeth from a non-fluoridated European population represented experimental and control groups. The tooth surfaces were cleaned and the enamel pellicle removed. Each tooth was scratched from the cementum enamel junction towards the occlusal surface and back, and from the mesial border of the lingual surface to the distal one using the tungsten burr loaded with 250 gr at a speed of 15 cm/sec. The specimens were prepared for SEM examination. Some of the teeth, fluorosed as well as controls, were first etched with 30% phosphoric acid for 60 seconds, then rinsed thoroughly with tap water.

For this research, the Peninj Australopithecus specimen was chosen because the preservation of the mandible and teeth allowed an accurate interpretation of enamel alterations to be made. The fossil, 1.5-2 million years old had been discovered in lacustrine deposits on the West bank of Lake Natron, Tanzania (Leakey and Leakey, 1964).

All the SEM micrographs were examined and ranked according to the classification system developed by Powers and Craig (1972) for recording surface failure on enamel with the addition of further evaluation criteria for sub-surface alterations. According to this classification: a ductil action is inferred from smooth furrows considered in Class 1; tensive cracks by cleaving and splits from Classes 2 and 3; and scars and chips cause "chevron" formations that are present in Classes 4 and 5.

Results

In rats, the SEM photographic analysis failed to distinguish between the experimental grooves cut in the enamel of the control (Fig. 1) and the fluoride-enriched teeth (Fig. 2). In all of the examined samples the tungsten burr produced a smooth groove corresponding to the tool's configuration. The subsurface displayed cracks (Fig. 3) originating from the detachment of enamel chips in the floor of the grooves (Fig. 4).

The fluorosed human teeth underwent a higher rate of abrasion than those in the control group. Frequent breaks along the top margins of the groove walls (Fig. 5) occurred whenever developmental fissures or preexisting enamel focal holes were encountered (Fejerskov et al., 1984). The scars were oriented opposite to the direction of movement of the burr and resembled "chevrons". At low magnification (100x), the bottom of the groove was apparently smooth, and it was only at 5000x magnification that its roughness, as well as the presence of cracks covered by small chips, became evident (Fig. 6). The fluorotic teeth pretreated with acid displayed an even higher rate of abrasion. The "chevron" formations were larger and more pronounced (Fig. 7) and the groove presented a granular surface ranked as Class 5. At magnifications of 1000x and 3000x, the defects seemed to result from tearing of tooth material. Etching of the control tooth surfaces raised the rate of wear only to Class 2 and 3 (Fig. 9). The defects were more apparent at a high magnification (Fig. 10).

The occlusal tooth surfaces of the Australopithecus boisei were covered by both older and more recent grooves and fissures visible at 100x (Fig. 11). Figure 12, at 300x magnification demonstrates the presence of "chevron" formations, the criterion for assigning it to Class 4.

Plate 1

Fig. 1. Groove in unfluoridated rat enamel (x1000).

Fig. 2. Groove in topically fluoridated rat enamel (x2000).

Fig. 3. In Class 2 cracks are present in the bottom of the groove; in Class 1 the furrow would be smooth (x5000).

Fig. 4. A rough aspect is present in the bottom of Class 3 groove (x5000).

Fig. 5. Broken borders of the groove (chevron formation) in fluorotic human enamel: Class 4 (x100).

Fig. 6. At a higher magnification (x5000) the bottom of the groove seen in Fig. 5 is rough.

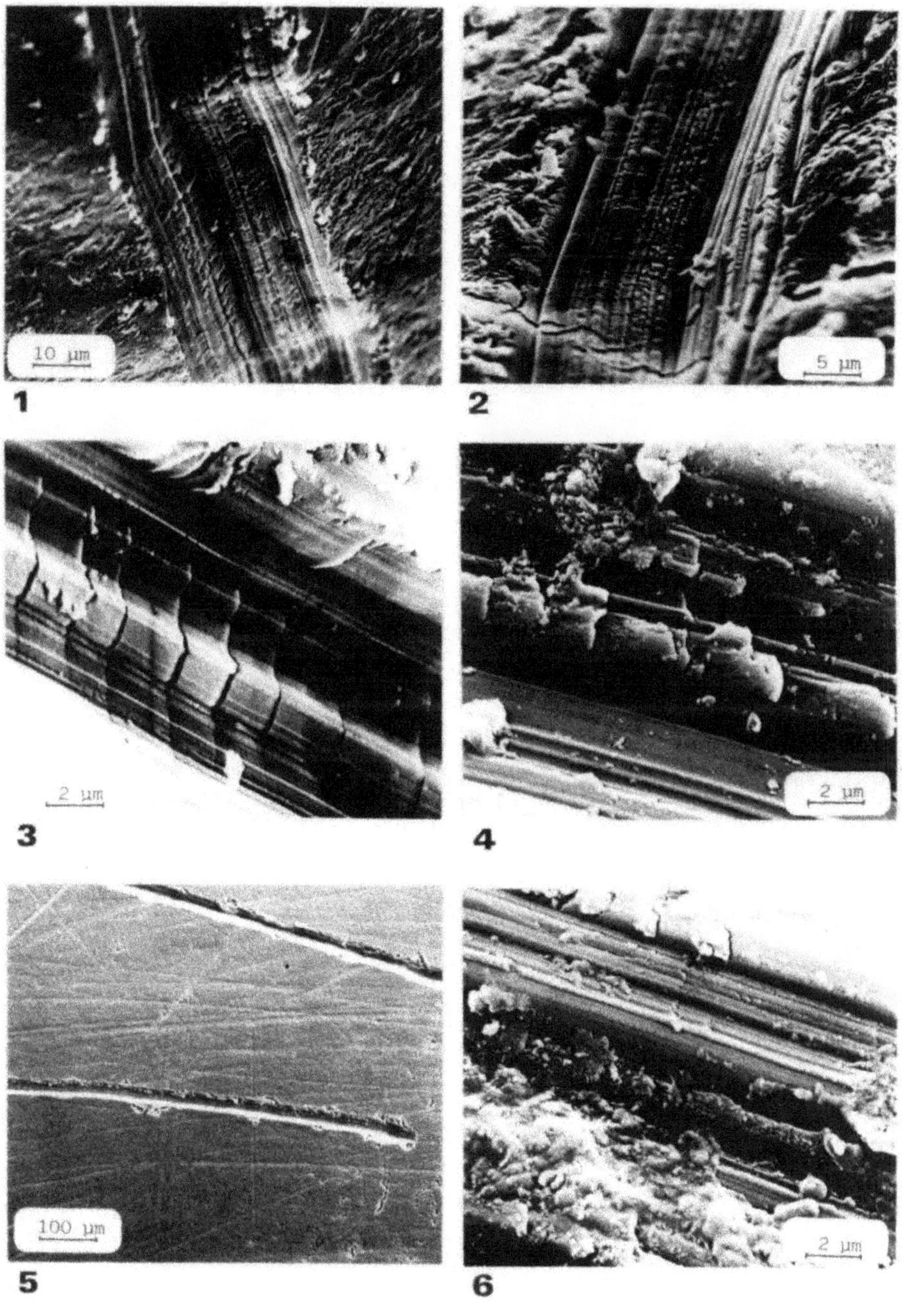
10 µm
5 µm
1
2
2 µm
2 µm
3
4
100 µm
2 µm
5
6

Plate 2

Fig. 7. The etched fluorotic human enamel appears to be more friable (x100).

Fig. 8. Examination of a higher magnification (x5000) demonstrated the porous aspect in the groove floor: Class 5.

Fig. 9. A better resistance to abrasion is found in the etched normal human enamel (x100), see also Fig. 10.

Fig. 10 At x1000, cracks appear in the smooth floor of the groove (Fig. 9) considered to be Class 2.

Fig. 11. Buccal border of the occlusal surface of the left mandibular second premolar of *Australopithecus boisei* from Peninj at x100.

Fig. 12. The same area shown in Fig. 11, at x300, gives evidence of chevron formation.

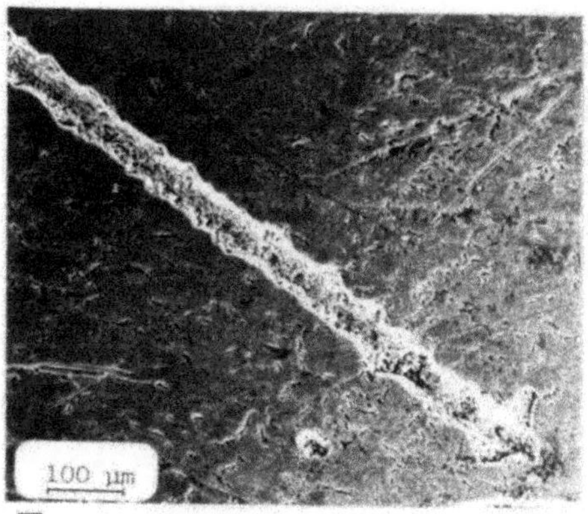

7

8

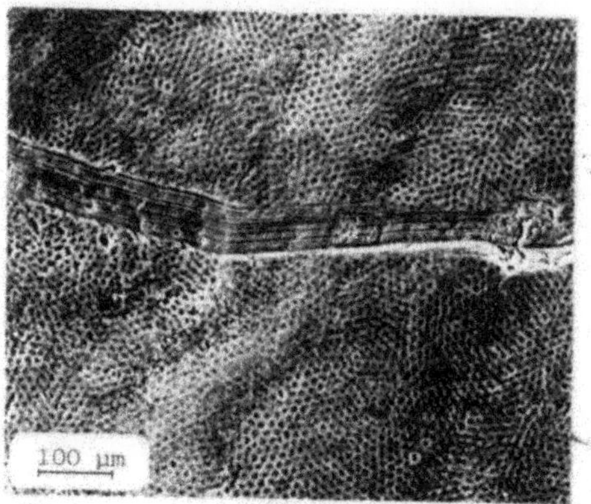

9

10

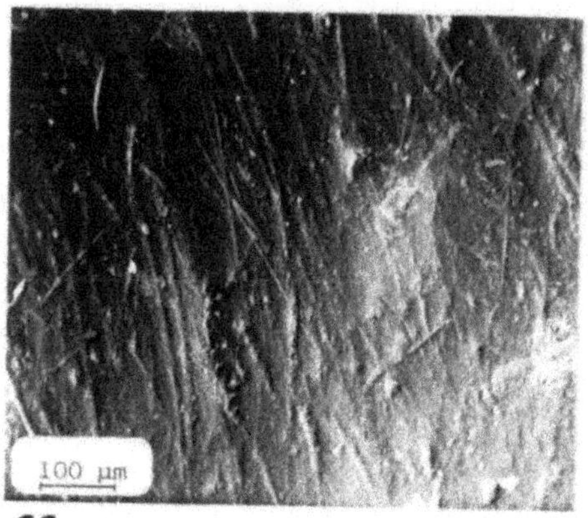

11

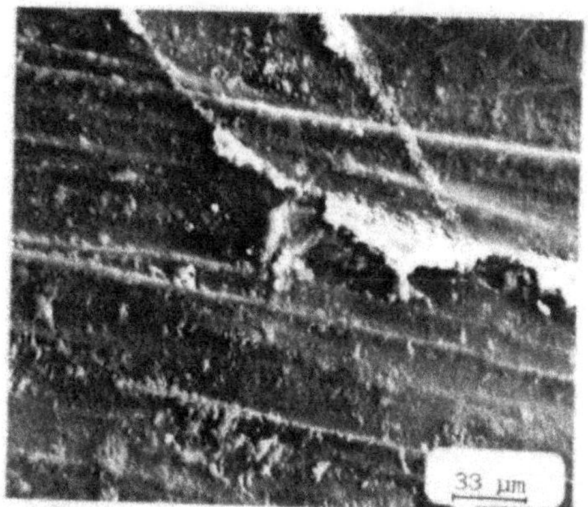

12

Discussion

The SEM examination of the Australopithecus boisei tooth enamel surface revealed the presence of "chevron" formations which probably resulted from a concurrently greater than normal hardness of the outer enamel layer and a more porous subsurface. A similar combination of surface and subsurface wear rate was found in fluorotic teeth of modern humans, but failed to be demonstrated in fluoride-enriched rat teeth (Fig. 2-3 and Fig. 7-9). In the present study only an acceptable classification for the assignment of enamel defects induced by a calibrated burr was delineated; quantification of the defects remained beyond the scope of this research. The Powers and Craig classification (1972) was enhanced by the addition of subsurface evaluation criteria with respect to changes occurring in the groove's depths. According to the proposed scoring system the severity of the damage was the criterion for assigning the observed defects to each of the described classes of surface failure. The characteristics of defect formation were different for each class.

Lingual surfaces rather than occlusal ones were examined, except in the case of the fossil tooth. The choice was made on the basis of the integrity of the lingual surfaces when compared to that of the masticatory surfaces, which were covered with many fissures, cracks, and furrows. The occlusal surfaces exposed permanently to abrasive contacts and direct pressures that cause severe damage to the integrity of the tooth material would also progressively develop different physico-chemical properties (Powers et al., 1973). Moreover, fluorosis has been accepted as a wear-enhancing factor by reducing enamel resiliency (Buckley, 1969).

In the 21-day-old rats, the ingestion of fluoridated water had only a topical effect (Markitziu et al., 1985), which did not result in increased enamel abrasion similar to that observed in human teeth originating in regions where endemic fluorosis occurs. The increased severity of the experimentally induced defects found by us in the etched fluorotic human teeth was probably the result of an enhanced friability of the subsurface layer (Figs. 7 and 8).

The experimentally induced abrasion on the lingual tooth surface presented herein simplifies the evaluation of the much more complex wear phenomenon that occurs naturally on the occlusal surfaces of teeth. The defects induced "in vivo" by the abrasive particles of food are filled with different alimentary deposits or smoothed by attrition, making it difficult to evaluate wear, whilst the experimental groove may be evaluated fairly accurately. Some problems have also been noted regarding the wear patterns on the Peninj teeth, which are different in appearance from other Australopithecine teeth. It has been suggested, for example, that there may be postexcavation damage from cleaning and preparation on the occlusal surfaces of the teeth (Alan Walker, pers. comm.).

References

Buckley, D.H. 1969. Effects of surface active media on friction deformation and fracture of calcium fluoride. NASA TN D-5580: Washington.

Fejerskow, O., Josephsen, K. and Nyvad, B. 1984. Surface ultrastructure of unerupted mature human enamel. Caries Research 18: 302-314.

Gantt, D.G. 1981. Enamel thickness and Neogene hominid evaluation. American Journal of Physical Anthropology 54: 222 (abstract).

Gantt, D.G. 1985. Enamel thickness and human evolution. American Journal of Physical Anthropology 66: 171 (abstract).

Horowitz, H.S., Heifetz, S.B., Driscoll, W.S., Kingman, A. and Meyers, R.J. 1984. A new method for assessing the prevalence of dental fluorosis - the Tooth Surface Index of Fluorosis (TSIF). Journal of the American Dental Association 109: 37.

Kay, R.F. 1981. The nut-crackers: a new theory of adaptation of the Ramapithecinae. American Journal of Physical Anthropology 56: 141-152.

Leakey, L.S.B. and Leakey, M.D. 1964. Recent discoveries of fossil hominids in Tanganyika: at Olduvai and near Lake Natron. Nature 202: 5-7.

Le Moan, G. 1984. Qu'est-ce que le fluor? Le fluor et la thérapeutique. In Goupil (ed.), Le Point Sur Le Fluor, pp. 19-30. Paris: SNPMD.

Markitziu, A., Salomon, I. and Gedalia, I. 1985. Tooth wear, solubility and fluoride concentration of molar-tooth surfaces in rats maintained on simultaneous or separate intake of food and fluoridated drinking water. Archives of Oral Biology 30: 167-169.

Powers, J.M. and Craig, R.G. 1972. Wear of fluorapatite single crystals: III. Classification of surface failure. Journal of Dental Research 51: 611-618.

Powers, J.M., Ludema, K.C. and Craig, R.G. 1973. Wear of fluorapatite single crystals: VI. Influence of multiple-pass sliding on surface failure. Journal of Dental Research 52: 1032-1040.

Puech, P.-F. 1985. Precisions sur la méthode dite des "répliques" pour l'étude des surfaces dentaires et osseuses. Bulletin de la Société Préhistorique Française 82: 72.

Simons, E.L. 1964. On the mandible of Ramapithecus. Proceedings of the National Academy of Sciences 51: 528-535.

Wood, B.A. 1981. Tooth size and shape and their relevance to studies of hominid evolution. Philosophical Transactions of the Royal Society of London 292: 65-76.

A SCANNING ELECTRON MICROSCOPE STUDY OF ARCHAEOLOGICAL DENTAL CALCULUS

Keith Dobney and Don Brothwell

Introduction

Dental calculus or 'tartar' as it is more commonly known, is one of the commonest types of ectopic concretion known to occur in man. Its precursor is dental plaque, a whitish-yellow, soft substance consisting basically of a network of non-mineralised micro-organisms which adhere to the tooth crown or root.

For reasons still not fully understood, the organic matrix begins to calcify and the resultant inorganic fraction consists mainly of combinations of calcium and phosphates (e.g. hydroxyapatite, whitlockite and brushite). Once calcified it is very durable and thus not surprisingly occurs commonly on a range of archaeological human and other mammal teeth (Figs. 1 and 2), some of great age. For example, several of the human neanderthal skulls from Shanidar cave in N.E. Iraq and from Krapina in Yugoslavia show quite severe deposits of supra-gingival calculus (Fig. 3). Various Pleistocene mammal teeth from a number of cave deposits in Britain also show varying amounts of dental calculus, as does a specimen of Anoplotherium modicum from the Upper Eocene (i.e. approximately 40 million years old) (Figs. 4-6).

During the course of an extensive research programme concerned with the potential value of archaeological dental calculus for providing direct information regarding past human and other mammalian diet (Dobney and Brothwell, 1986), the scanning electron microscope (SEM) was employed simply to locate microscopic fragments of food debris trapped within the calcified matrix. However, it soon became clear that a further rewarding avenue of research involving the SEM was emerging, namely the study of the ancient oral microflora.

Specimen Preparation

A number of specimens were selected from a variety of archaeological and modern samples, both human and other mammal. Some were viewed as removed from the tooth surface, while others were deliberately fractured or sectioned to reveal features deep within the matrix. Each specimen was cleaned by immersion first in alcohol and then briefly in an ultrasonic bath. When dried the specimens were left intact, fractured, or embedded in methyl methacrylate for sectioning. After preparation the specimens were mounted on Cambridge stubs, using carbon cement to secure them, and then coated with gold or carbon. Each was then viewed under the SEM at magnifications of between 60x to 15,000x, using a beam current of 15KV.

Results of Rapid Scanning

When scanning calculus specimens from a number of different sites and species, at magnifications of 2000x or more, it became apparent that large numbers of micro-organisms of various forms were present.

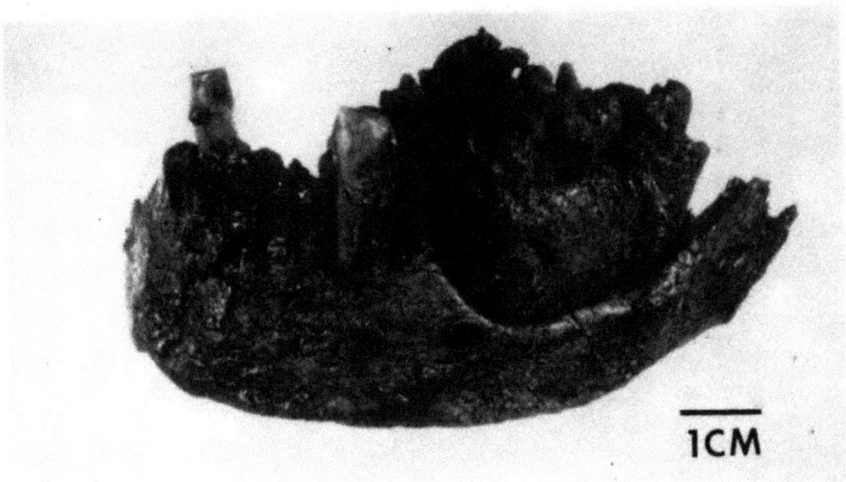

Fig. 1. Gross calculus deposit on an Anglo-Saxon human individual from the Mortimer Collection, Hull Museum. Note 3 maxillary teeth completely embedded and loss of occluding mandibular molars. Here the calculus actually forms a functional occlusal surface.

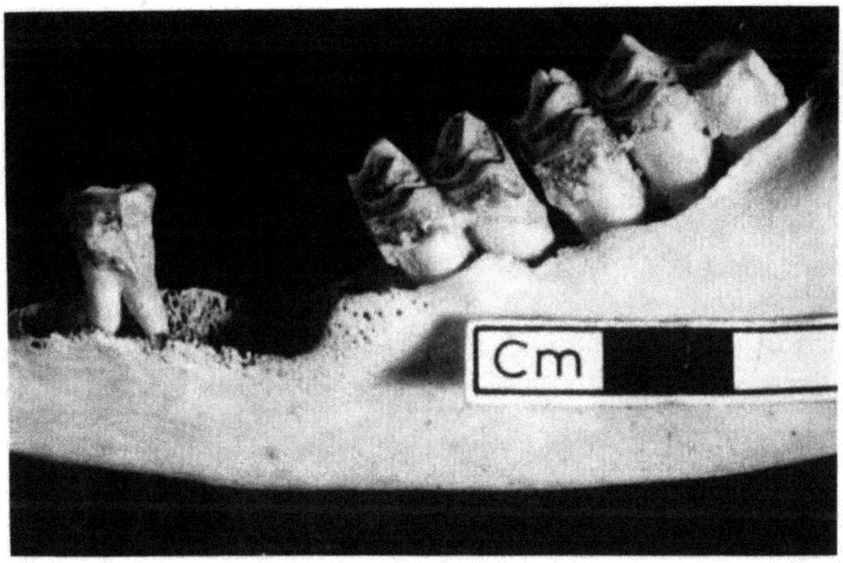

Fig. 2. Mild calculus deposit on the buccal surfaces of sheep dentition from Iron Age Danebury. Note also associated periodontal disease and ante mortem loss of P4 and M1.

Fig.3.

A Neanderthal jaw from Krapina displaying medium alveolar resorption and considerable calculus (After Kramberger 1906)

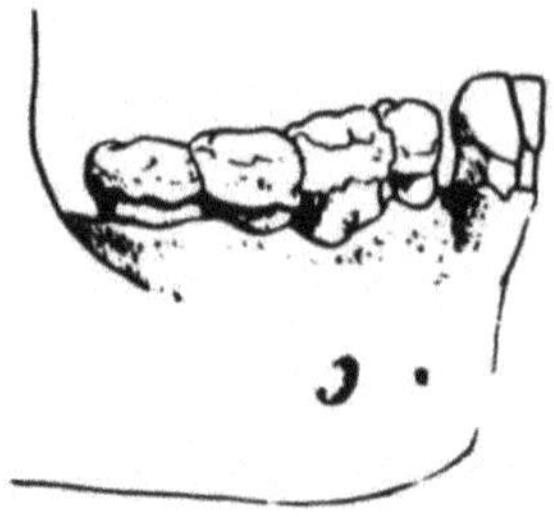

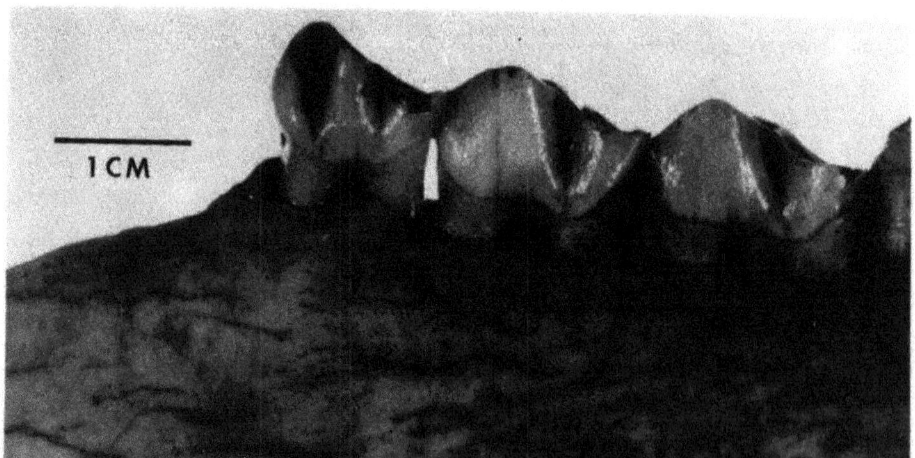

Fig. 4. Mildly eroded calculus deposits on buccal surface of Pleistocene reindeer dentition from Pinhole Cave.

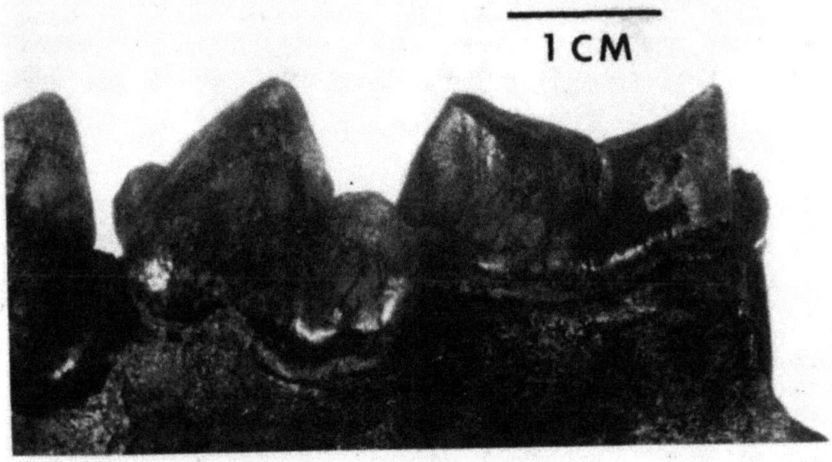

Fig. 5. Slight/mild calculus deposits on buccal and distal crown surfaces of Pleistocene hyena M1 from Pinhole Cave.

Fig. 6. Eroded mild/severe calculus deposits on buccal surface of P4, from an _Anoplotherium modicum_ specimen from Upper Eocene deposits.

Since each specimen had been thoroughly cleaned and the majority of these microbes seemed to be present deep within the calcified matrix, it can only be presumed that they represent the calcified remains of the original oral flora in a remarkable state of preservation. Hansen et al. (1986) observed and commented briefly on the same phenomenon in dental calculus from several frozen bodies from the coastal site of Qilakitsoq, Greenland. In most cases it is impossible to distinguish between these ancient calcified micro-organisms and scanning electon micrographs of their living counterparts isolated from the soft organic plaque.

The Microbiology of Dental Calculus

A limited amount of research has been undertaken on modern calculus using the SEM (Jones, 1972; Friskopp et al., 1980 and Lustmann et al., 1976) but detailed microbiological work has, for obvious reasons, been undertaken on living micro-organisms isolated from the dental plaque (Newman, 1974 and 1980; Poole and Newman, 1971a, b).

There are many different types of micro-organism which inhabit the oral cavity and they make up the normal indigenous or resident flora. Various different suites of micro-organisms inhabit the cheek mucosa, the tongue, the gingival crevice and the tooth surface itself. Each of these different micro-environments may encourage organisms with widely different nutritional requirements; for example the tooth surface encourages aerobic, anaerobic and facultative types of microbes, while those inhabiting the gingival crevice (the space between gum and tooth) tend mainly to be anaerobic and facultative types only.

Calculus formation appears to be associated with certain types of micro-organisms. At first, streptococcal forms predominate with filamentous types such as actinomyces present in low numbers. As the calculus develops there is an increase in filamentous forms at the expense of the coccal forms.

The Potential of Ancient Oral Ecology

As we have seen, locked within this calcified matrix lies a remarkable record of past oral ecology. What is of great interest to archaeologists and indeed palaeontologists, mainly with respect to man but also of potential importance in certain domestic animals, is the question: has the ecology of oral microbes always been as stable as modern studies might suggest or have there been changes brought about by marked dietary adaptations, hominid migration, disease and other environmental influences? It certainly appears to be established that influences of the physical and chemical constituents of the diet (Baer et al., 1961 and King and Glover, 1945), as well as variations in oral health may affect both the quantitative and qualitative relationships of oral micro-organisms (Nolte, ed., 1982).

The study of the ancient oral flora using the SEM does, however, present many problems, not least is the identification of the many different microbes. Modern workers studying oral microbiology deal with live bacteria grown in cultures and rely heavily on their physiology and growth for specific identification. It is obvious with these ancient bacteria that only gross morphological differences will enable us to distinguish between them. While it is true that many related species are very similar if not identical in appearance it is still, however, possible to classify them into general categories based on size and shape (Fig. 7). Thus, by using the SEM, it may be possible to gain an overall qualitative impression of the types of micro-organisms present in each individual. Such general differences can be noted in Figures 8 to 13. These represent a series of mammalian calculus specimens, both ancient and modern.

A further, more refined, evaluation, as yet unattempted on archaeological material, employs direct measurement of the micro-organisms from stereomicrographs to differentiate metrically-characterized oral bacteria. This has been attempted using components of the modern oral flora and has successfully demonstrated subtle differences between some of the more closely related species (Boyde and Williams, 1971). Thus, allowing for slight size changes brought about by the calcification process, groups of micro-organisms of similar size and morphology may be further classified.

Backscattered Electron Imaging

Using the backscattered electron image facility on the SEM, changes in density within the calculus matrix can be detected. By viewing polished sections, one can observe voids within the calculus where food debris and micro-organisms are likely to be trapped (Figs. 14 and 15). Patterns of calculus growth can also be investigated (as in Fig. 16) where growth and subsequent mineralisation appear to have occurred on a wide front away from the tooth surface in rhythmic cycles. A number of archaeological specimens show this phenomenon, which has also been noted in modern clinical studies (Spouge, 1973).

Each growth episode is so distinct that these phases of calcification can be readily counted (Fig. 17). By studying the frequency and intensity of these incremental lines, we may be able to gain some additional insight into the diet and environmental stress to which these various individuals were subjected during life.

Conclusion

The SEM has revolutionized the study of the many disciplines within archaeology as the breadth of the papers in this volume indicates. In the case of archaeological calculus the SEM has enabled a hitherto little known source of information to be realised and utilised in the study of past oral ecology.

As has been shown, many technical and interpretive problems will have to be overcome before the complex factors relating diet and disease to changes in the oral flora and subsequent growth rates of calculus are more fully understood. This means studying modern calculus and plaque where the micro-organisms can be more readily identified in order to construct a framework with which the

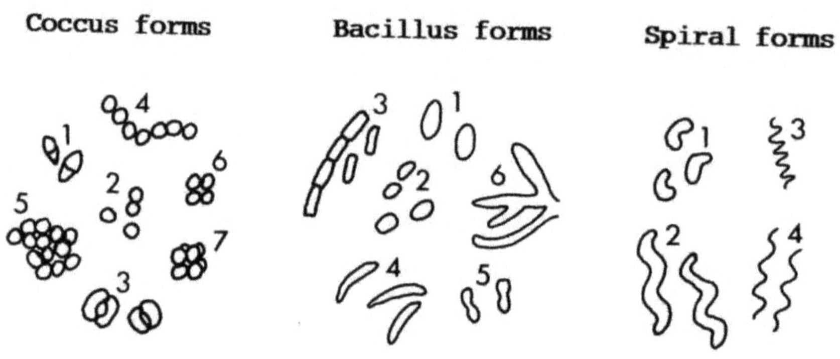

Coccus forms

1. Diplococci (lance-shaped)
2. Diplococci (round)
3. Diplococci (kidney-shaped)
4. Streptococci (cocci in chains)
5. Staphylococci (cocci in clusters)
6. Micrococci (cocci in tetrads)
7. Sarcinae (cocci in packets)

Bacillus forms

1. Rods (round-ended)
2. Rods (coccoid)
3. Rods (square-ended in chains)
4. Rods (spindle-shaped)
5. Rods (swollen-ended)
6. Filaments (Actinomycetes)

Spiral forms

1. Vibrios (rigid commas)
2. Spirilla (rigid spirals)
3. Treponemae* (flexible spirals)
4. Borreliae* (flexible spirals)

* referred to as Spirochetes

Fig. 7. General morphological classification of bacteria. (After Nolte 1982).

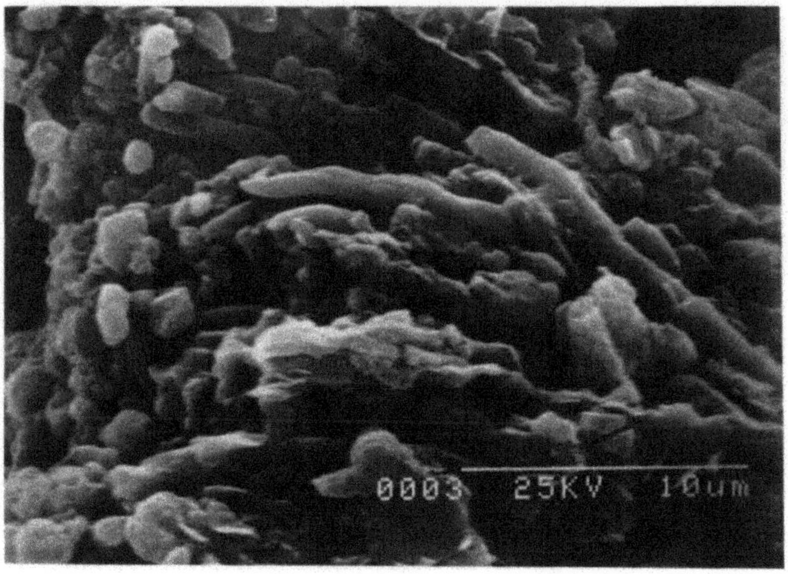

Fig. 8. Filamentous micro-organisms from Anglo-Saxon human calculus from Eccles.

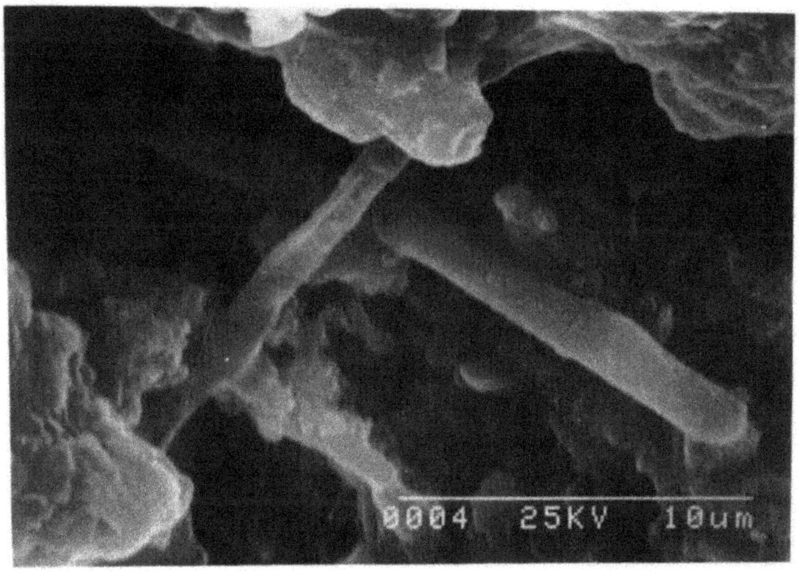

Fig. 9. Larger filamentous micro-organisms.

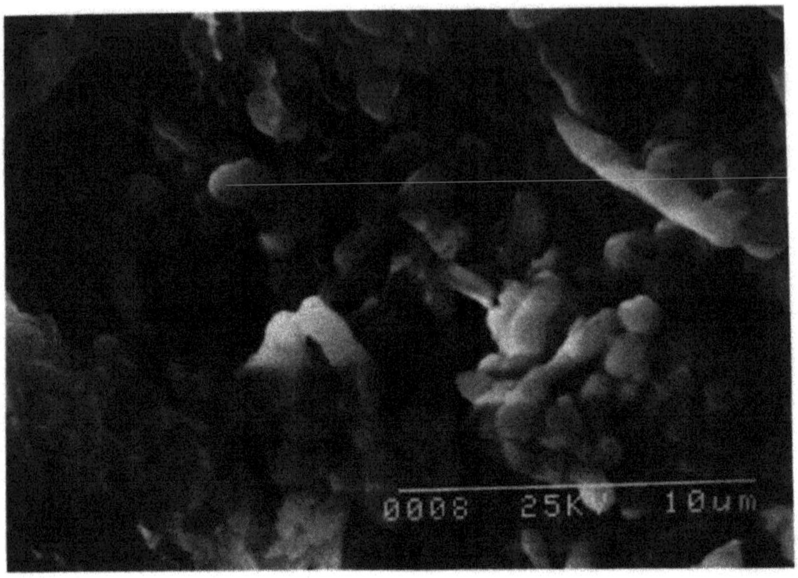

Fig. 10. Rod and coccal shaped micro-organisms (same specimen as in Fig. 7).

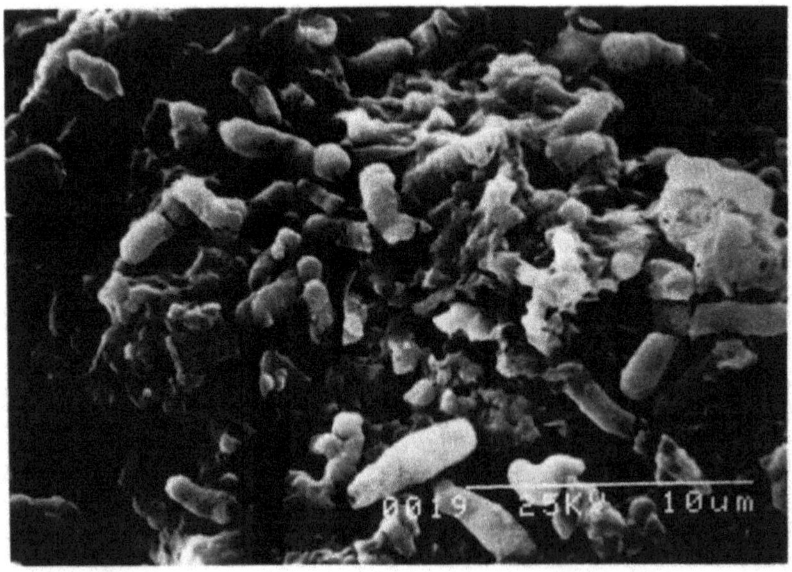

Fig. 11. Short rod-shaped micro-organisms from medieval human calculus, from Ilford, Essex.

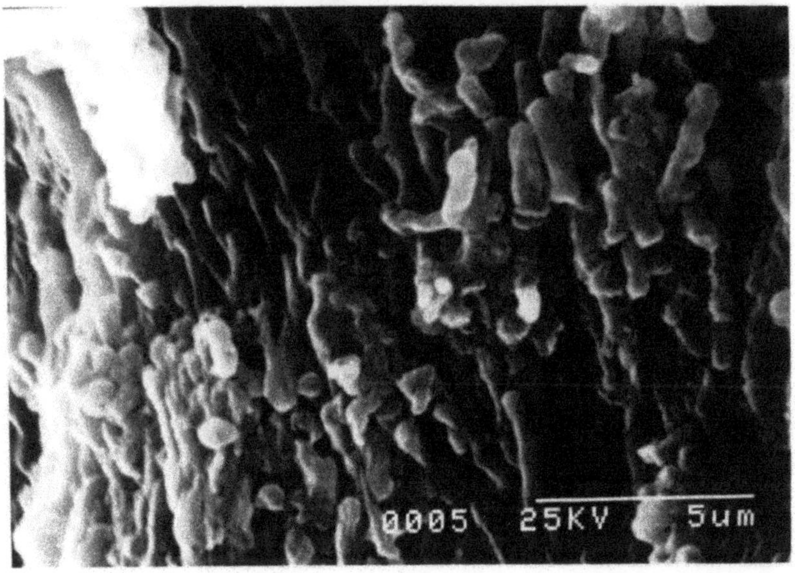

Fig. 12. Short, wavy, rod-shaped micro-organism isolated from modern North Ronaldsay sheep calculus. Note general alignment.

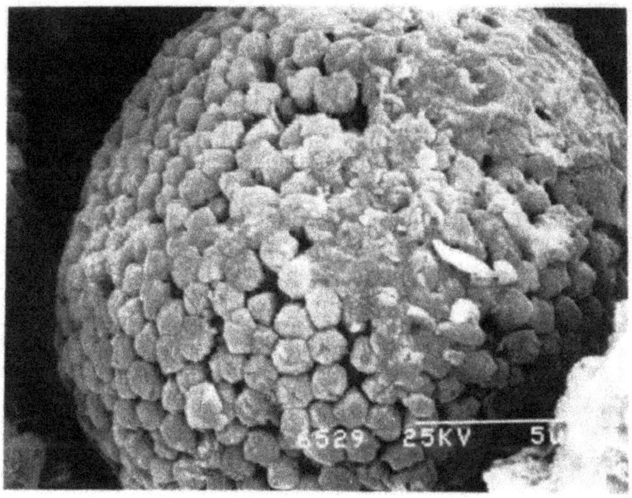

Fig. 13. Large unidentified (possible coccal) colony of rod-shaped micro-organisms from early medieval human calculus from York.

Fig. 14. Back-scattered electron image of polished human calculus from early medieval York. Note the numerous dark areas which represent voids within the calculus matrix.

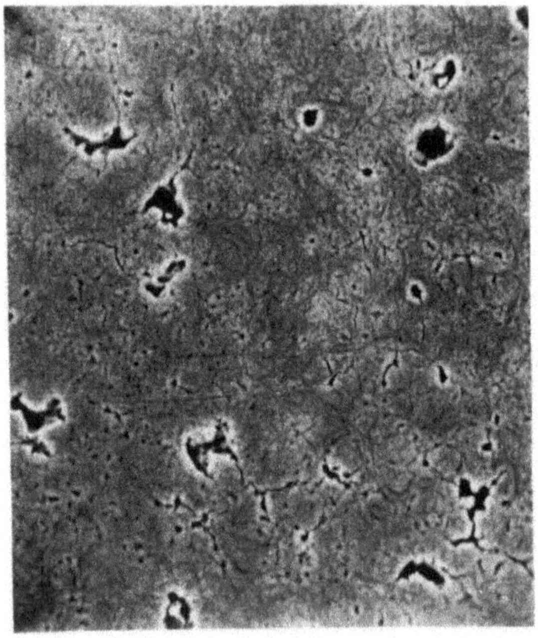

Fig. 15. Back-scattered electron image of polished cow calculus from medieval Canterbury. Note the nebulous appearance of the matrix, representing voids once occupied by living microbes.

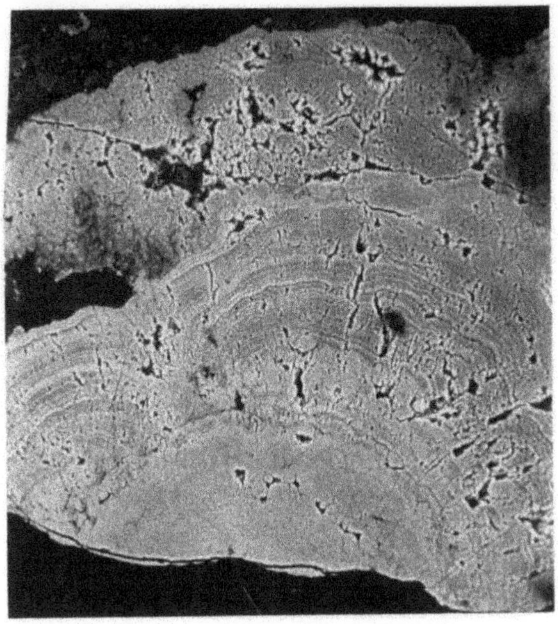

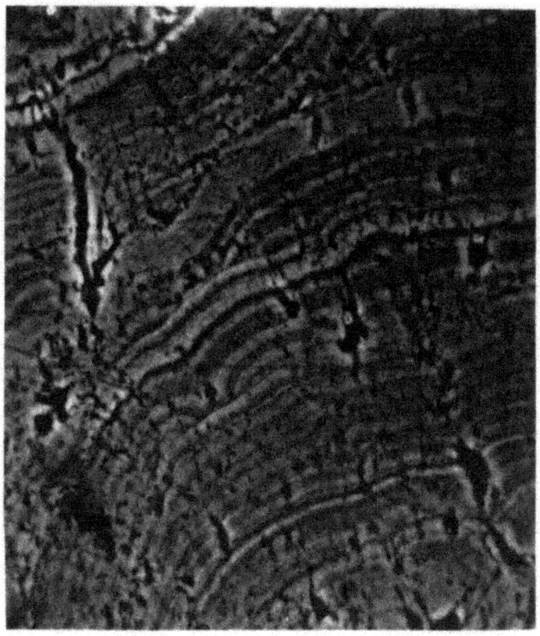

Fig. 16. Back-scattered electron image of polished human calculus from early medieval York. Note the many incremental lines radiating out from the surface nearest the tooth from right to left.

Fig. 17. Detail of Fig. 16.

archaeological data can be compared.

Using other supplementary techniques with the SEM, such as tandem scanning microscopy, micro-radiography and certain biochemical techniques to study the organic constituents of semi-calcified microbes, one might hope to collect a comprehensive body of data from contrasting archaeological samples. In this way the study of archaeological dental calculus could make a major contribution to the study of past diet and oral ecology.

Acknowledgements

The authors would like to thank: Professor Alan Boyde, Dr. Sheila Jones and Elaine Maconnachie for invaluable assistance. The project was funded by the British Science and Engineering Research Council.

References

Baer, P.N., Stephan, R.M. and White, G.L. 1961. Studies on experimental calculus formation in the rat. 1: effect of age, sex, strain, high carbohydrate, high protein diets. Journal of Periodontology 32: 190-196.

Boyde, A. and Williams, A. 1971. Estimation of the volumes of bacterial cells by scanning electron microscopy. Archives of Oral Biology 16: 259-267.

Dobney, K. and Brothwell, D.R. 1986. Dental calculus: the relevance to ancient diet and oral ecology. In Cruwys, E. and Foley, R. (eds.), Teeth and Anthropology, pp. 55-81. BAR International Series No. 291.

Friskopp, J. and Hammarstrom, L. 1980. A comparative scanning electron microscopic study of supra and sub-gingival calculus. Journal of Periodontology 51: 553-562.

Hansen, J.P., Meldgaard, J. and Nordqvist, J. 1986. Qilakitsoq: De Gronlandske Mummier Fra 1400-tallet. Kobenhavn: Glyndendals Bogklub.

Jones, S.J. 1972a. Calculus on human teeth. Apex 6: 2.

Jones, S.J. 1972b. Morphology of calculus formation on the human tooth surface. Proceedings of the Royal Society of Medicine 65: 903-905.

King, J.D. and Glover, R.E. 1945. The relative effects of dietary constituents and other factors upon calculus and gingival disease in the ferret. Journal of Pathological Bacteriology 57: 353-362.

Kramberger, K.G. 1986. Der Diluviale Mensch von Krapina in Kroatien. 277. Wiesbaden.

Lustmann, J., Lewin-Epstein, J. and Shteyer, A. 1976. Scanning electron microscopy of dental calculus. Calcified Tissue Research 21: 47-55.

Newman, H.N. 1974. Diet, attrition, plaque and dental disease. British Dental Journal 136: 491.

Newman, H.N. 1980. Dental Plaque: The Ecology of the Flora on Human Teeth. Illinois: Springfield.

Nolte, W.A. 1982. Oral microflora. In Nolte, W.A. (ed.), Oral Microbiology with Basic Microbiology and Immunology. 4th edition. London: C.V. Mosby.

Poole, D. and Newman, H.N. 1971a. Dental plaque and oral health. Nature 234: 329-331.

Poole, D. and Newman, H.N. 1971b. Topographical variations in dental plaque. Journal of Dental Research 50: 1175.

Spouge, J.D. 1973. Oral Pathology. St. Louis: C.V. Mosby.

BUCCAL STRIATIONS ON THE HOMINID ANTERIOR TEETH FROM ATAPUERCA (SPAIN)

Yolanda Fernández-Jalvo and José María Bermúdez de Castro

Introduction

Analysis of dental microwear in Primates and other mammals has been shown to be a useful indicator of the diets of fossil forms, the patterning of jaw movements during the chewing process and other specific oral activities (Walker 1976, 1977, 1979; Puech 1976, 1978, 1982; Puech et al., 1980; Ryan, 1979; Covert and Kay, 1981; Gordon, 1982 and others).

We have observed striations on the buccal surfaces of the anterior teeth of the middle Pleistocene fossils from Atapuerca. These fossils were recovered in 1976, 1984 and 1985 in the "Sima de los Huesos" (SH): a deep pit within the "Cueva-Mayor del Silo" system of the Sierra de Atapuerca near Ibeas de Juarros and Burgos, Spain (Fig. 1). The associated fauna of the SH site, the dates and the possible correlations with other archeological sites of the Sierra de Atapuerca, currently being excavated, are reported by Aguirre and Rosas (1985).

The fossil hominid remains recovered from the SH site deposits were included in the "anteneandertalians" group on the basis of their dental and skeletal morphology (Aguirre and de Lumley, 1977). The deposits are derived and most of the human remains, as well as the fossil carnivora recovered up to present, were found to have been relocated into secondary deposits by amateur excavators.

In 1973, de Lumley reported the presence of striations, similar to those we observed, on the buccal surface of the upper permanent incisors and canines from Hortus (de Lumley, 1973). Because these striations were oblique and coursed in the same direction, de Lumley suggested that these striations could be due to an intentionally directed motion. According to de Lumley, hominids from Hortus might have used a sharp flint tool to cut meat while holding the meat in their teeth, a task which could have scratched the enamel of their anterior teeth. This hypothesis is suggestive and we felt that it could be investigated with the Atapuerca hominid dental remains. In this report, we want to present the description and the distribution of the striations observed on the hominid anterior teeth from Atapuerca and discuss the implications of de Lumley's hypothesis.

Methods and Materials

In this study we looked at a collection of hominid and non-hominid teeth from the SH deposit in order to investigate the possibility that other taphonomic agents may have been responsible for their formation (Andrews and Cook, 1985; Behrensmeyer et al., 1986). If natural abrasive agents produced these striations, then both the human and non-human dental remains should have been affected in a similar way and on both the lingual and buccal surfaces.

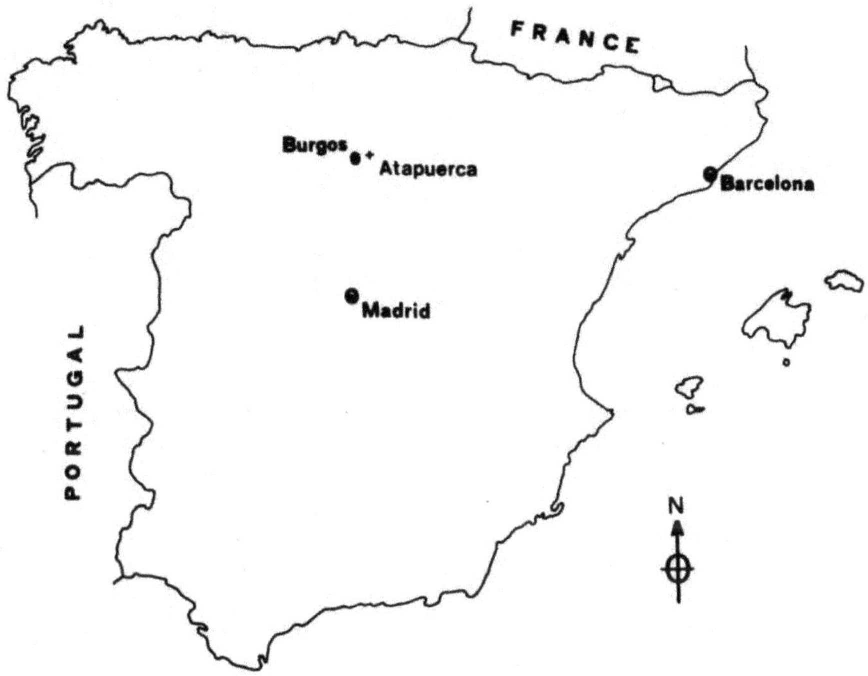

Fig. 1. Map of Spain showing the Atapuerca deposit location.

A total of 54 permanent teeth and one deciduous tooth from the SH site were recovered. Of these, 42 teeth were catalogued as isolated specimens, while the remaining 13 were preserved in mandibular remains. The identification and inventory numbers of the isolated teeth are shown in Table 1. The "in situ" teeth belong to the mandible AT-1 (with both complete molar series) and to the mandibular fragments AT-2 (C1-M1), AT-3 (P4) and AT-75 (M2-M3).

It was possible to estimate that at least nine individuals were represented among the human remains of the SH site of Ibeas (Table 2). This was derived from the data obtained on the mandibular remains and the isolated teeth. The proximal wear facets, the occlusal crown wear, as well as the morphology and size of the teeth were the criteria followed in order to determine the minimum number of individuals. At the moment, 21 isolated teeth cannot be accurately assigned to the nine individuals identified.

Using a Nikon SMZ10 Binocular microscope, we examined the total sample of hominid teeth and 233 non-human teeth from the SH deposit. Thirteen of the 16 incisors and canines from the hominid sample illustrated striations as described by de Lumley (1973). These are tabulated and presented with illustrations of the striations and their orientations (Table 3).

We have studied the striations on three of the Atapuerca hominid teeth with a scanning electron microscope (SEM): two incisor teeth (AT-42 and AT-104) and one molar tooth (AT-21). We also looked at two modern human teeth, which were manually scratched with two kinds of flint. One of the pieces of flint was calcareous and did not have a sharp edge, while the other was of crystalline nature and very sharp. Both were similar to the tools discovered at the Atapuerca sites.

These five specimens were indirectly observed by means of positive replicas (cast) made with Nural 23 epoxy-resin (Smooth-On, New Jersey) taken from a negative replica (mould) made with Exaflex Injection Type (G.-C. Dental Industrial Corp., Arizona) Vinyl Silicone Rubber Impression Material. The replication method used was an adaptation from Bromage (1985, in press).

The replicas were sputter-coated with gold-palladium and mounted on a SEM stub with conductive carbon paint. The analyses were performed with a JEOL JXA-840 SEM, in the secondary electron emission mode, at an accelerating voltage of 15KV. Kodak Plus-X-Pan 120 was used to photograph the specimens.

Results

Nature of the Striations. The first striking observation made on the hominid teeth studied was the fact that the striations were visible to the naked eye (Fig. 2). Further examination using a binocular microscope revealed these striations to have an oblique disposition and a recurrent orientation.

Table 1. Identification and inventory numbers of the isolated permanent teeth from the SH site of Ibeas. The only deciduous tooth (AT-90) is a lower right canine.

	MAXILLARY TEETH		MANDIBULAR TEETH	
	RIGHT	LEFT	RIGHT	LEFT
I 1	AT-42, AT-54	AT-8, AT-27		AT-104
I 2	AT-29	AT-7, AT-53	AT-5, AT-55	AT-103
C		AT-6, AT-44	AT-60	AT-67
P 3	AT-24	AT-41		AT-47, AT-64, AT-102
P 4	AT-68	AT-23	AT-9, AT-28	AT-74
M 1	AT-20, AT-26	AT-16	AT-101	AT-14, AT-22, AT-43, AT-21
M 2	AT-12, AT-46	AT-15		AT-11
M 3	AT-10		AT-30	AT-13, AT-100

Table 2. Inventory numbers of the teeth and mandibles from Atapuerca (SH, Cueva Mayor) assigned to the nine identified individuals. At present, AT-13 and AT-21 cannot be assigned to any particular individual.

Individuals								
I	II	III	IV	V	VI	VII	VIII	IX
AT-1	AT-2	AT-3	AT-14	AT-11	AT-75	AT-16	AT-15	AT-90
AT-60	AT-27	AT-28	AT-74		AT-22	AT-23		
	AT-42	AT-47	AT-100		AT-101	AT-46		
	AT-43	AT-67	AT-102					
	AT-55	AT-103						
		AT-104						

INDIVIDUAL	INVENTORY NUMBER	TOOTH	FREQUENCY	DIRECTION	SKETCH
I	AT-60	C, R	o	——	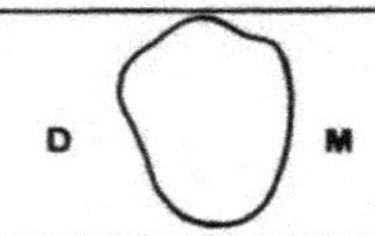
II	AT-42	I^1 R	+ + + +	D / M	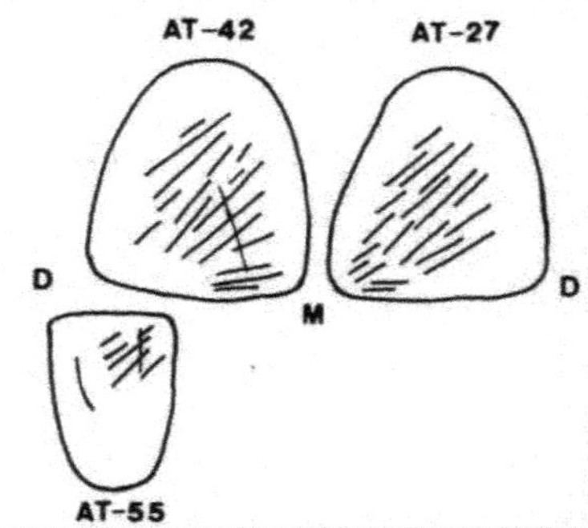
	AT-27	I^1 L	+ + + +	M / D	
	AT-55	I_2 R	+ +	D / M	
III	AT-104	I_1 L	+ + + +	M / D	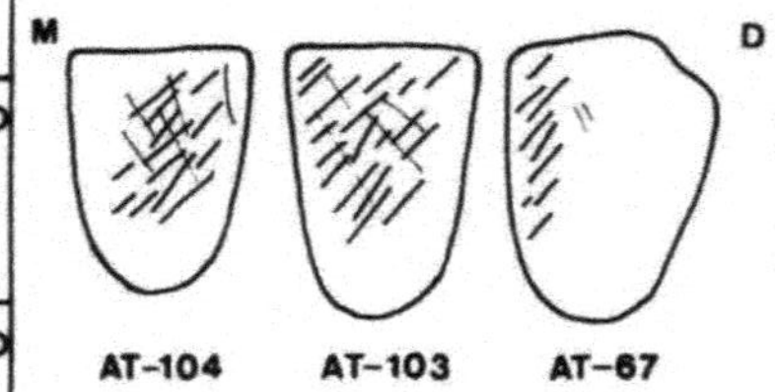
	AT-103	I_2 L	+ + + +	M / D	
	AT-67	C, L	+ + +	M / D	
IX	AT-90	dC, R	+ + +	D / M	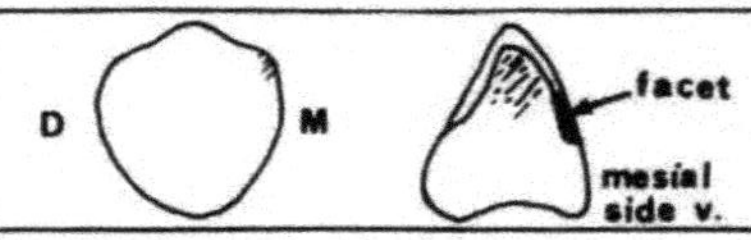

Table 3. Key data for the striations on hominid incisors and canines from the SH site.

INDIVIDUAL	INVENTORY NUMBER	TOOTH	FREQUENCY	DIRECTION	SKETCH
—	AT-54	I^1 R	+	D–M	D M
—	AT-29	I^2 R	+ + +	D–M	D M
—	AT-8	I^1 L	+ + +	M–D	M D
—	AT-7	I^2 L	+ +	M–D	M D
—	AT-53	I^2 L	o	—	M D very damaged
—	AT-6	C' L	+	M–D	M D damaged
—	AT-44	C' L	o	—	M D damaged
—	AT-5	I_2 R	+ +	D–M	D M

Table 3 (con't.).

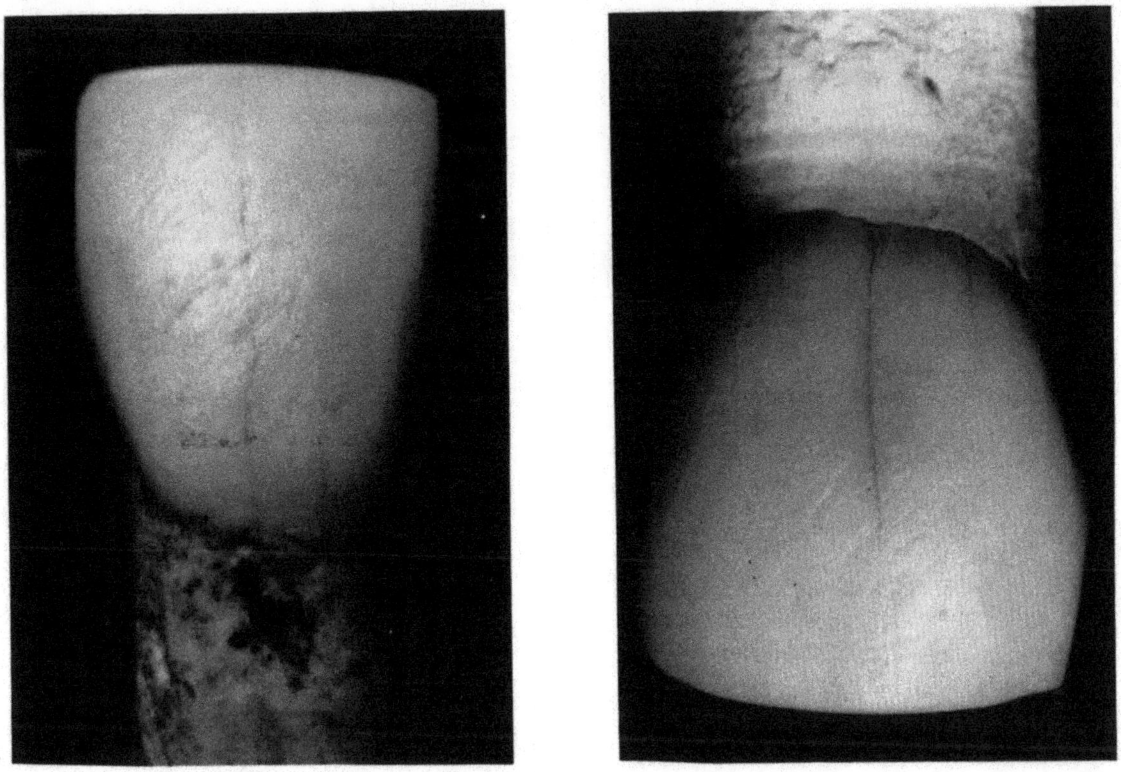

Fig. 2. Buccal view of AT-104 (I_1 L) (left photo) and AT-42 (I^1 R) (right photo) showing the striations reported in this work.

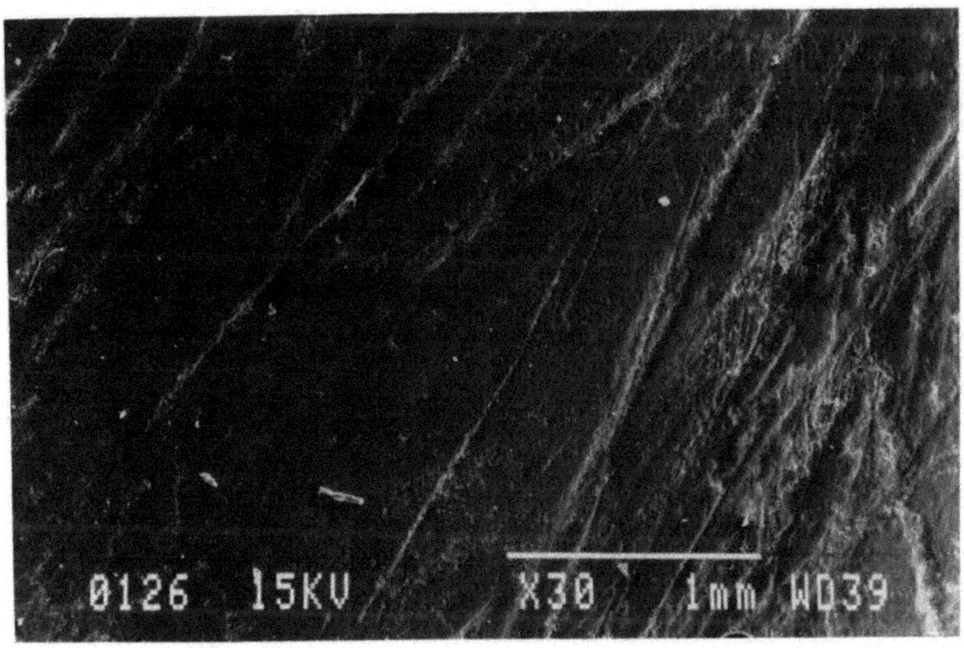

Fig. 3. SEM photo of AT-42 replica. Overall view showing the orientation and the obliquity of the striations. The occlusal surface is at the top of the photo. Note the shrinkage of the silicon rubber mould, causing a wrinkled appearance of the surface.

Next, we looked for similar striations on the teeth of *Ursus deningeri*, the most abundant species at the SH site (Torres, 1984). About 3000 skeletal remains and teeth accounting for at least 99 individuals were recovered (Ceballos, 1986). Out of a total of 585 teeth collected, 57 incisors, 26 canines, 40 premolars and 110 molars were studied using a binocular microscope. We examined the buccal (or labial), lingual, mesial and distal faces of these teeth.

We could not find any striation similar to those found on the human anterior teeth on the non-human teeth studied. Instead, we observed other striations which could be attributed to abrasion and particle action during the chewing process.

Location of the Striations. On the hominid teeth studied with the binocular microscope, we observed prominent striations with a preferential distribution and orientation on the buccal faces of the incisors and some canines. Neither lingual faces of these nor the premolars and molars exhibit such striations.

Table 3 shows the list of all the hominid incisors and canines of the SH site with the most relevant data concerning the striations on these teeth. Out of the total sample of anterior teeth, AT-44 and AT-60, upper and lower canines respectively, are the only ones which do not show striations on their buccal faces. AT-53 has no visible striations, but this incisor is heavily damaged.

The number of striations on the buccal surfaces of the teeth is highly variable; some teeth show only a few striae, whereas others exhibit many (AT-42, AT-27, AT-103 and AT-104). Most of the striations are oblique and run from the upper left part of the buccal surface down to the lower right part, as the tooth is oriented in the mouth.

The striations are observed to be more frequent on the central area, are less common toward the lateral areas, and are more scarce on distal than mesial aspects. They are also absent from the mesial and distal margins near the interproximal contact with the adjacent tooth.

Some interesting observations were made on the special distribution of the scratches on AT-67 and AT-90. AT-67, the left lower canine of the Atapuerca III individual, shows striations on the mesio-buccal aspect, while no striations can be found on the disto-buccal aspect. The right lower deciduous canine AT-90 of the Atapuerca IX individual shows an interproximal facet located on its lingual face produced by the contact with the lateral incisor. The canine seems to have a "rotated eruption" with respect to its usual position in the jaw. In this way, the mesial side of the AT-90 is located at the same level as the buccal face of the incisors. Striations exhibiting an oblique disposition are exclusively present on the mesial side.

We have also observed some heavily worn teeth (AT-54, AT-29, AT-8, AT-7 and AT-5), which show striations near their cervical margin whereas teeth with less wear tend to have striations occurring further up the crown.

SEM Results. Figure 3 clearly illustrates the characteristic striations on the replica of an anterior tooth of Atapuerca II individual (AT-42). The first outstanding observation is the remarkable width of the striations (Fig. 4), which are distinguishable from the fine abrasion observed on the buccal faces of the postcanine teeth. Puech (1982) has reported that the phytoliths contained on certain plants scratch the enamel, leaving striations with width variation of 0.2 to 1 micron.

The striations on the anterior hominid teeth from the SH site often exhibit a range of widths from about 20 microns up to, in some cases, 100 microns (Fig. 5: AT-104). This variation in striation width is very similar to that observed on the modern human teeth altered experimentally by us using the two kinds of flint.

It is also possible to differentiate these striations on the buccal surfaces of the anterior teeth from other narrower (less than 10 microns) striae that have been observed on the archaeological teeth. We believe that this finer abrasion is the result of mastication.

The striations produced experimentally presented another interesting characteristic: there are fine, parallel lines on the walls and floors of the grooves (Fig. 6) which are similar to those in cut-marks on bone (Shipman and Rose, 1983). We have also found similar internal structures in some of the striations on the teeth of the Atapuerca hominids (Figs. 7 and 8), despite the fact that most of the specimens were too damaged by sedimentary abrasion to permit us to study their internal structure in any detail.

Discussion

We have compared hominid and non-hominid teeth in order to investigate the nature and origin of the striations observed on the anterior teeth of the middle Pleistocene hominids from Atapuerca. The striations on the hominid teeth proved to be of oblique disposition with a preferential orientation. On the non-hominid teeth, there were no such characteristic striations. This indicates that natural agents such as transport, could not have been the cause of the striations.

De Lumley (1973) hypothesized that the striations she had observed on the buccal surfaces of the upper incisors and canines from Hortus might have originated from an intentional and directed motion between a stone tool and teeth in order to cut meat. If this hypothesis is true, then the striations should only be found on the anterior teeth and not on the postcanine teeth. Furthermore, only the buccal faces of the anterior teeth should present the characteristic striations with an oblique disposition and preferential orientation.

Our results support the hypothesis in so far as we have only observed these characteristic striations on the buccal faces of the incisors and some canines, noting a complete absence of these striae on the lingual faces of the Atapuerca hominid teeth. Therefore, our results tend to confirm the hypothesis posed by de Lumley (1973).

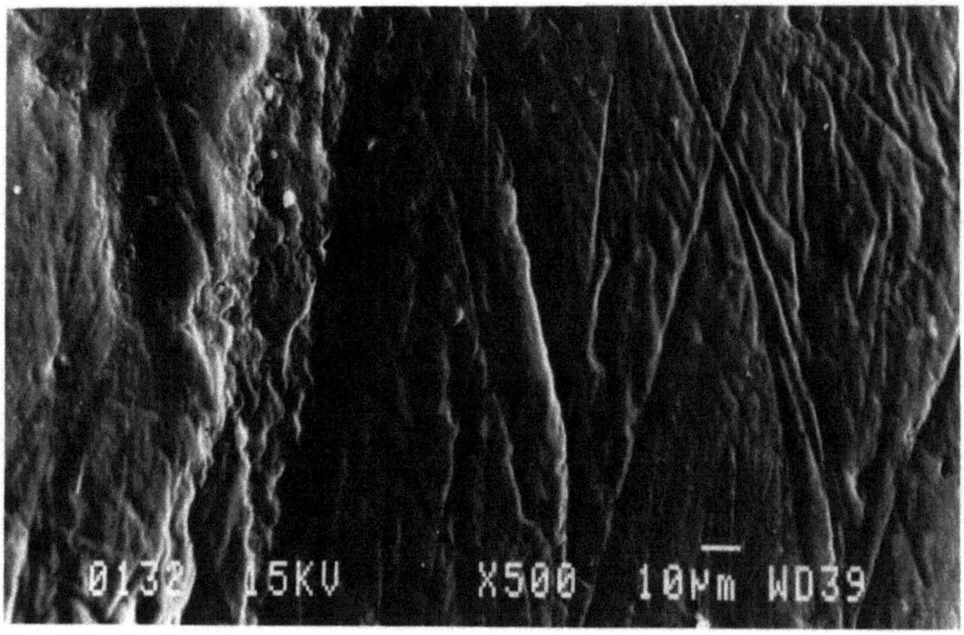

Fig. 4. SEM photo of AT-104 replica. Comparison between one of the striations (left photo) and microwear produced by mastication (right photo). The occlusal surface is at the top of the photo.

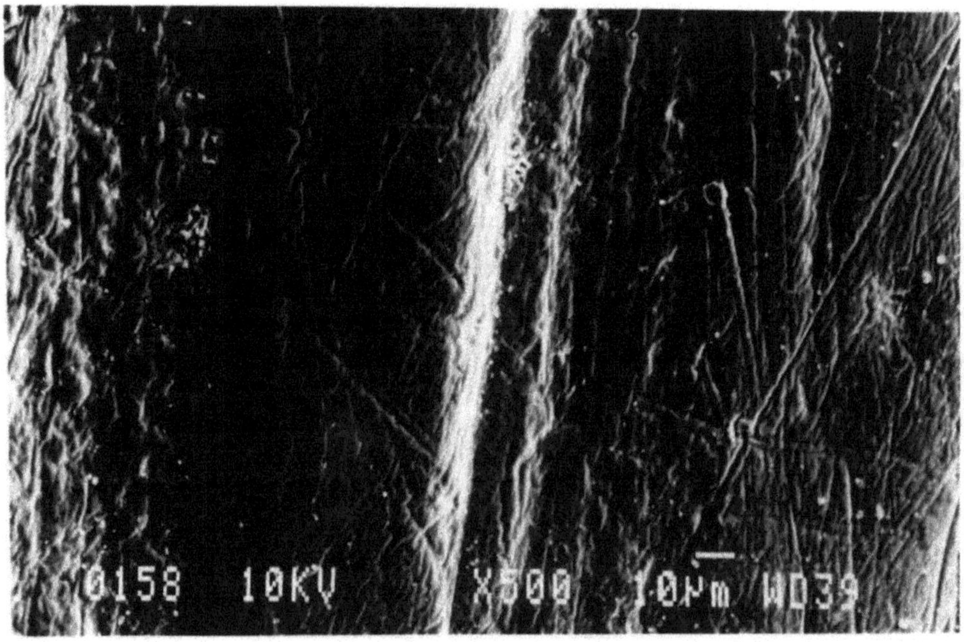

Fig. 5. SEM photo of AT-104 replica. View showing the different widths of the striations. The occlusal surface is at the top of the photo).

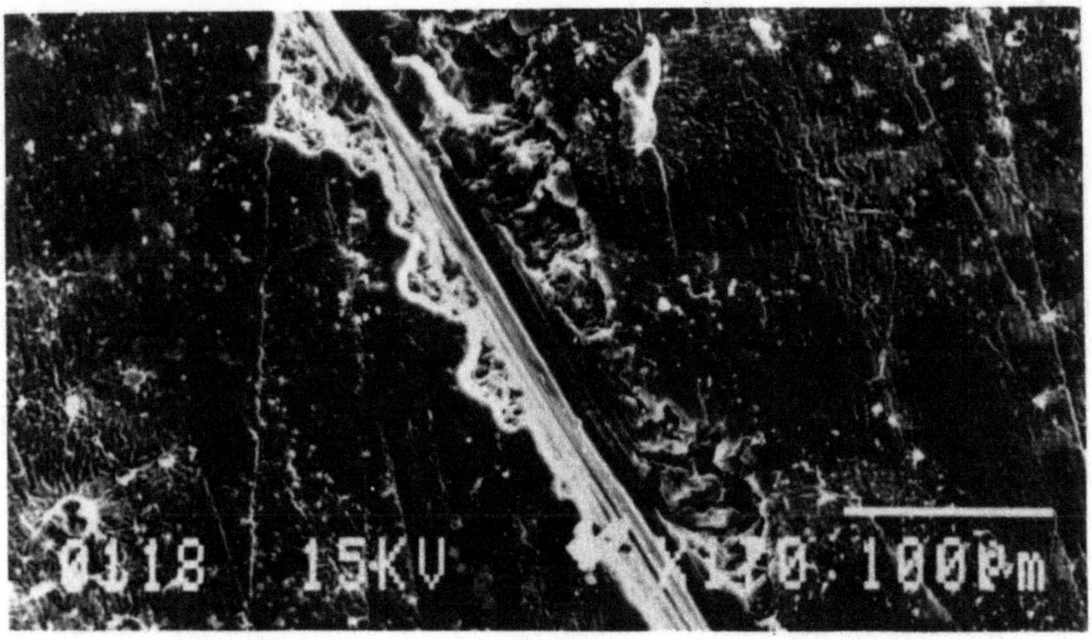

Fig. 6. SEM photo of a modern human tooth showing the internal structures of fine parallel lines within striations produced with a flint tool. The occlusal surface is at the top of the photo.

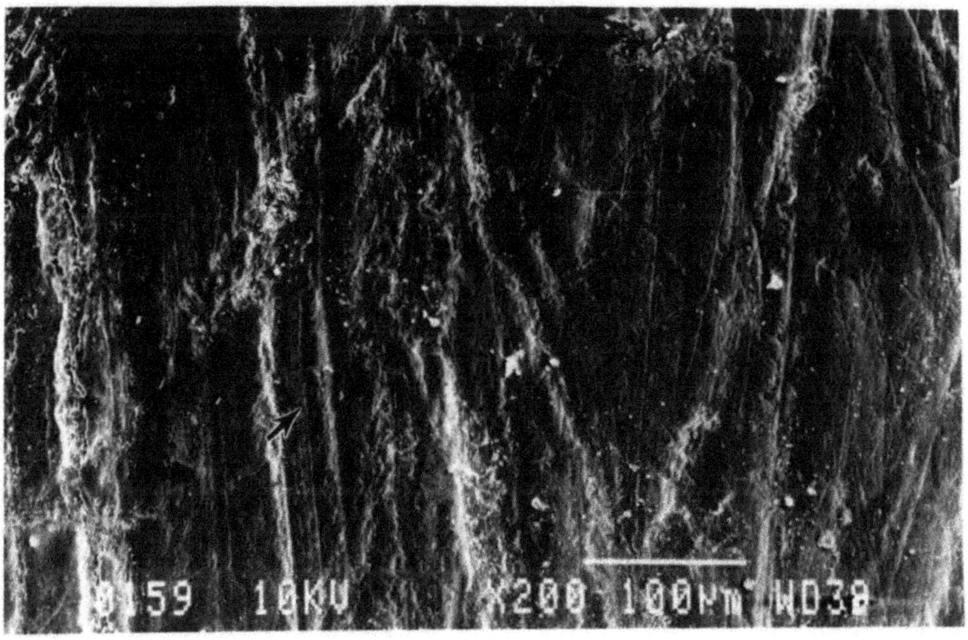

Fig. 7. SEM photo of AT-104 replica. Internal structure of the ancient striation. The occlusal surface is at the top of the photo.

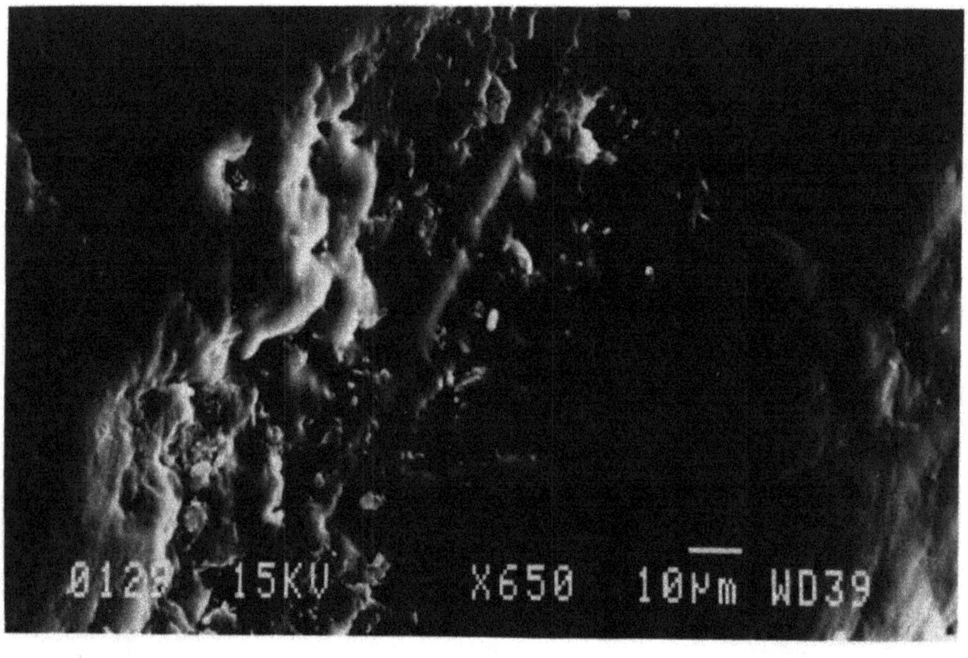

Fig. 8. SEM photo of AT-104 replica. Traces of internal structure within a striation with postdepositional damage. The occlusal surface is at the top of the photo.

The SEM analysis of the Pleistocene hominid teeth and our experimentally scratched modern teeth demonstrates that the characteristic striations observed on them are very similar. However, the distinct internal structures of the striations seen on the modern teeth are not very well-defined on the teeth from Atapuerca due to postdepositional abrasion.

Finally, it was possible to determine the directionality of the intentional striations on both the ancient and the modern human teeth using the criteria outlined by Gordon (1984). This was useful because these teeth provide a unique opportunity to examine in vivo surface damage made by the intentional use of a hand-held stone tool. Directionality was not the focus of the present study, but it was nevertheless apparent that within a group of oblique striations showing a preferential orientation, striae coursed in both directions (e.g. up and down). This suggest that the cutting action was more like a sawing motion rather than an abrupt, sequential unidirectional cutting movement. Furthermore, using the criteria set out by Bromage (in press) and Bermudez de Castro, et al. (in prep) it was possible to say that hominid II of Atapuerca preferentially used his right hand when making the striae.

Conclusions

Striations with a characteristic distribution and orientation were found on specific teeth from the SH site of Atapuerca. This has led us to conclude that these striations could not have been produced randomly by natural, post-mortem processes, but rather were the result of an "ante-mortem" process.

We have also concluded that there is a similarity between those characteristic striations observed on the anterior teeth of Atapuerca hominids and those produced experimentally on modern human teeth with a stone tool.

We hope that our future research will further define the origin of this special kind of striations.

Acknowledgements

We want to thank Timothy Bromage and Hediye Erdjument for their valuable help in carefully reading the manuscript, providing helpful comments and ideas, and especially for their friendship. Sincere thanks are given to Paloma Adeva and Manolo Gimenez for technical advice on the operation of the SEM in the "Centro Nacional de Investigaciones Metalurgicas". Our thanks also to the "Escuela Superior de Ingenieros Aeronauticos", where the specimens were coated. We wish to extend our gratitude to Antonio Sanchez, Antonio Rosas, Josefa Enamorado, Carlos Diez, Marta Ceballos and Begona Sanchez for their suggestions and critical comments concerning with this work. Finally, we are grateful to Dr. Jose Luis Monreal Lago for the experimental material provided and especially to the photography service of the "Museo Nacional de Ciencias Naturales" for its professional assistance. This research was supported by a grant from the "Comision Asesora de Investigacion Cientifica y Tecnica" in relation to the Atapuerca 2 Project (n. 1849/82).

Addendum

A more detailed study of the lingual striations observed on the Atapuerca-Ibeas anterior teeth, as well as those of other fossil hominids (La Quina 5, France, and Cova Negra, I^1, Spain) may be found in: Bermúdez de Castro, J.M., Bromage, T.G. and Fernández Jalvo, Y. (1988, in press). In this paper, we conclude that the macro- and microscopic studies of the striations, together with those striations produced in an experimental study, suggest that scratching of these hominid anterior teeth was indeed the result of **ante-mortem** tool use, producing striations indicating right- (normally) or left-handedness (rarely).

At the present time, 19 anterior teeth from Atapuerca-Ibeas are known to possess these buccal striations and the minimum number of affected individuals is 11.

References

Andrews, P. and Cook, J. 1985. Natural modifications to bones in a temperate setting. Man 20: 675-691.

Aguirre, E. and de Lumley, M.A. 1977. Fossil men from Atapuerca, Spain: their bearing on human evolution in the Middle Pleistocene. Journal of Human Evolution 6: 681-688.

Aguirre, E. and Rosas, A. 1985. Fossil man from Cueva Mayor, Ibeas, Spain: new findings and taxonomic discussion. In: Hominid Evolution: Past, Present and Future. Tobias, P.V. (ed.) Alan R. Liss, Inc., New York: 319-328.

Behrensmeyer, A.K., Gordon, K.D. and Yanagi, T.G. 1986. Trampling as a cause of bone surface damage and pseudo-cutmarks. Nature 319: 768-771.

Bermúdez de Castro, J.M., Bromage, T.G. and Fernández Jalvo, Y. (1988, in press) Buccal striations on fossil human anterior teeth: evidence of handedness in the Middle and early Upper Pleistocene. Journal of Human Evolution 17.

Bromage, T.G. 1985. Systematic inquiry in tests of negative/positive replica combinations for SEM. Journal of Microscopy 137: 209-216.

Bromage, T.G. (in press) Scanning electron microscopy/replica technique and recent applications to the study of fossil bone. Scanning Electron Microscopy.

Ceballos, M. 1986. Estudio biométrico y morfológico de una población de Ursus deningeri del Pleistoceno medio de la Sierra de Atapuerca (Burgos). Memoria de Licenciatura. Facultad de Ciencias Biológicas de la Universidad Complutense de Madrid.

Covert, H.H. and Kay, R.F. 1981. Dental microwear and diet: implications for determining the feeding behaviors of extinct primates with a comment on the dietary pattern of Sivapithecus. American Journal of Physical Anthropology 55: 331-336.

De Lumley, M.A. 1973. Anténéandertaliens et Néandertaliens du Basin Mediterranéen Occidental Européen. Études Quaternaires, Memoire 2, Université de Provence.

Gordon, K.D. 1982. A study of microwear on chimpanzee molars: implications for dental microwear analysis. American Journal of Physical Anthropology 59: 195-215.

Gordon, K.D. 1984. Microfracture patterns of abrasive wear striations on teeth indicate directionality. American Journal of Physical Anthropology 63: 315-322.

Puech, P.-F. 1976. Recherche sur la mode d'alimentation des hommes du Paléolithique par l'etude microscopique des couronnes dentaires. In La Prèhistoire Française I, de Lumley, H. (ed.) C.N.R.S. Paris: 708-709.

Puech, P.-F. 1978. L'alimentation de l'homme prèhistorique. La Recherche 9: 1029-1031.

Puech, P.-F. 1982. L'usure dentaire de l'homme de Tautavel. In L'Homo erectus et la place de l'homme de Tautavel parmi les hominides fossiles. Congrés International de Paléontologie Humaine, 1 Congrés, Nice. C.N.R.S. Pretirage, Tome 1: 249-275.

Puech, P.-F., Prone, A. and Kraatz, R. 1980. Microscopie de l'usure dentaire chez l'homme fossile; bol alimentaire et environment. Comptes Rendus des Seances de l'Academie des Sciences, Paris 290: serie D: 1413-1416.

Ryan, A. 1979. Wear striation direction on primate teeth: a scanning electron microscope examination. American Journal of Physical Anthropology 50: 155-168.

Shipman, P. and Rose, J.J. 1983. Early hominid hunting, butchering and carcass-processing behaviors: approaches to the fossil record. Journal of Anthropological Archeology 2: 57-98.

Torres, T.J. 1984. Ursidos del Pleistoceno-Holoceno de la Peninsula Iberica. Ph.D. Thesis. Universidad Politecnica. Madrid.

Walker, A.C. 1976. Wear striations on the incisors of cercopithecoid monkeys as an index of diet and habitat preferences. American Journal of Physical Anthropology 45: 299-308.

Walker, A.C. 1977. Scanning electron microscopy of early hominid teeth. American Journal of Physical Anthropology 47: 376.

Walker, A.C. 1979. SEM analysis of microwear and its correlation with dietary patterns. American Journal of Physical Anthropology 50:489.

THE OCCURRENCE OF INTERPROXIMAL ABRASION GROOVES IN SOME IRISH POPULATIONS

C. Power and V.R. O'Sullivan

Introduction

Interproximal abrasion grooves are horizontal grooves found beneath the interproximal contact area on the mesial or distal surfaces of the crowns or cervical regions of teeth (Fig. 1). Tiny spicules of bone, and rough particles of sand in food and drinking water can cause abrasion. Interproximal abrasion grooves have occasionally been found on the teeth of Australian Aborigines (Heithersay, 1959), Bushmen (Drennan, 1929), American Indians (Schulz, 1977), and in fossil hominids including Neanderthals (Martin, 1929) and Homo erectus (Weidenreich, 1937). These people subsisted on gritty, abrasive diets. Such grooves are not found in individuals eating soft foods relatively free of abrasives such as the diet of modern western society (Wallace, 1974), and a separate study of over 1200 modern Irish teeth has produced no evidence of interproximal abrasion grooves (Power and O'Sullivan, 1986).

The purpose of this study is to examine interproximal abrasion grooves found in the human dentitions from some Irish archaeological sites. Groove morphology was examined using the scanning electron microscope (S.E.M.).

Materials

Thirty known examples of interproximal abrasion grooves which had been reported in human dentitions from four Irish archaeological sites were examined in this study. These were as follows:

1. Ten teeth from two inhumated individuals from the neolithic passage grave at Newgrange, County Meath (O'Sullivan, 1982).
2. Two teeth from two subjects from a total of 45 individuals from Wood Quay, Dublin City. These remains date to the 11th/12th century (Power, 1984).
3. Nine teeth from six individuals from the Cistercian Abbey of Tintern Minor in County Wexford. The skeletal remains of 50 individuals were uncovered in the 1982 season of excavation. These remains date to the 16th century and probably represent the remains of local peasantry (Power, 1984).
4. Nine teeth from one individual from 19th century Cork City.

Methods

All the teeth examined were initially cleaned using acetone and a no. 6 soft sable brush to flick away extraneous material adjacent to the tooth surface. The teeth were examined using a Jeol JSM-35 SEM without sputter coating, at an operating voltage of 10 kV. Electron micrographs were taken of the interproximal abrasion grooves at various magnifications from x10 to x900.

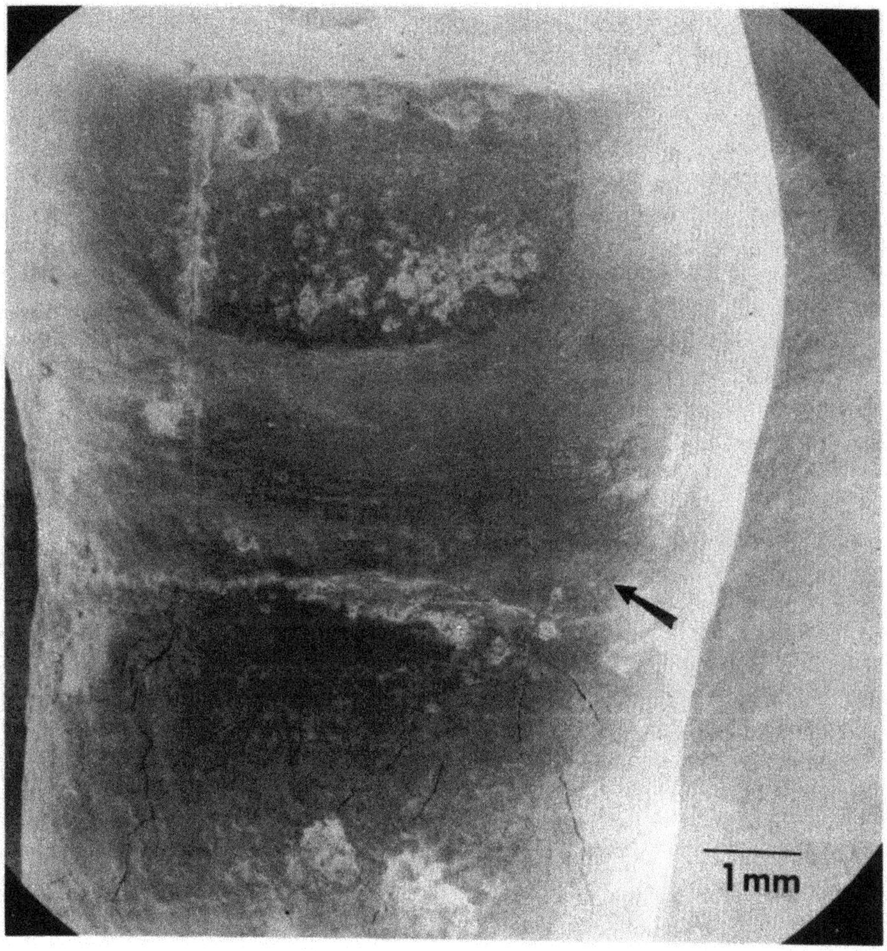

Fig. 1. An interproximal abrasion groove is present beneath a large interproximal contact area on the distal surface of a lower left first premolar from Tintern, County Wexford, Ireland.

Results

A total of 30 teeth with interproximal abrasion grooves was available for examination. Of these, 10 had a groove on the mesial surface only, 10 had a groove on the distal surface only, and 10 had a groove on each of the interproximal surfaces. Macroscopic examination showed that most of the grooves were confined to the crown (Fig. 1). Some of the grooves were rounded and shallow, and did not penetrate the enamel. Some of the larger grooves in the cervical region involved parts of the crown and root. A small number of grooves were deep and trench-like with angular profiles.

SEM examination showed that dentine was exposed in the floor of some abrasion grooves and the floor of the interproximal abrasion groove contained a large number of parallel scratch marks (Fig. 2). The scratch marks vary considerably in size. Two groups of scratch marks on the lower left first permanent premolar formed a criss-cross pattern (Fig. 3). Twelve of the grooves examined were present on the teeth of the single individual from Cork City. Three were deep, angular grooves and the floor of one of these showed a V-shaped pattern of parallel scratch marks. The remaining grooves on this individual and on the teeth from Newgrange, Wood Quay and Tintern showed rounded shallow interproximal abrasion grooves.

Discussion

The findings illustrate that interproximal abrasion grooves can occur in several different forms. Some grooves are confined to the enamel and others can involve part of the root, as well (Fig. 1). They can be shallow or deep, rounded or angular and trench-like. SEM revealed a large number of parallel scratch marks in the interproximal abrasion grooves (Figs. 2-3). The criss-cross pattern seen in one groove may reflect a change in the direction of the current of fluid moving between the teeth, perhaps as a result of diverse cleansing or chewing habits. It has been suggested that some grooves are caused by toothpicks (Ubelaker et al., 1969). The V-shaped pattern of scratch marks in the floor of one angular, trench-like groove in the 19th century individual from Cork City suggests the expected pattern produced by the insertion and withdrawal of a toothpick. Some studies have attempted to show a correlation between the angle of the groove to the curve of the dental arch and the most probable path of insertion of a toothpick (Berryman et al., 1979; Ubelaker et al., 1969). Berryman et al. (ibid.) suggested "a probe such as a toothpick for therapeutic use" as one of the possible causes for these grooves in the dentitions of prehistoric Arikara Indians from Larson, South Dakota. Grooves can also be caused by the use of dental floss or the insertion of vegetable material between the teeth. Dental plant fibre-processing (for craft functions) and the archaeological occurrence of abraded grooves has been reported by Cybulski (1975) in an ethnographic study from British Columbia. Wallace (1974) proposed that grooving results from hydraulic patterns of grit-laden saliva forced between the teeth after gingival recession. However, Gordon (pers. comm.) believes that grit in saliva would not transmit enough force to scratch tooth surfaces.

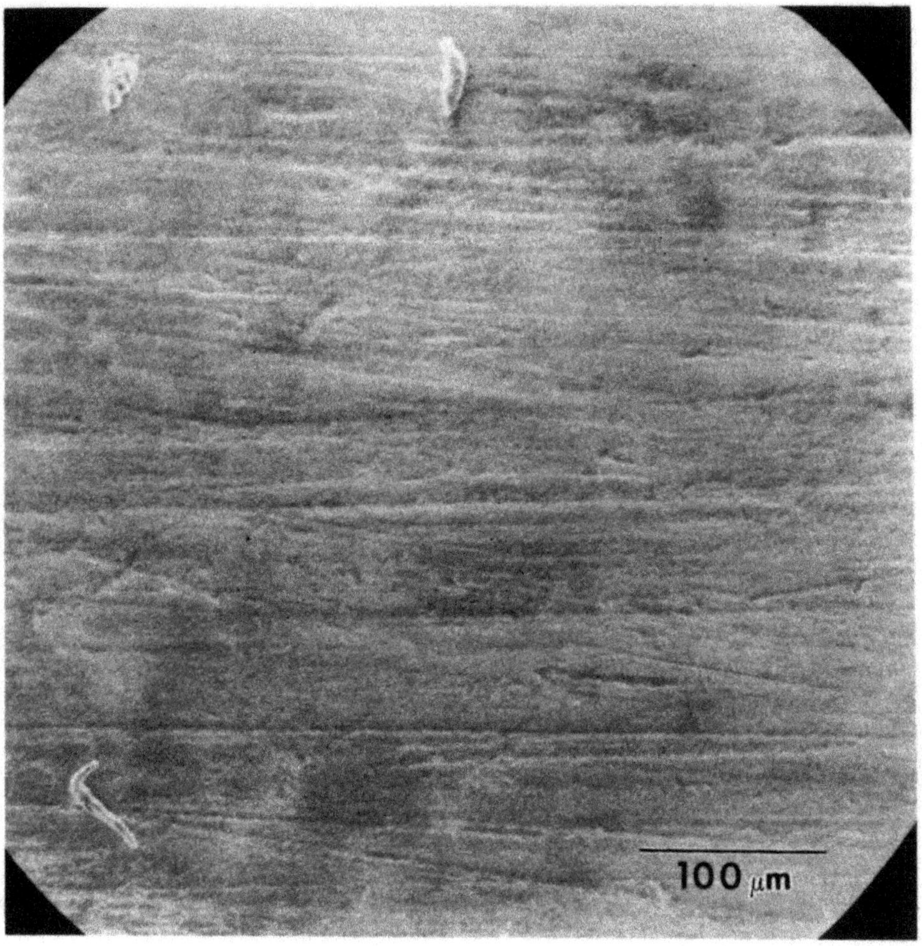

Fig. 2. Sets of parallel scratch marks in the floor of an interproximal abrasion groove on the mesial surface of a lower left second premolar from Tintern, County Wexford. The numerous scratch marks are orientated in the same general direction.

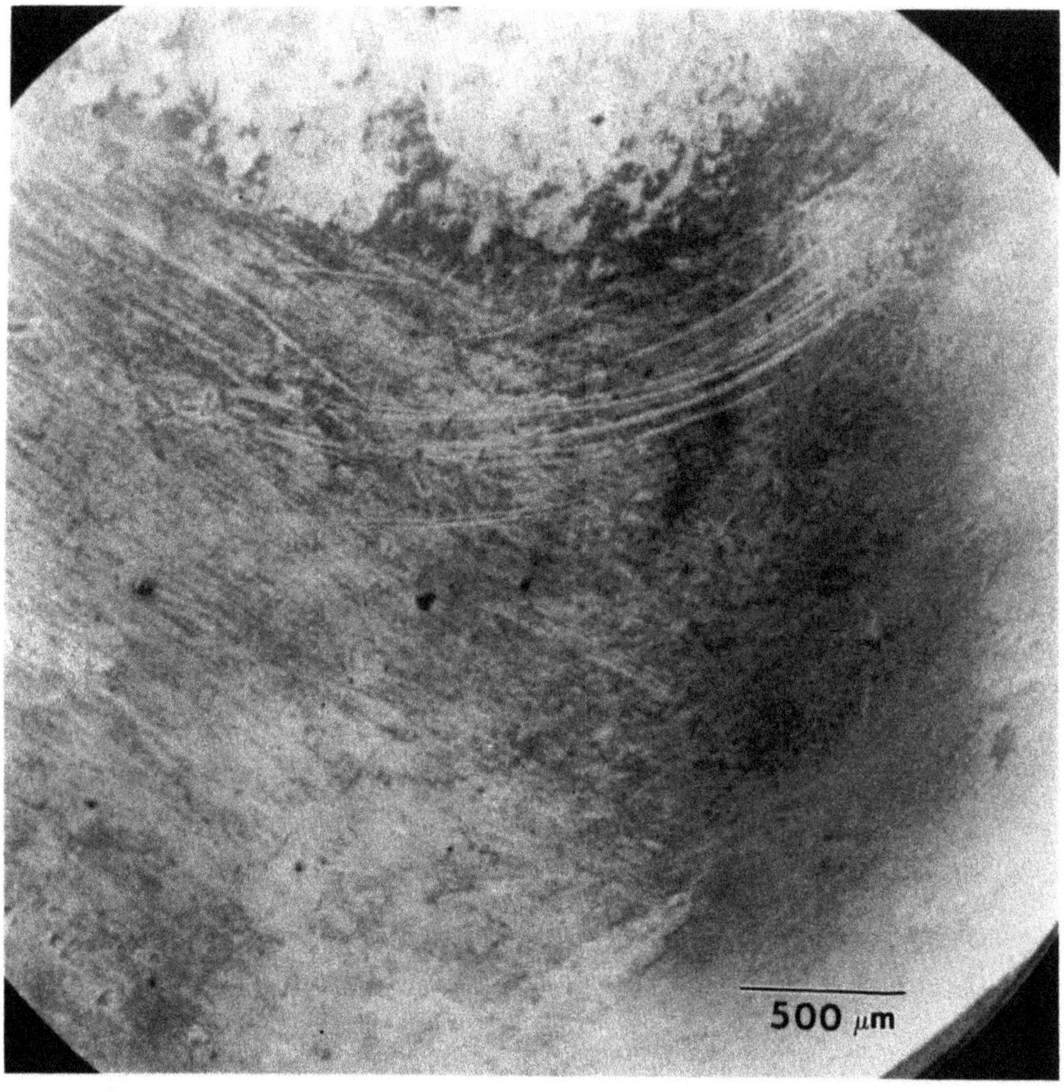

Fig. 3. Grooving in the form of concentric arcs and parallel lines is evident at the edge of the interproximal abrasion groove on the mesial surface of a lower left first premolar from Tintern, County Wexford.

Interproximal abrasion grooves have been found in non-human dentitions (Von Koenigswald, 1972; O'Sullivan et al., 1986), which indicates that such grooves can be produced without the use of toothpicks or similar agents.

Conclusions

A variety of potential causes exist which are capable of producing abrasion grooves on the interproximal surfaces of teeth. Such grooves may be produced without the use of a toothpick, but it has yet to be determined whether or not toothpicks were actually used to produce the sort of interproximal abrasion grooves seen in the cases discussed above.

References

Berryman, H.E., Owsley, D.W. and Henderson, A.M. 1979. Non-carious interproximal grooves in Arikara Indian dentitions. American Journal of Physical Anthropology 50: 2: 209-212.

Cybulski, J.S. 1975. Tooth wear and material culture: pre contact patterns in the Tsimshian area, British Columbia. Syesis 7: 31-35.

Drennan, M.R. 1929. The dentition of a Bushman tribe. Annals of the South Africa Museum 24: 61-87.

Gordon, K. 1984. The Johns Hopkins University School of Medicine, Baltimore, Maryland, personal communication.

Heithersay, G.S. 1959. A dental survey of the Aborigines at Haast's Bluff, Central Australia. Medical Journal of Australia 1: 721-729.

Martin, H. 1923. L'homme fossile de La Quina. Archives de Morphologie Facs 15.

O'Sullivan, V.R. 1982. The human dental remains. In O'Kelly, M.J. (ed.), Newgrange-Archaeology, Art and Legend, pp. 205-213. London.

O'Sullivan, V.R., Fisher, E., Kelly, M. and Clifford, H.S. 1986. Mammalian interproximal abrasion grooves. Journal of Dental Research (in press).

Power, C. 1984. Anthropological Studies on the Dental Remains from some Irish Archaeological Sites. Unpublished Master's Thesis, University College, Cork, Ireland.

Power, C. and O'Sullivan, V.R. 1986. The occurrence of interproximal abrasion grooves in Irish populations. Journal of Dental Research, in press.

Schulz, P.D. 1977. Task activity and anterior tooth grooving in prehistoric California Indians. American Journal of Physical Anthropology 46: 1: 87-91.

Ubelaker, D.J., Phenice, T.W. and Bass, W.M. 1969. Artificial interproximal grooving of the teeth in American Indians. American Journal of Physical Anthropology 30: 145-150.

Von Koenigswald, G.H.R. 1972. Tooth wear and culture: a survey of tooth functions among some prehistoric populations. In Molnar, S. (ed.), Current Anthropology 13: 5: 511-515.

Wallace, J.A. 1974. Approximal grooving of teeth. American Journal of Physical Anthropology 40: 385-390.

Weidenreich, F. 1937. Sinanthropus dentition. Palaeontologia Sinica 101: 1-180.

www.ingramcontent.com/pod-product-compliance
Lightning Source LLC
LaVergne TN
LVHW071203210926
848482LV00015B/712